Advances in Terrestrial Drilling

Advances in Terrestrial Drilling

Ground, Ice, and Underwater

VOLUME I

Advances in Terrestrial and Extraterrestrial Drilling:
Ground, Ice, and Underwater

Edited by
Yoseph Bar-Cohen
Kris Zacny

CRC Press
Taylor & Francis Group
Boca Raton London New York

CRC Press is an imprint of the
Taylor & Francis Group, an **informa** business

First edition published 2021
by CRC Press
6000 Broken Sound Parkway NW, Suite 300, Boca Raton, FL 33487-2742

and by CRC Press
2 Park Square, Milton Park, Abingdon, Oxon, OX14 4RN

CRC Press is an imprint of Taylor & Francis Group, LLC

Library of Congress Cataloging-in-Publication Data
Names: Bar-Cohen, Yoseph, editor. | Zacny, Kris, editor.
Title: Advances in terrestrial and extraterrestrial drilling : ground, ice, and underwater / Yoseph Bar-Cohen, Kris Zacny.
Description: Boca Raton, FL : CRC Press/Taylor & Francis Group, LLC, 2021. | Includes bibliographical references and index. | Summary: "This 2 volume set will include the latest principles behind the processes of drilling and excavation on Earth and other planets. It will cover the categories of drills, the history of drilling and excavation, various drilling techniques and associated issues, including rock coring as well as unconsolidated soil drilling and borehole stability"-- Provided by publisher.
Identifiers: LCCN 2020037780 (print) | LCCN 2020037781 (ebook) | ISBN 9780367653460 (v. 1 ; hardback) | ISBN 9780367653477 (v. 2 ; hardback) | ISBN 9780367674861 (v. 1 ; ebook) | ISBN 9781003131519 (v. 2 ; ebook)
Subjects: LCSH: Boring. | Space probes--Equipment and supplies. | Drilling and boring machinery.
Classification: LCC TN281 .A234 2021 (print) | LCC TN281 (ebook) | DDC 622/.3381--dc23
LC record available at https://lccn.loc.gov/2020037780
LC ebook record available at https://lccn.loc.gov/2020037781

ISBN: 978-0-367-65346-0 (hbk)
ISBN: 978-0-367-67486-1 (ebk)

Typeset in Times
by SPi Global, India

Contents

Preface

Drilling can simply be defined as the action of penetrating into solid media. For terrestrial applications, drilling technology is well established with a range of commercial tools that are readily available. There are also significant research and development efforts that enhance the drilling capability for much harsher conditions while reducing operational cost and increasing profit margins. A number of terrestrial efforts are focused on enabling faster and cheaper drilling and the move to robotic operations is for human safety. While autonomous robotic drilling is also a major requirement for planetary efforts, there are a number of unique challenges that need to be addressed. Drilling or coring on other planets requires a highly detailed understanding of drilling processes and development of new capabilities for operation under extreme temperature, pressure, as well as low gravity. Some of these conditions are also applicable on Earth and will be key in reaching extreme depths to tap into new oil reservoirs, drilling geothermal wells in very hot rocks, and/or exploring ice layers in the Antarctic. This book has been compiled to cover the latest scope of knowledge in drilling as provided by leading scientists and engineers around the world.

We focused on drilling with emphasis on penetration of the ground and various subsurface materials, including rocks, permafrost, ice, soil, and regolith, to name a few. We have covered a range of mechanical and other drilling techniques as well as related issues including cuttings transportation and their disposal, borehole stability, the current and future levels of drilling autonomy. We have also covered the need for sample acquisition, caching and transport and restoration of in situ conditions necessary for eventual integrated science instruments and data interpretation. We describe the drilling process from basic science and the associated process of breaking and penetrating various media, and the required hardware and the process of excavation and analysis of the sampled media.

The first chapter covers the various drilling techniques and unique applications, in addition to directions of the evolution of technology. Chapter 2 covers modeling and analysis of vibro impact drilling systems. Chapter 3 covers subtractive and additive manufacturing applied to drilling systems and describes the use of such processes as 3D printing. Chapter 4 covers on-shore drilling while Chapter 5 covers offshore deepwater drilling. Chapter 6 covers recent innovations in drilling in ice while Chapter 7 covers environmental drilling/sampling and offshore modeling systems. And Chapter 8 covers drilling automation. Chapter 9 covers specialized drilling techniques for medical applications.

Yoseph Bar-Cohen
Pasadena, CA

Kris Zacny
Altadena, CA
July 2020

Acknowledgments

The editors would like to acknowledge and express their appreciation to the following individuals who took the time to review various book chapters. Their contributions are greatly appreciated and helped make this book of significantly greater value to readers. The individuals who served as reviewers of chapters in this book are as follows:

Chapter 1
Jared Atkinson, Honeybee Robotics, Altadena, CA
Gregory H. Peters, Jet Propulsion Laboratory (JPL)/California Institute of Technology (Caltech), Pasadena, CA

Chapter 2
Boleslaw Mellerowicz, Honeybee Robotics Ltd., Altadena, California
Patrick Harkness, University of Glasgow, Scotland
Ramesh Malla, Department of Civil and Environmental Engineering, University of Connecticut, CT

Chapter 3
Sathish Nammi, Brunel University, London, UK
John Paul C. Borgonia, Jet Propulsion Laboratory (JPL)/California Institute of Technology (Caltech), Pasadena, CA

Chapter 4
Christopher Yahnker, Jet Propulsion Laboratory (JPL)/California Institute of Technology (Caltech), Pasadena, CA
Frederick Bruce Growcock, Occidental Petroleum Corporation, Houston, TX
Arthur Hale, Aramco Services Company, Houston, TX

Chapter 5
Heitor Lima, Texas A&M University, College Station, TX;
João Carlos Ribeiro Plácido, Mechanical Engineering Department, DEM PUC-Rio, Brazil
Shiniti Ohara, Vice President Operations, Barra Energia, Rio de Janeiro, RJ, Brazil

Chapter 6
Victor Zagorodnov, Cryosphere Research Solutions LLC, Columbus, OH
Matthias Huether and Jan Tell Alfred Wegener Institute, Bremerhaven, Germany
Boleslaw Mellerowicz, Honeybee Robotics, Altadena, CA

Chapter 7
Alfred William (Bill) Eustes III, Petroleum Engineering Department at the Colorado School of Mines, Golden, CO
Gary Covatch, Petroleum Engineer, National Energy Technology Laboratory (NETL), US Department of Energy (DOE), Morgantown, WV

Chapter 8
Gokturk Tunc, Schlumberger Limited, Schlumberger Limited, Houston, TX
Richard Harmer, Schlumberger Limited, Houston, TX
Rodrigo Gallo Covarrubias, Schlumberger Limited, Houston, TX

Chapter 9
Mark Mewissen, Interventional Radiologist, SLH Milwaukee, WI;
Ron Waxman Cardiology Director, Cardiovascular Research and Advanced Education, MedStar Heart and Vascular Institute, MedStar Washington Hospital Center, Washington DC
Craig Ford, President, SterileBits, Inc., Marina Del Rey, CA

Editors

Dr. Yoseph Bar-Cohen is the Supervisor of the Electroactive Technologies Group (http://ndeaa.jpl.nasa.gov/) and a Senior Research Scientist at the Jet Propulsion Lab/Caltech, Pasadena, CA. In 1979, he received his PhD in Physics from the Hebrew University, Jerusalem, Israel. His research is focused on electro-mechanics including planetary sample handling mechanisms, novel actuators that are driven by such materials as piezoelectric and EAP (also known as artificial muscles), and biomimetics.

Dr. Kris Zacny is a Senior Scientist and Vice President of Exploration Systems at Honeybee Robotics, Altadena, CA. His expertise includes space mining, sample handling, soil and rock mechanics, extraterrestrial drilling, and In Situ Resource Utilization (ISRU). Dr. Zacny received his PhD from UC Berkeley (2005) in Geotechnical Engineering with focus on planetary drilling and space mining, ME from UC Berkeley (2001) in Petroleum Engineering, and BSc cum laude from University of Cape Town (1997) in Mechanical Engineering. He spent several years working in South African mines and tested space drills in Antarctica, Arctic, Greenland, and the Atacama. Dr. Kris Zacny is a Senior Scientist and Vice President of Exploration Systems at Honeybee Robotics, Altadena, CA. His expertise includes space mining, sample handling, soil and rock mechanics, extraterrestrial drilling, and In Situ Resource Utilization (ISRU). Dr. Zacny received his PhD from UC Berkeley (2005) in Geotechnical Engineering with focus on planetary drilling and space mining, ME from UC Berkeley (2001) in Petroleum Engineering, and BSc cum laude from University of Cape Town (1997) in Mechanical Engineering. He spent several years working in South African mines and tested space drills in Antarctica, Arctic, Greenland, and the Atacama.

Contributors

Mary R. Albert
Thayer School of Engineering
Dartmouth College
Hanover, New Hampshire

Saleh Alhaidari
Petroleum Engineering Department
Colorado School of Mines
Golden, Colorado

Ahmed Amer
Petroleum Engineering Department
Colorado School of Mines
Golden, Colorado

Mircea Badescu
Jet Propulsion Laboratory (JPL)
California Institute of Technology (Caltech)
Pasadena, California

Xiaoqi Bao
Jet Propulsion Laboratory (JPL)
California Institute of Technology (Caltech)
Pasadena, California

Yoseph Bar-Cohen
Jet Propulsion Laboratory (JPL)
California Institute of Technology (Caltech)
Pasadena, California

Grant V. Boeckmann
Space Science and Engineering Center
University of Wisconsin
Madison, Wisconsin

Nicole Bourdon
Petroleum Engineering Department
Colorado School of Mines
Golden, Colorado

Javaid Butt
Anglia Ruskin University
Chelmsford, UK

Alfred Eustes
Petroleum Engineering Department
Colorado School of Mines
Golden, Colorado

Chris J. Gibson
Space Science and Engineering Center
University of Wisconsin
Madison, Wisconsin

Jay A. Johnson
Space Science and Engineering Center
University of Wisconsin
Madison, Wisconsin

Deep Joshi
Petroleum Engineering Department
Colorado School of Mines
Golden, Colorado

Hyeong Jae Lee
Jet Propulsion Laboratory (JPL)
California Institute of Technology (Caltech)
Pasadena, California

Roy Long
DOE/FE National Energy Technology Laboratory
Houston, Texas

Peter Lucon
Montana Technological University
Butte, Montana

Ernie Majer
Lawrence Berkeley National Laboratory
Berkeley, California

Keith Makinson
British Antarctic Survey
Cambridge, UK

Kirtland McKenna
Petroleum Engineering Department
Colorado School of Mines
Golden, Colorado

Richard Meehan
Schlumberger Limited
Houston, Texas

José Ricardo Pelaquim Mendes
Faculty of Mechanical Engineering
University of Campinas
Campinas, São Paulo, Brazil

Kazuo Miura
Faculty of Mechanical Engineering
University of Campinas
Campinas, São Paulo, Brazil

Vahaj Mohaghegh
Anglia Ruskin University
Chelmsford, UK

Julius Rix
British Antarctic Survey
Cambridge, UK

Kelly Rose
DOE/FE National Energy Technology Laboratory
Houston, Texas

Shabnam Sadeghi-Esfahlani
Anglia Ruskin University
Cambridge, UK

Yoseph Shalev
Interventional Cardiologist and Medical Director
Placidus Medical
Fox Point, Wisconsin

Stewart Sherrit
Jet Propulsion Laboratory (JPL)
California Institute of Technology (Caltech)
Pasadena, California

Hassan Shirvani
Anglia Ruskin University
Chelmsford, UK

Kristina R. Slawny
Space Science and Engineering Center
University of Wisconsin
Madison, Wisconsin

Ozan Uzun
Petroleum Engineering Department
Colorado School of Mines
Golden, Colorado

Marcio Yamamoto
National Maritime Research Institute
Mitaka, Tokyo, Japan

Kris Zacny
Honeybee Robotics Spacecraft Mechanisms Corporation
Pasadena, California

Zachary Zody
Petroleum Engineering Department
Colorado School of Mines
Golden, Colorado

1 Introduction
Drilling as Means of Penetrating Solids

Yoseph Bar-Cohen and
Jet Propulsion Laboratory (JPL)/California Institute of Technology (Caltech), Pasadena, CA
Kris Zacny
Honeybee Robotics Spacecraft Mechanisms Corporation, Altadena, CA

CONTENTS

1.1 THE EVOLUTION OF DRILLING TECHNOLOGY

Drilling is a process of penetrating solid objects by creating holes using various tools or approaches. When applying rotation to create a hole, the bit has a circular cross section and the produced borehole has a round shape. Alternatively, percussive or hammering drills can create non-round boreholes. Mechanisms of penetrating objects and formations, i.e., drills, are widely used for many applications including making holes, sampling, exploration, and excavation (Bar-Cohen and Zacny, 2009; Clark, 1987; Hossain and Al-Majed, 2015; Rollins, 2010; Zacny et al., 2008). While the field is quite well established, there are still many challenges including operation at extreme conditions as well as drilling extremely hard materials. Drilling on other extraterrestrial bodies also poses significant challenges resulting from the limited resources that are available and the difficult conditions involved. The earliest successful planetary drilling was performed on the moon in 1970 by the Soviets' robotic Luna 16 lander (Bar Cohen and Zacny, 2009). In recent years, missions have been increasingly launched to drill on Mars and penetrate the surfaces of other extraterrestrial bodies.

These exploration missions are seeking to investigate the history of our solar system and better understand our planet Earth (Bar-Cohen and Zacny, 2009; Zacny et al., 2008, 2013a). To penetrate a large variety of materials on Earth, scientists and engineers have developed many types of drills, of which mechanical drills are the most common. These drills employ a bit with a tip that interacts with the drilled material and employs large shear and/or impact stresses over a small area to cut or break the formation. Such bits are widely used and are available commercially at local hardware stores.

For millions of years, animals and insects have been penetrating and removing cuttings from solid objects made of soil, rocks, and wood. Termites (Figure 1.1), earthworms, squirrels (Figure 1.2), rodents, woodpeckers, and many other living creatures create holes and tunnels for habitation and food extraction (Bar-Cohen, 2005, 2011). Roots of plants also have enormous capabilities to penetrate rocks and hard soil as well as lifting heavy curb structures and concrete slabs (Figure 1.3).

Digging, excavation, and ground penetration have been done by humans since ancient times for many purposes, including digging water wells, supporting columns and structures, burying objects, mining for resources, searching for food (such as plant bulbs and roots), among others. The capability to penetrate solid materials and formations has improved from simple digging by hand to the sophisticated methods used today. Technology has advanced so much that we can easily make holes as shallow as one micron or less to several kilometers deep. The advances in developing penetration tools followed the improvements in techniques of effective fabrication of strong materials and methods of processing and machining, as well as the increased capability to leverage forces using powerful mechanical, electrical, pneumatic, and hydraulic actuators (Dansgaard, 2004; Karanam and Misra, 1998).

The need for strong, hard, and sharp tools has been recognized by humans in the search for improving the ability to dig and drill. The advances in using metals have made major contributions to addressing the required capability and may have begun in the Bronze Age. As far back as 3150 BC, the ancient Egyptians have been attributed to using metallic tools for drilling. In addition, archeological findings indicate that around 2500 BC the Egyptians have used diamond drilling tools for the construction of the pyramids. Moreover, around 600 BC, the Chinese drilled holes of about 35 cm (14 in.) diameter to as deep as 600 m (~2,000 ft). Around the 11th century, Carthusian monks used a percussive technique to drill in search for water and were able to reach depths of about 300 m (1,000 ft) (De Villiers, 2001).

Advances in actuation mechanisms have made the most impact on the capability to drill. The invention of the steam engine in the 18th century greatly improved the capability to drill on large scales and with it came a surge in demand for coal to fuel steam engines. The pioneering of this technology can be attributed to the invention of the steam-driven rotary drill in 1813 by Richard Trevithick, England, (Encyclopedia Britannica, 1986). Over centuries, coal has been the most extracted material, being also the most abundant fossil fuel on Earth. Its predominant use has been for producing energy in the form of heat. In the 18th and 19th centuries, coal was the most important source for energy that fueled the Industrial Revolution and was a major driver in the development

FIGURE 1.1 Termites make holes in wood.

FIGURE 1.2 (a) A squirrel tunnel. (b) The squirrel next to the tunnel.

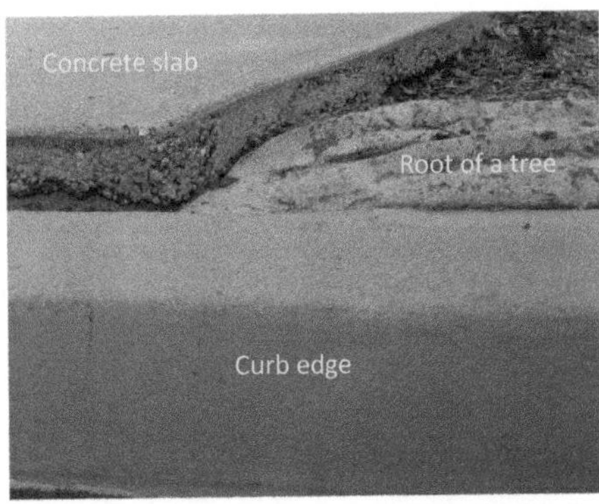

FIGURE 1.3 Roots of trees are capable of lifting heavy concrete slabs, curbs, and structures.

of tools for large-scale applications. Other notable steam driven drills were invented by Joseph W. Fowle (1849–51) and Cavé, Paris (1851). Even though these drills found many surface penetration applications, they were not suitable for underground drilling where most drilling for mineral excavation had been taking place (Rollins, 2010). In the early stages, long hoses were not available and therefore the steam machinery had to be kept close to the boiler. In 1951 Joseph W. Fowle patented the first use of a flexible steam hose for operating at great distances from the location of the boiler. Fowle's invention became the basis for the design and development of modern rock drills.

The capability to mine hard rocks and the significant reduction in the cost and time required for excavation were a direct result of using mechanical drills powered by compressed air, i.e. the pneumatic hammers that are also known today as "jackhammers." The first patented pneumatic drill was invented in 1849 by Jonathan Couch, Philadelphia, USA (Encyclopedia Britannica, 1986). This pneumatic drill is of a hammering type that delivers impact blows via a metal rod into rocks. The 1871 invention of Samuel Ingersoll pioneered pneumatic drilling while electric drills emerged in the 20th century and revolutionized the on-demand capability to penetrate tough materials. In 1852, the physicist Jean-Daniel Colladon used compressed air for drilling blast holes to create the Mont Cenis tunnel in the Western Alps. Using compressed air to drive the drilling tools was also explored by Germain Sommeiller and others between 1852 and 1860 (Peele, 1920). The use of compressed air offered the following three major advantages:

1. Low losses in the transmission line.
2. Leakage is safer than in steam systems, which pose a burn hazard to the operators.
3. The released compressed gas provides additional ventilation in the confined underground work areas.

In 1890, the first hammering drill using an air-leg feed (Bar-Cohen and Zacny, 2009) was developed by C. H. Shaw (Encyclopedia Britannica, 1986) and it represented a major contribution to pneumatic drilling technology. The air-leg feeder consisted of an air-ram assembly that was attached at one end to the rear of the drill system and the other to a support/anchoring structure. When charged with compressed air, the air-leg, reacting against the support structure, pushes the drill deeper into the borehole. Following this development, in 1896, J. George Leyner invented the hollow drill bit, which signified another significant improvement in drilling technology. Leyner's contribution involved blowing air through the center of the drill and its hollow bit. This way, air removed the cuttings from the drilled hole as well as cooled the drill bit and the drilled formation, thus improving the drilling efficiency.

Advances of the Industrial Revolution brought about the use of compressed air as a working fluid. However, compressed air was not the only working fluid used to improve drilling processes. Water was the first drill-fluid and its use is documented among ancient Egyptian and Chinese cultures (Barrett, 2011). Even into the 1800s single-jack and double-jack hard rock drilling used water as working fluid. Double-jack drilling involved one or two drillers hitting a steel rod with heavy sledgehammers while a holder turned the rod after each blow. The holder also had the responsibility of pouring water into the borehole periodically. Leftover percussion energy shot water and cuttings from the borehole at each blow of the hammer.

Rotary drills in oilfields were first introduced in 1884 (Lee et al., 1988; Maurer, 1968) and initially utilized water as a working fluid. In the early 1900s as oil exploration expanded, drillers began noticing that their drilling efficiency dramatically increased when drill-water mixed with the natural clays of some of the formations that they were drilling through. Drilling into and through clay-rich rock formations, water/clay mixtures were inadvertently transformed into viscous working fluids. This quickly led to the development of engineered drill-mud mix designs (Barrett, 2011). An example of a drilling rig that uses an air/mud rotary combination is shown in Figure 1.4.

Handheld jackhammers, developed in the 1940s, operate on the same principle as double-jack (percussive) drilling. However, instead of operating via human power, a flexible attachment to an

FIGURE 1.4 A drilling rig system (STAR 50K, Enid, OK) using air/mud rotary combination. *Source:* Photographed at Jet Propulsion Lab (JPL) by Yoseph Bar-Cohen, the lead author of this chapter.

air cylinder and a valve body operates a piston, which serves as the hammer acting on the drill rod. Where two sledgehammer wielding drillers could deliver percussive blows to a drill rod at a rate of around one hammer-blow per second (1 Hz), a pneumatically actuated jackhammer will operate at much higher percussion rates, substantially increasing the drilling speed and enabling a single driller to drill in any direction without mounting the drill on a support structure.

Hydraulic mechanisms, rather than compressed air, as direct-drill drives were introduced in the 1970s. This technology has evolved to the level that allows for significantly quieter operation with noise levels at 10-m distance that are less than 85 dB (made by Atlas Copco).

The first electric rotary drill was patented in 1889 by James Arnot, who designed it primarily to penetrate rocks and coal. About six years afterwards, the invention of the portable drill was patented by Wilhelm Fein. A major milestone took place in 1917 by Black & Decker with the invention of the trigger switch that is mounted on the drill pistol-grip handle (Black and Decker, 1917). Using batteries to drive drills significantly contributed to the portability of drills and, in 1961, the first cordless electric drill was introduced by Black & Decker. For planetary exploration cordless technology has been a critical technology—it enabled astronauts on the Apollo 15, 16, and 17 missions to drill on the moon and to reach as deep as 3 m (Heiken and Jones, 2007) in lunar regolith. The A. L. Hawkesworth's invention of the replaceable bit in 1918 greatly reduced the cost of drills and improved the efficiency of drilling since it was no longer necessary to replace an entire drill when the bit was worn out.

The significant demand for mining resources first led to exploitation of the easy-to-mine areas and then to the significant growth in drilling technologies for extremely challenging conditions with much greater requirements for speed, safety, durability, and cost. Advances in metallurgy, including the 1890s emergence of heat-treated steel alloys led to significantly more deformation-resistant drill bits as well as improved hammer drills that are faster, lighter, and much more effective. In addition, bits made of harder materials, such as tungsten carbide, were produced. The use of diamonds on drill bits was first introduced in 1869, providing greater performance and durability in cutting hard materials. To address health hazards from fine dust in mines and to enhance drill performance,

J. George Leyner invented the fluid-tight swiveling coupler that introduced drilling fluids into difficult places to drill (Leyner, 1914).

The demand for oil led to the development of the directional drilling that requires drilling off-vertical or horizontally. Such drilling was performed as early as the 1920s where rigs on one property were used to tap into reservoirs in neighboring properties. A key issue related to this technology was the ability to track the movement of the drill bit with respect to the subsurface geological formation and overall structure.

As improvements have been introduced, greater drilling challenges have been addressed including drilling at extreme conditions and reaching great depths with performance that is superior to the existing technology. The former Soviet Union took on the goal to drill to the greatest reachable depth on Earth; to as deep as 15,000 m (49,210 ft). This purely scientific goal (learning about the properties of subsurface rocks) was first proposed in 1962 and after about 8 years of preparations, the actual drilling began at Kola Peninsula. In 1989, the deepest drilling depth of 12,261 m had been reached beyond which it was not feasible to continue due to the rock plasticity. The plasticity was the result of the high temperatures; at that depth, the rock temperature was about 180°C (356°F). The acquired cores are continuing to be analyzed to this day and they have led to many discoveries and scientific publications (e.g., MacDonald, 1988). The discoveries include the fact that below the 3–6 km granite layer there is a metamorphic layer. In addition, microscopic fossils were found at as deep as 6.7 km below the ground surface and analysis of recovered rock samples has shown them to be 2.7 billion years old (Kola Superdeep Borehole, 2018). In 2008, the dismantling of this drill system has begun but it is still one of the greatest drilling achievements ever.

In 1961, also seeking to reach an extensive depth, the US attempted a parallel initiative to the Russian Kola project. The US project, which was called Mohole, was led by the American Miscellaneous Society and funded by the National Science Foundation (NSF) (Burleson, 1998). Mohole's goal was to drill through the ocean floor and reach the boundary between the Earth's crust and the mantle. The project took place off the coast of Guadalupe, Mexico. Drilling was done through 3,500 m (11,700 ft) of deep water and through the seafloor to a depth of 183 m (601 ft). However, in 1967, the project was canceled by the US Congress.

These deep-drilling initiatives illustrate how extreme drilling conditions can drive drilling technology. Where human-powered drills gave way to steam-powered drills, which gave way to pneumatic drills, and so on, each new challenge requires better methods or new applications of the old methods. In no place are drilling challenges more substantial than on extraterrestrial planetary bodies. Planetary missions have great limitations on the available mass and power, and they require autonomous operation due to the communication delays (approaching 20 min each way with Mars) making remote control operation impractical (Zacny et al., 2008; Zacny and Cooper, 2006). For Mars missions, once a command is sent out, it takes more than 40 min to learn whether the drill indeed started to drill or not. Communications relaying problems with drilling would be not be received by operators on Earth until about 20 min after they had initially been sent. Corrective commands sent back by the operators could not be received by the drill system for yet another 20 min.

Of course, this all assumes that there is a direct communications link between the landed systems on Mars and the operators on Earth, which is not the case. Communications links to landed systems are more tortuous than that. On Mars for instance, landed systems (rovers, landers, instruments, drill systems, etc.) rely on periodic passes from spacecraft orbiting Mars to receive commands from Earth (uplink), as well as to send their data back to Earth (downlink). For missions during which drilling is performed, sequences that the drill is directed to run are first uplinked to a Mars orbiter. These drill sequences cannot be run until that orbiter passes over the drilling system to upload the sequences. The drill system will run these sequences, and another orbital pass will be needed to downlink the drill data. Due to the need of at least two orbital passes, plus the time it takes to run the drill sequence, plus the delay in communications, any problems encountered during the drilling operation will not be received on Earth for 10s of hours to days.

Robotic drill systems that are sent to just about anywhere in the solar system are going to need autonomous control. The exception to this rule may be Earth's moon. With its close proximity to Earth, the problems associated with great distances are mitigated. Earth's moon may be a good place to develop drill system autonomy regardless. The moon is only three days' journey away, it offers an extreme low-pressure environment, both high and low temperature extremes, and it's close enough that humans may remain in the control loop to assist and correct autonomous operations.

Communications and automation are not the only challenges associated with extraterrestrial drilling. Limitations on mass and power, plus the factors associated with operating in extreme environments make planetary drilling tasks far more complex and demanding than similar operations on Earth. Extraterrestrial subsurface formation may be unknown and possibly consist of significant geological uncertainties and challenges. The local gravity, which provides the necessary preload from the support platform (rover or lander), may be very limited. In addition, the environment may be very cold or hot and the pressure may be high, low, or vacuum. Examples include Venus with surface temperature as high as 465°C and pressure over 92 times higher than Earth atmospheric pressure (Zacny et al., 2015). The other neighboring body, Mars, has gravity of about one-third of Earth, whereas the gravity on asteroids and comets is many hundreds or thousands times smaller.

On many planetary bodies, oxidation and irradiation by cosmic rays may have affected the content of the subsurface making it essential to drill below several meters to reach locations where the effects have been minimal. The Icebreaker drill (see Figure 1.5) has been specifically designed and tested under Mars conditions to demonstrate that drilling a few meters below the surface at Mars would be feasible. These effects suggest the possibility that finding evidence of putative extraterrestrial life may require reaching depths of meters and beyond. Either the acquired samples can be analyzed on the surface using a suite of in situ instruments or, ideally, returned to Earth for a thorough analysis via equipment that is more sophisticated (McKay, 1997).

FIGURE 1.5 Icebreaker drill undergoing testing in a Mars chamber. *Source:* Courtesy Honeybee Robotics.

1.2 DRILLING METHODS

There are four basic methods of fracturing/penetrating natural material (rocks, soils, etc.): mechanical, thermal, jetting, and chemical. The use of explosives is also widely applied on large-scale excavations (Ostrovskii, 1960, Rollins, 2010). The focus of this book is on the methods of drilling that use mechanical forces to penetrate objects. However, for the sake of completeness this section provides an overview of both mechanical and alternative drilling techniques.

1.2.1 Mechanical Techniques

Mechanical techniques apply stresses that exceed the compressive, tensile, or shear strength of the penetrated solid, causing brittle failure or plastic yielding. These techniques are the most widely used drilling mechanisms and they consist of rotation, cutting, or shearing the medium, and/or applying repeated impacts to produce finely crushed rock under the bit surface. High-impact stresses cause the formation of cracks and fragments (Maurer, 1968) and the cuttings are removed along the flutes of a rotating bit.

We tend to think of rocks as being very hard. However, under mechanical loading, most rocks behave as elasto-plastic (or semi-brittle) materials. Many rocks have a tendency to yield under stress without breaking. The deformation under stress prior to failure is referred to as the elastic strain. However, when the stress level exceeds the rock's yield stress, permanent deformation takes place. The stress (pressure) at which a rock becomes permanently deformed (i.e. it has yielded) is referred to as its yield strength. Yielding is not the same thing as fracturing. Brittle to semi-brittle rocks will fracture as more stress is applied. However, rocks that are more ductile tend to deform as more stress is applied. The mechanical method best employed to drill into a solid material is directly tied to the material's toughness. Toughness relates to where materials reside on the strength vs. hardness matrix. Rock toughness is a better determining factor for what method should be employed when drilling. Percussive techniques with relatively dull, chisel-like bits are more efficient for drilling semi-brittle rocks, whereas rotary-only drilling with a sharp bit is more efficiently used to drill into ductile materials. Materials that are very strong and very hard may require grinding methods for the creation of a borehole.

Most, but not all rocks reside in the semi-brittle regime. Cyclic loading of semi-brittle rocks can cause fatigue failure with formation of miniature cracks that grow with time until they reach a critical length that causes breakage. For breaking rocks, they can be subjected to tensile, shear and pure compressive stresses from a bit and all the three stress types can be combined too. The relative level of stresses of loading a rock is affected not only by the loading characteristics but also by the drill bit design. The latter affects the geometry of the produced hole and the dynamic stress wave reflections from the drilled medium. While pure compression causes crushing and breakage of grains, tension and shear cause breakage of the inter-granular bonds.

The challenges to drilling operation significantly increase when penetrating deep into the subsurface. An example includes drilling soil (on Earth) or regolith (on extraterrestrial bodies) where encasing the borehole wall is required to avoid borehole collapse. Another issue that needs attention is the increase in the borehole wall pressure as the depth increases. If the drilling medium consists of ice, refreezing of melted ice can jam the bit.

1.2.2 Thermal Techniques

Rocks can be penetrated by the use of heat causing local fractures or melting and vaporization. The related principal methods are thermal-spalling and thermal-melting, respectively. While the thermal-spalling occurs at about 400°C–600°C, the thermal-melting is performed in the range of 1,100°C–2,200°C (Maurer, 1980).

- **Thermal-spalling**: Thermal-spalling is caused by stresses generated from mismatch in thermal expansion of constituents and grains within the structure of rocks (Just, 1963). Such breakage is a natural process that is also known as "onion weathering" or "exfoliation." The thermal stresses cause fracture of rocks in the form of fragment flaking and the severity is a function of the thermal gradients that are introduced into the rock. This method has limited applicability and requires significant heterogeneity to cause spalling. The process is similar to exfoliation decomposition of rocks, which has been documented in many desert environments (Blackwelder, 1925). During thermal exfoliation, the surfaces of rocks are heated and expanded during the day time, while at night the surface is cooled and contracts. This cyclic process leads to small cracks that grow until entire layers are peeled off. The process can be accelerated when water within the cracks turns into ice when the temperature falls below the freezing—as ice expands it causes cracking and fractures. Heating, particularly non-uniform heating, causes degradation and loss of strength in most rocks and can cause them to crumble.

- **Melting and vaporization**: Heating of solids (including metals, rocks, etc.) at very high temperatures causes melting and vaporization (Ready, 1997). This method requires very high power and it is not practical for sampling and analysis of rocks due to the damage caused to the material structure. For ice, the use of melting is quite simple and can be used to reach great depths (Zacny et al., 2016a). This method is frequently used in Antarctic drilling and allows rates as fast as 100 m/h. Drilling using an ice-melting probe for planetary exploration is currently being considered for reaching kilometer-plus depths on Europa, a moon of Jupiter (Rapp, 2007; Smith et al., 2006; Zacny et al., 2016b, 2018a).

Heat can be introduced by various methods including the use of electric resistance (Maurer, 1968), laser, plasma, electric discharge, and microwave, just to name a few. The use of laser as a heating source provides significant precision in making holes that can be microns in size and have any desired shape. The use of microwaves as a form of heating has been investigated, as an example, by scientists at Tel Aviv University, Israel, performed drilling by applying microwaves directly to the test medium via dielectric losses (Jerby et al., 2002). For this purpose, a coaxial near-field radiator was fed by a conventional microwave source and acted as a bit. The drill bit operated as an antenna that focused microwave energy onto the drilled medium and was capable of melting any nonconductive materials along its path. Temperatures as high as 1500°C have been reached and, using a 2-mm diameter bit, holes in concrete as deep as 2 cm have been produced in about a minute. This method has the advantage that it does not require a rotating bit, does not produce dust, and is very quiet. Disadvantages for in situ planetary exploration include the need for very high power and the process causes thermal damage to the drilled material making the extracted material unusable for scientific analysis.

Plasma cutting is a heating process that involves using an accelerated jet of hot plasma for cutting electrically conductive materials. These materials include metals such as aluminum, brass, copper, and stainless steel. Plasma cutting consists of creating an electrical channel of superheated and electrically ionized gas. It provides high precision and speedy cutting at relatively low cost and, therefore, it is finding applications in computer-controlled machines at a wide range of scales.

Another method of cutting that involves heating is Electrical Discharge Machining (EDM) (Figure 1.6). It involves using a wire electrode to form various desired shapes. The process consists of bringing an electrode close to the workpiece that is being cut where a dielectric liquid fills the gap between the two electrodes (the workpiece serves as the second electrode). The workpiece is a metallic material and it is cut by electric discharge that leads to erosion. The electric field between the two electrodes is dictated by the applied voltage, and the higher the voltage the greater the field and its related drilling action. The liquid enables the removal of the formed debris and it is needed to assure insulation between the two electrodes.

1.2.3 Water Jetting

A high pressure water jet uses abrasives for erosion drilling and machining. The method is very efficient since the abrasive material can be filtered out and reused. Water jetting has been used as far

FIGURE 1.6 An electrical discharge machine. *Source*: This photo was downloaded from Wikipedia and it has been released into the public domain by its author https://commons.wikimedia.org/wiki/File:Electrical-discharge-machine.jpg

FIGURE 1.7 An abrasive water-jet cutter being used to cut a metal tool. *Source:* Courtesy the US Air Force. https://commons.wikimedia.org/wiki/File:Water_jet_cutter_tool.jpg

back as the mid-1800s, while in the 1930s the use of narrow jets appeared in industrial applications. The use of an abrasive water-jet cutter is shown in Figure 1.7. Today, ultrahigh pressures at the levels of 30–90 ksi (210–620 MPa) are used with abrasive particles. A significant advantage of the water jetting is that it does not cause heat damage to the processed parts.

1.2.4 Chemical Techniques

Penetration of rocks can be performed chemically by dissolution using such strong solvents as fluorine and halogens (Maurer, 1968). This method can be quite effective in penetrating specific types of rocks (McGee, 1955) but it involves a violent reaction and may even cause fire as well as pose health hazards to its users. Another chemical method that can be used in drilling ice could include salt, such as NaCl. Since salt depresses freezing point, adding salt would assist in locally melting ice at temperatures that are several degrees below freezing.

1.3 TYPES OF MECHANICAL DRILLS

Mechanical drills are the most widely used penetration mechanisms and are manually, electrically, hydraulically, or pneumatically powered (Astakhov, 2014). They range in size from as compact as a battery-driven hand-held drill to as large as oil rigs that reach kilometers in depth. The drill types include:

- **Rotary drills**: These are the most widely used drills and include portable and fixture mountable versions. The fixture mountable drills are mounted on a stand or bolted to the floor or workbench and are often known as a drill press, pedestal drill, pillar drill, or bench drill. These drills are advantageous over hand-held drills since using a lever allows moving the chuck and spindle in a controlled manner. They also require less effort to apply preload and they offer greater position precision. The drill is typically driven by an electric motor and consists of a base, column (or pillar), table, spindle (or quill), and drill head.
- **Hammer drills**: Hammer drills are widely used to penetrate semi-brittle materials, such as concrete, and employ impacts at various frequencies. Large hammer drills (particularly pneumatic ones) act crudely with highly varying energy delivered in each stroke. The impacts may cause collateral damage to the drilled medium and its surroundings. The Special Direct System (SDS) drills (developed in the 1970s by Bosch) introduce high power hammering action that can gently pulverize the drilled material with less damage to the borehole.
- **Rotary-hammer drills**: These drills combine rotation and hammering and are highly effective in drilling concrete, masonry, and other hard and brittle materials. These type of drills are also known as roto-hammer or masonry drills; they introduce impact forces to cause breakage and bit rotation to remove the cuttings. The combined action significantly enhances the drill operation over the use of either rotation or hammering only.

1.3.1 Drill Bits—The End-Effector of Mechanical Drills

Mechanical drills are effectively a drill bit with an actuator that either rotates or hammers a bit or both. In selecting a bit, the characteristics of the required borehole or sample need to be taken into account. If a sample is not required or cuttings are deemed sufficient for analysis, a straight borehole can be drilled. If geologists require a core, a coring bit is needed. Generally, there are many types, shapes, dimensions, and material composition of bits as well as tip coatings.

Twist drill bits (invented in 1861 by Steven A. Morse, East Bridgewater, Massachusetts) are the most widely used type of bits. They are applied to make a hole in effectively every solid material including metal, concrete, rocks, plastic, and wood. Originally, this type of bit has been produced by cutting two grooves on opposite sides of a metal rod and twisting the rod to form helical flutes. Nowadays, twist drill bits are produced by rotating a rod while cutting flutes using a grinding wheel, drawing a hot rod through a die, or using 3D printing techniques. The geometry and shape of the bit cutting edges dictate its performance; as the sharpness degrades with use, the bit loses efficiency. Besides the cutting edges of the bit, the geometry of the spiral twist and their tip angle are important factors in the performance of drilling operation. The spiral twist is optimized between the speeds of cuttings removal and drilling rate as well as the material that is being drilled. The cutting angle controls the bit walk and chatter (uncontrolled off axis movement of the bit) as well as the wear rate of the bit. Under a given axial load, a higher angle leads to more aggressive cutting, but it can cause extensive wear, binding, and even drill bit failure. A cutter rake angle is designed based on what material type (hard or soft) the drill bit will be penetrating. The most commonly used cutting angle is 118° (from the vertical), about 90° for very soft plastics, while a shallow angle of about 160° is used to drill tough metals (e.g., steel alloys).

Gun drill bits are used for drilling long straight holes in metals, wood, and some plastics. This type of bit is so named because they are also used to produce gun barrels and firearms by making very straight and accurately sized holes. In order to cool the bit while drilling, a hole is provided through the bit core allowing for flow of compressed air or coolant liquid. Besides lubricating and cooling the cutting edges of the bit, the coolant also removes the cuttings. For long life and ability to drill hard materials, bits are generally made with a carbide tip.

For starting holes and preventing drill walk or wander, centering and spotting bits are used to initiate the borehole in the form of a conical indentation into the surface. Spotting drill bits are designed with an angle that is the same or greater than the bit to be used. Thus, stresses on the bit corners are minimized, reducing potential premature failure of the bit or damage to the borehole quality. Alternatively, centering punch tools are used to produce a pilot hole. In addition, there are carbide drills that are specifically designed to start their own hole.

1.3.2 Material Makeup of Bits

The material makeup of drill bits is critical to its performance, durability, and cost, as well as its applicability. Bit materials include soft low carbon steel for drilling wood while hardened and tempered high carbon steel are used for drilling wood and metals. For high speed and hard materials drilling, tool steel, cobalt steel, and tungsten carbide are used. Due to the high cost of tungsten carbide, it is used in small pieces that are screwed or brazed onto the tip of the bit. Excessive heating causes bit softening and loss of tempering and, therefore, subsequent dulling of the bit cutting edge.

Bits with diamond crystals embedded in the tip are used to provide strong abrasion resistance for hard materials (Figure 1.8). Such bits are more expensive but often last longer. Generally, diamond crystals have a large strength anisotropy and can be very brittle when impacted at some angles (Denning, 1953; Zacny 2011). To maximize the benefit of these crystals and to address the significant anisotropy, they are used by fusing many of them together to create a polycrystalline diamond. Such polycrystalline diamonds (also known as PCD) exhibit weaker average hardness, but they are much stronger than the weakest diamond. Polycrystalline diamonds that are used for bits are made synthetically and are much harder and tougher, i.e. have greater resistance to brittle fracture (Zacny and Cooper, 2001). Generally, polycrystalline diamonds on the cutting surface are bonded to a tungsten carbide substrate layer, where the diamond abrades the rock by shearing, whereas the tungsten carbide layer provides mechanical support and impact resistance. Such bits are sensitive to heat, making them unsuitable for drilling ferrous materials including steel. This limitation is due to the fact that, in the presence of such metals such as iron or cobalt, diamonds heated to as low as 500°C revert to graphite and oxidize (Evans and Phaal, 1962).

Alternatives that are less effective, but less expensive, than diamonds include such crystals as Cubic Boron Nitride (CBN). In addition, various bit coatings are used for drilling hard materials such as Black Oxide (FeO), Titanium Nitride, Titanium Aluminum Nitride, Titanium Carbon Nitride (TiCN), and Zirconium Nitride.

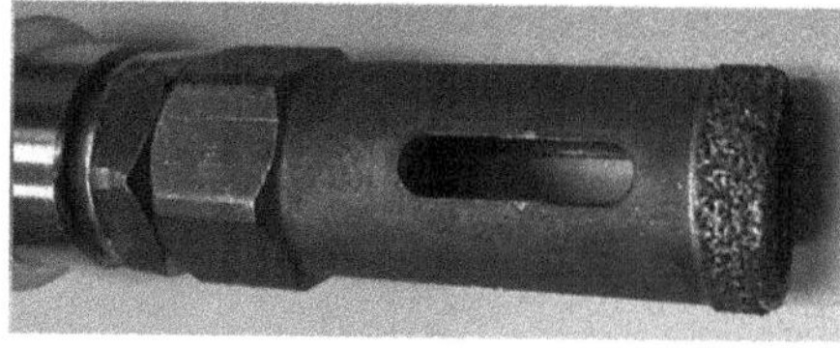

FIGURE 1.8 A coring bit with diamond-coated tip section.

1.3.3 Additive and Subtractive Manufacturing of Drill Bits

3D printing has revolutionized production processes allowing fabrication of parts with minimal loss of material (Bar-Cohen, 2018). Besides the ability to produce complex shapes that are dictated by CAD programs, parts can be made with hybrid compositions and predictable performance. The example of an auger bit with fluted shape shown in Figure 1.9 (produced at the author's lab using a 3D printer) demonstrates the possibilities of making bits with complex structures. This helicoid with ridges along the edge has been incorporated into the auger as a means of removing a greater amount of cuttings during the drilling process. Machining such a configuration is extremely difficult, while for 3D printing it is as easy as making a simple structure.

Compared to machining, the ability to produce complexly shaped metallic parts using 3D printing is a very important benefit to drilling technology. An example of making a complex shape stainless steel, 316L, end-effector of a drill has been prototyped for use in a future potential NASA mission to Europa (Figure 1.10). Machining such drill bit parts is very difficult while 3D printing is significantly easier.

3D printed drill bits have been successfully used in the permafrost of Antarctica, drilling at rates of 1 m/h with only 100 W of power (Zacny et al., 2013a).

1.4 DRILLING APPLICATIONS

Drilling is widely used in resource exploration, including petroleum, gold, coal for geological studies, construction, and many other applications. Extracted cores (Figure 1.11) can be used to assess the grade of minerals within a rock matrix, as well as the economic feasibility of the related mining. For geological studies, careful extraction of samples from a core is used to assess the strength of the subsurface formation in order to determine if the ground is strong enough to support a planned building or road. In addition, cores are used to learn the geological history of planet Earth. For example, ice cores acquired from deep in the Antarctic ice sheets were used to determine that they are in the age range of 800,000 years. These ice cores were also used to identify the chronology of

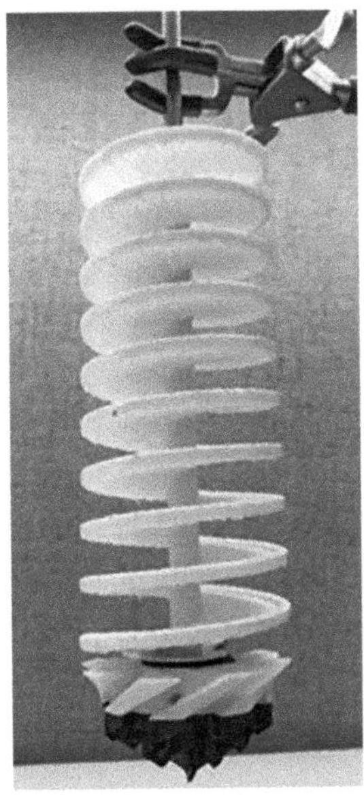

FIGURE 1.9 A fluted bit auger produced by 3D printing. *Source:* Courtesy NASA JPL.

FIGURE 1.10 End-effector of a drill bit for application for potential drilling on Europa, which is a moon of Jupiter. *Source:* Courtesy NASA JPL.

FIGURE 1.11 Auto-Gopher 1 acquired >3 m of rock cores during robotic drilling. *Source:* Courtesy of Zacny et al. (2013b).

previous ice ages and interglacial periods, map historical variations in Earth's climate, and determine the atmospheres during those times. Other applications are discussed in the following sections.

1.4.1 Small-Scale Drilling

The most widely applied drilling mechanisms are small scale, such as dental drilling (Christensen, 2006; Simonsen, 1989). The dental drill is a hand-held mechanical instrument that is used to perform

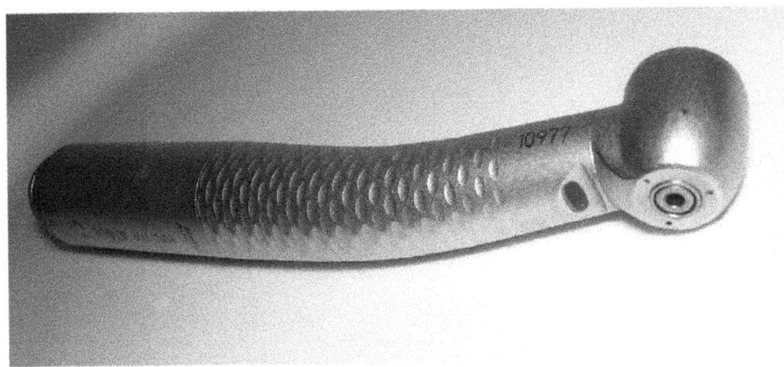

FIGURE 1.12 A dental handpiece. The hole on the right is the place where the dental burrs are inserted. *Source*: This photo is from the Wikimedia Commons, which is from the Wikipedia freely licensed media file repository. https://en.wikipedia.org/wiki/File:Dentalhandpiece0111-26-05.jpg

a wide variety of dental procedures, including removing of decay prior to filling, polishing fillings, as well as modifying prostheses. The handpiece contains the internal mechanical components that causes the rotation of the inserted bit, known as dental burr (Figure 1.12). Some handpieces are equipped with a light source as well as a water-spray cooling system allowing for increased visibility as well as accuracy of the procedure and patient comfort. The drill bit of current drills operates at speeds of hundreds of thousands of revolutions per minute.

Physically, as the drilling scale becomes very small as in the case of microelectronic boards, it is not feasible to drill such boards via mechanical tools. Therefore, laser as a thermal drilling technique is widely used where controlled and precise heating is applied. This non-contact process addresses the issues that are involved with need to avoid damage to the electronic parts. The process is done automatically and it maintains the original quality of the electronic components.

1.4.2 Large-Scale Excavation

Large-scale excavation requires powerful drilling tools and examples include mining and tunneling. Such drills are increasingly being produced as automatic systems. This way, the labor cost is reduced and the health hazard is removed. In these fully autonomous rigs, such as the Atlas Copco family of SmartRigs™, the drilling process starts with programming the location, depth, and inclination of the required holes. Then, the data is uploaded into the drill-rig computer and, from then on, the operator observes the drill process safely and comfortably from the cabin. The drill rig is equipped with the necessary sensors for positioning the drill at the required location and allowing maintenance of the programmed penetration rate. The drill-rig control system makes Measurement While Drilling (MWD) and it records in real time the drilling telemetry including the weight on bit, rotational speed, and penetration rate.

Mining is another form of large-scale excavation and it is the extraction of natural resources. The early humans mined resources by picking up loose rocks and using them as tools to enhance their manual excavating of soft ores. The use of metallic tools significantly enhanced humans' capability and it started with the use of iron in the Iron Age (during the second millennium BC). The development of iron-made tools enormously improved people's lives, allowing production of tools that were used to break very hard ores. To enhance rocks' breaking capability, fire has been used to introduce fractures through thermal-spallation. In the 1880s, the use of fire in Japan allowed the production of long tunnels. The most significant impact on mining technology has been made in the late 1500s with the introduction of explosions using gunpowder. Initially, existing cracks in rocks were used for inserting the gunpowder (Ostrovskii, 1960). Later, it was realized that one could use iron tools to create deep holes for more effective explosions and excavations. The first recorded use of drilling

FIGURE 1.13 RedWater is using coiled-tubing drilling (CTD) for penetrating through Martian regolith to reach subsurface ice and produce water. *Source:* Courtesy of Zacny et al. (2018b).

and blasting for mining was by Martin Weigel in 1613 at Freiberg, Germany. Improvements in metallurgy increased the strength and durability of the mining and excavation tools and they led to enormous improvement of the tunneling capabilities.

Other natural resources that are extracted by drilling are oil and gas, and this process began in the 19th century (Hunt et al., 2002). For at least several decades, oil is expected to continue to be one of the most important energy sources but it is also used for making petroleum products. Drilling for oil is done anywhere oil exists, including schoolyards in the Los Angeles area. Drilling oil and gas wells used to be done mostly via a rotating string of rigid steel pipe sections and an end-effector at the tip. Sections are added to the string as the penetration progresses. In recent years, flexible tubing is used in coiled-tubing drilling (CTD) that allows penetration as deep as 3 km (10,000 ft) or more where a hydraulic motor pushes the tubing while fluid is pressurized through the tubing to rotate the drill bit. The drill bit and the hydraulic motor are lowered into the borehole while unspooling the coiled tubing. The drill bit is raised and lowered inside the borehole at rates up to ten times faster than what is possible with conventional rotary drills. This particular approach is currently being developed for water extraction on Mars (Figure 1.13).

Another large-scale drilling application is the formation of tunnels (Hemphill, 2012; Kolymbas, 2005). Generally, as commonly done by the oil and gas industry, the process involves digging and insertion of a shield as a protection of the wall from collapse. The first tunneling shield was conceived in 1825 by Sir Marc Isambard Brunel for the excavation of an underwater Thames Tunnel in London, England. The first boring machine, called *Mountain Slicer*, was built by Henri-Joseph in 1845 to dig the Fréjus Rail Tunnel between France and Italy through the Alps. Today, tunnel boring machines (TBM) with diameters as large as about 18 m are being used.

1.4.3 Off-shore Drilling

Advances in drilling technology enabled performing off-shore drilling and have addressed significant challenges. Today, drilling can be performed from ships that dynamically hold their position above the seafloor. Further, drilling to depths of over 2,000 m is possible. Drilling projects with the objective of improving the understanding of the origins of earthquakes and of Earth history occur extensively, thanks to funds from the US National Science Foundation (NSF). The tasks that are performed include coring of unstable thick sediment sections and investigating the fundamentals of ocean crust formation. Demand for energy is leading oil and gas exploration companies to implement numerous platforms at various sea locations of the world, including the Gulf of Mexico and the North Slope of Alaska.

1.4.4 Ice Drilling

Drilling ice requires special care because of the fact that the ice hardness increases quite significantly as temperature drops (Bar-Cohen, 2016; Durham et al., 1992; Mellor, 1971; Atkinson et al., 2018). In addition, it is essential to address the fact that melted ice can refreeze and may result in jamming of the bit. Such jamming occurred during the drilling test of the JPL's Ultrasonic/Sonic Ice Gopher at Lake Vida where, after reaching a depth of 176 cm using the percussive actuated drill, the drill froze and required alternative removal method (Bar-Cohen et al., 2007, 2018). A photo of the Ultrasonic/Sonic Ice Gopher inside the drilled hole is shown in Figure 1.14.

1.4.5 Planetary Drilling and Sampling

NASA exploration missions are increasingly involving in situ drilling and sample acquisition. The challenges are significantly more complex than drilling on Earth (Bar-Cohen and Zacny, 2009; Briggs and Gross, 2002; Zacny et al., 2008). Increased complexity is the result of limited resources, unknown properties of the drilled medium, extreme temperatures and pressures, and low gravity provides low axial load (Bar-Cohen, 2014, 2016). Since extraterrestrial drilling offers the potential of major scientific discoveries, it provides significant incentives for the planetary science and engineering community to tackle the related challenges.

Drilling on other bodies in the solar system dates back to the 1970s. The Apollo 15 astronauts drilled as deep as 3 m on the moon and recovered core samples. The Apollo Lunar Surface Drill (ALSD), which was a 450-W rotary-percussive drill, was used with a coring bit that acquired a continuous ~2-cm diameter core to as deep as about 3 m and had an auger for carrying cuttings to the surface.

The search for evidence that life may have developed elsewhere in our solar system is one of the major goals of NASA's planetary exploration programs. Samples acquired from the subsurface and analyzed at the site or returned for analysis back to Earth provide opportunity to address the question if life has ever formed elsewhere in the universe. Mars is one of the major exploration targets but its surface today is a very hostile environment in terms of radiation and oxidation (Cockell and Barlow, 2002). Therefore, efforts are made to drill the subsurface for acquiring pristine samples.

FIGURE 1.14 Drilling ice at Lake Vida, Antarctica, using the Ultrasonic/Sonic Ice Gopher.

In addition to the search for life, the science goals of NASA's exploration mission include studies of the geological record of planets. For this purpose, the stratigraphic record from core samples provides information about historical events that are buried beneath more recently emplaced material. Data from drill telemetry can be helpful for material characterization and to identify changes in stratigraphy in order to identify regions that have scientific value for further investigation. In addition, efforts are underway to aid the potential of future human exploration of the solar system. For this purpose, resources are sought that can be used to "live off the land" by means of In-Situ Resource Utilization (ISRU) (Badescu and Zacny, 2018). Moreover, knowledge of the mechanical properties of the soil would be necessary for construction of habitats and assessment of the mobility in a human outpost (Atkinson and Zacny, 2018).

1.5 SUMMARY/CONCLUSIONS

Penetrating solids using drilling techniques provides enormous benefits in a broad range of applications allowing understanding, controlling, and exploring our world and the solar system. Advances in materials and other technologies contributed significantly to the ability to drill to great depths, using a wide range of drill sizes and the ability to drill very soft to very hard objects. We are now able to reach thousands of meters deep in the subsurface of Earth as well as to great depths under the seafloor. These technological advances allowed the building of the enormous infrastructure of our cities including freeways, tunnels, sewage and water systems, and the many utilities that we have. Exploring the subsurface of other planetary bodies in the universe began with the moon, where science fiction has been turned into engineering reality. Seeking to make drills with high efficiency led to modeling the physics of drilling as well as developing analytical approaches and innovation in making end-effector systems (Karanam and Misra, 1998; Wyllie, 1999). The specific requirements may depend on the user objectives. For example, the petroleum industry seeks to reduce the human labor by automating the drilling process in order to increase the profitability while extraterrestrial drilling involves long communication delays and mass and power limitations. The results are leading to longer-lasting drills, improved penetration rates and much higher levels of autonomy.

There are many drilling mechanisms; the key criteria for selecting these depend on specific applications or focus such as safety, power, and cost. There have been many advances over the last decade and one of them includes the introduction of 3D manufacturing as industrial tools for fabricating complex metallic tools and drill bits. Improvements in the technology of drilling is going to continue to be made and the challenges that will be addressed include drilling to greater depths, operating at hard-to-access locations, operating at extreme temperatures, autonomous drilling in hazardous and/or environmentally sensitive areas, and many others.

ACKNOWLEDGMENTS

Some of the research reported in this chapter was conducted at the Jet Propulsion Laboratory (JPL), California Institute of Technology, and Honeybee Robotics Spacecraft Mechanisms Corporation under a contract with the National Aeronautics and Space Administration (NASA). The authors would like to thank Jared Atkinson, Honeybee Robotics; and Greg Peters, Jet Propulsion Laboratory (JPL), California Institute of Technology, for reviewing this chapter and providing valuable technical comments and suggestions.

REFERENCES

Astakhov V. P., (April 8, 2014), *Drills: Science and Technology of Advanced Operations (Manufacturing Design and Technology)*, CRC Press, Boca Raton, FL, 888.

Atkinson J., and K. Zacny, (2018), "*Mechanical Properties of Icy Lunar Regolith: Application to ISRU on the Moon and Mars*," *ASCE Earth and Space 2018 Conference*, Cleveland, OH, April 10–12, 2018.

Atkinson J., W. B. Durham, and S. Seager, (2018), *The Strength of Ice-saturated Extraterrestrial Rock Analogs*, Elsevier, Icarus, 315, 61–68.

Badescu V., and K. Zacny, (2018), *Outer Solar System, Prospective Energy and Materials Resources*, Springer.

Bar-Cohen Y., (Ed.), (November 2005), *Biomimetics—Biologically Inspired Technologies*, CRC Press, Boca Raton, FL, 1–527, ISBN 0849331633.

Bar-Cohen Y. (Ed.), (September 2011), *Biomimetics: Nature-Based Innovation*, CRC Press, Taylor & Francis Group, Boca Raton, FL, 788.

Bar-Cohen Y. (Ed.), (March 2014), *High Temperature Materials and Mechanisms*, CRC Press, Taylor & Francis Group, Boca Raton, FL, 1–551, ISBN 13: 9781466566453, ISBN 10: 1466566450 http://www.crcpress.com/product/isbn/9781466566453.

Bar-Cohen Y. (Ed.), (July 2016), *Low Temperature Materials and Mechanisms*, CRC Press, Taylor & Francis Group, Boca Raton, FL, 500, ISBN-10: 1498700381, ISBN-13: 978-1498700382, http://www.crcpress.com/product/isbn/9781498700382.

Bar-Cohen Y. (Ed.), (September 2018), *Advances in Manufacturing and Processing of Materials and Structures*, CRC Press, Taylor & Francis Group, Boca Raton, FL, 1–550.

Bar-Cohen Y., and K. Zacny (Eds.), (2009), *Drilling in Extreme Environments—Penetration and Sampling on Earth and Other Planets*, Wiley-VCH, Hoboken, NJ, 827, ISBN-10: 3527408525, ISBN-13: 9783527408528.

Bar-Cohen Y., S. Sherrit, X. Bao, M. Badescu, J. Aldrich, and Z. Chang, (2007), "*Ultrasonic/ Sonic Driller/ Corer (USDC) as a Subsurface Sampler and Sensors Platform for Planetary Exploration Applications*," *Proceedings of the NASA Science Technology Conference (NSTC-07)*, University of Maryland Conference Center, June 19–21, 2007, 7.

Bar-Cohen Y., S. Sherrit, M. Badescu, H. J. Lee, and X. Bao, (2018), "Drilling Mechanisms Using Piezoelectric Actuators Developed at Jet Propulsion Laboratory," Chapter 6 in V. Badescu and K. Zacny (Eds.), *Outer Solar System. Prospective Energy and Material Resources*, Springer-Verlag, 181–259, ISBN 978-3-319-73845-1.

Barrett M. L., (2011), "Drilling Mud: A 20th Century History," *Oil-Industry History*, 12(1), 161–168.

Black S. D. and A. G. Decker, "Electrically Driven Tool," US Patent 1,245,860, issued on November 6, 1917.

Blackwelder E., (1925), "Exfoliation as a Phase of Rock Weathering," *The Journal of Geology*, 33(8), 793–806.

Briggs G. and A. Gross, (2002), "Technical Challenges of Drilling on Mars," American Institute of Aeronautics and Astronautics, Report No. 0469.

Burleson C. W., (1998), *Deep Challenge: Our Quest for Energy Beneath the Sea*, Gulf Professional Publishing, Houston, TX, ISBN-10: 0884152197, ISBN-13: 978-0884152194.

Christensen G. J., (2006), "The 'New' Operative Dentistry," *Journal of American Dentists Association*, 137(4), 531–533.

Clark G. B., (1987), *Principles of Rock Fragmentation*, John Wiley & Sons, New York, 622, ISBN-10: 0471888540, ISBN-13: 978-0471888543.

Cockell C. S. and N. G. Barlow, (February 2002), "Impact Excavation and the Search for Subsurface Life on Mars," *Icarus*, 155(2), 340–349.

Dansgaard W., (2004), *Frozen Annals—Greenland Ice Sheet Research*, The Niels Bohr Institute, University of Copenhagen, Denmark, ISBN: 87-990078-0-0.

De Villiers M., (2001), *Water: The Fate of Our Most Precious Resource*, Mariner Books, Wilmington, MA, 368, ISBN-13: 9780618127443.

Denning R. M., (1953), "Directional Grinding Hardness in Diamond," *American Mineralogist*, 38, 108–117.

Durham W. B., Stephen H. Kirby, and Laura A. Stern, (December 25, 1992), "Effects of Dispersed Particulates on the Rheology of Water Ice at Planetary Conditions," *Journal of Geophysical Research*, 97(E12), 20883–20897.

Encyclopedia Britannica, (1986), Chicago, IL, Vol. 4, pp. 227, https://www.britannica.com/technology/drilling-machinery, accessed June 24, 2019.

Evans, S., and C. Phaal (Eds.), (1962), *Proceedings of the 5th Conference on Carbon*, Pergamon, New York.

Heiken, G., and E. Jones, 2007, *On the Moon: The Apollo Journals*, Springer, New York

Hemphill G. B., (November 13, 2012), *Practical Tunnel Construction*, Wiley, Brentwood, CA, 304, ISBN-10: 0470641975; ISBN-13: 978-0470641972.

Hossain M. E. and A. A. Al-Majed, (March 2, 2015), *Fundamentals of Sustainable Drilling Engineering*, Wiley-Scrivener, 786, ISBN-10: 0470878177, ISBN-13: 978-0470878170.

Hunt J. M., R. P. Philp, and K. A. Kvenvolden, (September 2002), "Early Developments in Petroleum Geochemistry," *Organic Geochemistry*, 33(9), 1025–1052.

Jerby E., V. Dikhtyar, and O. Grosglick, (October 18, 2002), "The Microwave Drill," *Science*, 298(5593), 587–589.

Just G. D., (June 1963), "The Jet Piercing Process," *Quarry Managers' Journal Institute of Quarrying Transactions*, 219–226.

Karanam R. U. M., and B. Misra, (1998), *Principles of Rock Drilling*, Balkema Press, Rotterdam, Netherlands, 1–265.

Kola Superdeep Borehole, (2018), http://en.wikipedia.org/wiki/Kola_Superdeep_Borehole#_note-0, accessed on July 30, 2018.

Kolymbas D., (2005), *Tunneling and Tunnel Mechanics: A Rational Approach to Tunneling*, Springer-Verlag, New York, 438, ISBN-10: 3540251960, ISBN-13: 978-3540251965.

Lee T. H., H. R. Linden, D. A. Dreyfus, and T. Vasko, (1988), *The Methane Age*, Springer, New York, ISBN-10: 9027727457, ISBN-13: 978-9027727459.

Leyner J. G., (1914), U.S. Patent No. 1,096,436. Washington, DC: U.S. Patent and Trademark Office.

MacDonald G. J., (1988), "Major Questions About Deep Continental Structures," In A. Bodén and K.G. Eriksson (Eds.), *Deep Drilling in Crystalline Bedrock*, Springer-Verlag, Berlin, Germany, Vol. 1, 28–48, ISBN-10: 3540189955; ISBN-13: 9783540189954.

Maurer W. C., (1968), *Novel Drilling Techniques*, Pergamon Press, New York, NY

Maurer W. C., (1980), *Advanced Drilling Techniques*, Petroleum Publishing Company, Tulsa, OK.

McGee E., (August 8, 1955), "New Down-hole Tool," *Oil and Gas Journal*, 54, 67.

McKay, C., (1997), "The Search for Life on Mars," *Origins of Life and Evolution of Biosphere*, 27(1–3), 263–289.

Mellor M., (May 1971), "Strength and Deformability of Rocks at Low Temperatures," Report No. AD726372, Cold Regions Research and Engineering Laboratory, Corps of Engineers, U.S. Army, Hanover, New Hampshire.

Ostrovskii N. P., (1960), *Deep-hole Drilling with Explosives*, Gostoptekhia 'dat Moscow, Translated by Consultants Bureau Enterprises, Inc., New York.

Peele R., (1920), *Compressed Air Plant: The Production, Transmission and Use of Compressed Air*, 4th Edition, John Wiley & Sons, New York.

Rapp D., (2007), *Human Missions to Mars: Enabling Technologies for Exploring the Red Planet*, Springer/Praxis Publishing, Chichester, UK, ISBN: 3-540-72938-9, Appendix C, "Water on Mars."

Ready J. F., (1997), *Industrial Applications of Lasers*, 2nd Edition, Academic Press, New York, ISBN-10: 0125839618, ISBN-13: 978-0125839617.

Rollins J., (2010), *Excavation*, Harper, New York, 525, ISBN-10: 0061965812, ISBN-13: 978-0061965814.

Simonsen R., (1989), *Dentistry in the 21st Century—A Global Perspective*, Quintessence Publishing Co., Hanover Park, IL.

Smith M., G. Cardell, R. Kowalczyk, and M. H. Hecht, (2006), "The Chronos Thermal Drill and Sample Handling Technology," *The 4th International Conference on Mars Polar Science and Exploration*, Davos, Switzerland, held on October 2–6, 2006, Abstract 8095.

Wyllie D. C., (1999), *Foundations on Rock*, 2nd Edition, Taylor and Francis Press, Boca Raton, FL, 345–347, ISBN-10: 0419232109, ISBN-13: 978-0419232100.

Zacny K., (2011), "Fracture and Fatigue of Polycrystalline Diamond Compacts," *SPE Drilling & Completion, 2011*, https://doi.org/10.2118/150001-PA. Document ID SPE-150001-PA0.

Zacny, K. et al., (2016a), "Drilling and Breaking Ice," Chp. 10 in Bar-Cohen Y. (Ed.), *Low Temperature Materials and Mechanisms*, CRC Press.

Zacny K., and G. Cooper, (2001), "Evaluation of a New Thermally Stable Polycrystalline Diamond Material," *SPE 68784; Proceedings of Society of Petroleum Engineers Western Regional Meeting*, Bakersfield, California, March 26–30, 2001.

Zacny K., and G. Cooper (2006) "Considerations, Constraints and Strategies for Drilling on Mars," *Planetary and Space Science Journal*, 54(4), 345–356, doi:10.1016/j.pss.2005.12.003.

Zacny K., Y. Bar-Cohen, D. Boucher, M. Brennan, G. Briggs, G. Cooper, K. Davis, B. Dolgin, D. Glaser, B. Glass, S. Gorevan, J. Guerrero, G. Paulsen, S. Stanley, and C. Stoker, (2008), "Drilling Systems for Extraterrestrial Subsurface Exploration," *Special Paper in the Journal of Astrobiology*, 8(3), 665–706.

Zacny K., G. Paulsen, C. P. McKay, B. Glass, A. Dave, A. Davila, M. Marinova, B. Mellerowicz, J. Heldmann, C. Stoker, N. Cabrol, M. Hedlund, and J. Craft, (2013a), "Reaching 1 m Deep on Mars: The Icebreaker Drill," *Astrobiology*, 13, 1166–1198, 10.1089/ast.2013.1038.

Zacny K., G. Paulsen, B. Mellerowicz, Y. Bar-Cohen, L. Beegle, S. Sherrit, M. Badescu, F. Corsetti, J. Craft, Y. Ibarra, X. Bao, and H. J. Lee, (2013b), "Wireline Deep Drill for Exploration of Mars, Europa, and Enceladus," *2013 IEEE Aerospace Conference*, Big Sky, Montana, March 2–9, 2013.

Zacny K., M. Shara, G. Paulsen, B. Mellerowicz, J. Spring, A. Ridilla, H. Nguyen, K. Ridilla, M. Hedlund, R. Sharpe, J. Bowsher, N. Hoisington, S. Gorevan, J. Abrashkin, Lou Cubrich, and Mark Reichenbach, (2016b), "Development of a Planetary Deep Drill," *ASCE Earth and Space Conference*, Orlando, FL, April 11–15, 2016.

Zacny, K., J. Spring, G. Paulsen, S. Ford, P. Chu, and S. Kondos, (2015), "Pneumatic Drilling and Excavation in Support of Venus Science and Exploration," Chp 8 in Badescu, Viorel, and Zacny, Kris (Eds.), *Inner Solar System: Prospective Energy and Material Resources*, Springer, New York, NY.

Zacny K. et al., (2018a), "SLUSH: Europa Hybrid Deep Drill," *IEEE Aerospace Conf*, Big Sky, MT, March 1–5, 2018.

Zacny K., P. van Susante, T. Putzig, M. Hecht, and D. Sabahi, (2018b), "RedWater: Extraction of Water from Mars Ice Deposits, Space Resources Roundtable," *Colorado School of Mines*, Golden, CO, June 12–14, 2018.

2 Design, Modeling, and Testing of Piezoelectric Actuated Percussive Drills

Hyeong Jae Lee, Xiaoqi Bao, Mircea Badescu, Stewart Sherrit, and Yoseph Bar-Cohen
Jet Propulsion Laboratory (JPL)/California Institute of Technology (Caltech), Pasadena, CA

CONTENTS

2.1 INTRODUCTION

The acquisition of rock and soil samples plays an important role in NASA's space exploration missions. Tools capable of coring, drilling, and abrading are necessary for exposing fresh rock sample or sampling for in situ instruments or sample return. Due to the energy/power and mass constraints of planetary missions, high power consumption as is found in conventional drills is extremely undesirable. In order to address these contraints a piezoelectric-driven Ultrasonic/Sonic Driller/Corer (USDC) mechanism has been developed for planetary sampling. The USDC is a solid-state hammering mechanism that uses ultrasonic displacements to create larger-stroke percussive displacements. Inside the USDC, a piezoelectric stack generates vibrations that propagate and are amplified via a horn and the vibrations impact a free-flying mass. The mass then impacts a drill bit introducing stress pulses onto the bit/rock interface, which produce stress pulses that exceed the ultimate strain of the rock that it is in contact with and fractures. The USDC's key benefits include being lightweight, requiring low axial preload, and its ability to be configured as an in situ analyzer (Bao et al., 2003; Bar-Cohen and Zacny, 2009).

Since the initial development of the USDC, significant advances have been made improving its percussive drilling capability. A wire-line USDC drill design combined with rotary actuation, called the Auto-Gopher, is one of the improved versions that has been developed allowing it to reach depths greater than the drill length without adding drill segments. Originally, the USDC mechanism was developed using a free-flying mass, but in an effort to enable well-controlled impact frequency and impulses as well as improving the target fracturing capability, a constrained flexured striker mechanism has also been developed (Badescu et al., 2017).

This chapter presents an overview of the USDC mechanism and its finite element modeling. This includes modal, harmonic, and the piezoelectric horn interactions with the flexured confined striker.

The main focus of this chapter is on the performance analyis and demonstration of the new designs using modeling, simulation, and testing. The general background of the piezoelectric horn actuator is presented, followed by the concept design, modeling, and simulation results of the piezoelectric actuator and the flexured striker. The impact forces generated by the piezoelectric actuator were also investigated using transient dynamic modeling of the horn/impactor/drill bit. In addition, preliminary results obtained from testing a flexured striker actuator prototype are also reported in this chapter.

2.2 USDC PIEZOELECTRIC ACTUATOR

Under an electric field, piezoelectric materials inherently produce low strain. Therefore, to make them effective actuators they require design approaches that amplify their strokes. Fortunately, piezoelectric materials generate very large stresses per volt. In order to turn their small displacements that are on the order of nano- or picometers per volt, one can trade some of the stress for increased stroke. In addition, the resonance frequency of a piezoelectric material is generally limited to the range of hundreds of kHz and above, which is not practical for use in drilling applications. To address both of these limitations, a variety of actuator configurations have been used that involve combining a piezoelectric multilayer stack with a stepped-horn or a flex-tensional piezoelectric actuator and a longitudinal-torsional horn (Bar-Cohen and Zacny, 2009). These actuators have the capability to produce high force with high stroke at reasonable operating frequency range (10–40 kHz).

In many of the different piezoelectric actuator design applications, the stepped-horn actuator has been used for the implementation of the USDC mechanism. An ultrasonic stepped-horn design (shown in Figure 2.1) is an actuator design that allows for the generation of large vibration levels and high acoustic power into drilling target materials. Its design consists of three major parts: stack of piezoelectric rings, ultrasonic horn, and metallic backing. The piezoelectric stack consists of a number of thin alternately poled piezoelectric layers, connected mechanically in series and electrically in parallel. It provides increased vibration amplitude for a given voltage that is proportional to the number (n) of the piezoelectric stack elements. In comparison with a single piece of thick piezoelectric element, the effective piezoelectric strain coefficient (d_{33}), and capacitance (C) of stacks are proportional to the number of the piezoelectric stack elements (i.e., $d^*_{33} = n.d_{33}$, $C^* = n^2.C$), meaning that the required voltage can be reduced by a factor of the number of piezoelectric elements for a given required displacement.

The presence of the horn amplifies the displacement that is generated by the piezoelectric stack. The horn is usually designed to have a higher displacement/velocity at the tip so as to provide sufficient displacement for drilling. This amplification of the displacement can be achieved by reducing the cross section along the length of the horn. The materials used in ultrasonic horns are aluminum, titanium, and stainless steel, as they have high mechanical quaility factor. Among them, titanium is the most suitable as it has a high-yield strength, comparable to stainless steel, and low density.

Under dynamic operation at resonance, the displacement of the horn is amplified by the square of the diameter ratio of the horn end and face (Sherrit et al., 2002). The role of the backing is to

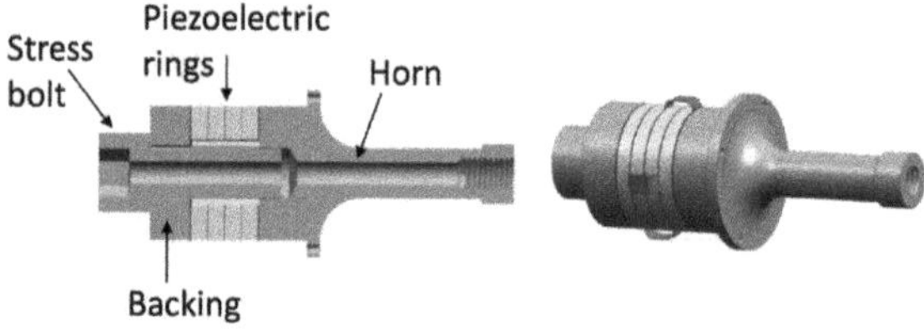

FIGURE 2.1 Actuator based on a piezoelectric transducer with a stepped-horn. The horn design shown is a dog-bone horn and it includes threads for attaching a drilling-wire.

direct energy out the front of the piezoelectrics and eliminate energy loss from the backside of the piezoelectrics.

In the USDC, the high-impact hammering mechanism is realized by the use of a free-mass to generate high-stress pulses, which increases the efficiency of rock breaking. Over the years, various novel prototypes have been conceived and developed including the Ultrasonic/Sonic Gopher (USG), the Ultrasonic Rock Abrasion Tool (URAT), the Ultrasonic/Sonic Ice-Gopher and the Auto-Gopher (Badescu et al., 2008; Bar-Cohen and Zacny, 2009; Bar-Cohen et al., 2012; Zacny et al., 2013). A schematic representation and an example of a CAD model of a USDC-type actuator are shown in Figure 2.2 and Figure 2.3. In this design, the bit is connected with a soft spring allowing contact between the bit and free-mass after impact. The basic mechanism of this type of drilling mechanism is that, once excitated, the free-mass inside the USDC bounces back and forth between the horn and the drill bit. The excited mass impacts the bit at much lower frequency (<1 kHz) and much higher velocity. This allows for effective conversion of high frequency and large force vibration of the horn tip into low frequency and large displacement hammering. The advantage of this method compared to ultrasonic drilling is the realization of high impact stress pulses onto the target providing effective rock breaking.

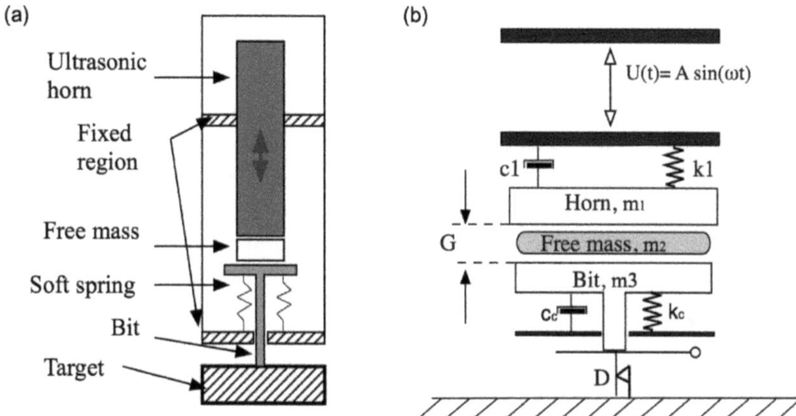

FIGURE 2.2 Schematic diagram (a) and 1-DOF dynamic model (b) of a USDC configuration. m_1, m_2, and m_3 are the masses of the ultrasonic horn, free-mass, and bit, respectively, whereas D is the resistive force that represents the resistance of the drilling surface.

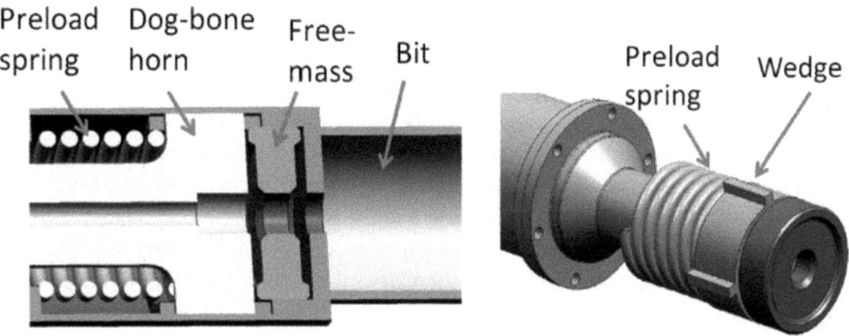

FIGURE 2.3 Cross-section view (left) and solid model (right) of the USDC mechanism showing the free-mass with preload spring and dog-bone horn.

The kinetic characteristics of the impact momentum and the energy transfer between the ultrasonic horn and the impactor and between the impactor and the drill bit can be determined based on the conservation of momentum and energy with appropriate assumptions (Bao et al., 2003; Vila and Malla, 2016). For the sake of simplicity, if we assume that all collisions are perfectly elastic (restitution coefficient = 1), the energy loss of the impact is negligible. The post-collision velocities, when the overall force developed in the spring is less than the dry friction force D, can be calculated as follows:

$$m_1 v_{i1} + m_2 v_{i2} = m_1 v_{f1} + m_2 v_{f2} \tag{2.1}$$

$$\frac{1}{2} m_1 v_{i1}^2 + \frac{1}{2} m_2 v_{i2}^2 = \frac{m_1 {v_{f1}}^2}{2} + \frac{m_2 v_{f2}^2}{2} \tag{2.2}$$

The equations lead to the post-collision velocities by solving for v_{f2}:

$$v_{f2} = \frac{m_1 \left(2v_{i1} - v_{i2}\right) + m_2 v_{i2}}{m_1 + m_2} \tag{2.3}$$

where v_{f2} is post-collision velocities of impactor, and v_{i1} and v_{i2} are pre-collision velocities of horn and impactor, respectively.

Assuming that the mass of the horn (m_1) is much higher than the mass of impactor (m_2), i.e., $m_1 >> m_2$, the post collision velocity of impactor after hitting the horn can be simplified as $v_{f2} \approx -v_{i2} + 2v_{i1}$ (i.e., v_{f2} = post impactor velocity, v_{i1} = horn velocity, and v_{i2} = pre-impactor velocity). For the case of impact between drill bit and impactor, assuming that the drill bit is not moving (no break, v_{bit} = 0 and $m_{bit} >> m_2$), the post-velocity of impactor (v_{f2}) is $v_{f2} \approx -v_{i2}$, by applying $v_{bit} = v_{il}$ and $m_{bit} = m_1$ on Eq. (2.3). Thus, the velocity $v_f(n)$, position $p_f(n)$ of impactor, and the time (t_c) of the collision between the horn and the impactor or between the impactor and the bit can be expressed as follows:

$$v_f(n) = \begin{cases} 2v_h t_c(n-1) - v_f(n-1) & if \quad p_f(n) = p_h(n) \\ -v_f(n-1) & if \quad p_f(n) = 0 \end{cases}$$

$$p_f(n) = p_f\left(0\right) + \sum_{n=1}^{N} v_f(t_c(n) - t_c \tag{2.4}$$

$$t_c(n) = \sum_{n=1}^{N} \frac{d - (u_h \times t_c(n-1))}{v_f(n)}$$

where $v_f(0) = 0$, $p_f(0) = d$, and $t_c(0) = 0$. p_h is position of horn, d is initial position of horn, and N is the total number of collisions. $v_h(t_c)$ *is* the velocity of horn at t_c.

The position of each component (horn, impactor, and bit) can be obtained by continuous solving of the equations by iterating the calculated time between the horn and the impactor and between the impactor and the drill bit and the velocity parameters. Figure 2.4 shows the example of velocity and position profile as a function of time after solving the Eq. (2.4), with the assumptions that there are no losses and gravity effects during collision events. Note that the post velocity of the impactor depends on the specific situation of the collision between horn and impactor. If the velocities of horn and impactor are in opposite directions, the velocity of the impactor is increased; however if the velocities of the horn and the impactor are in the same direction of horn velocity, the velocity of the impactor is slowed down. In addition, depending on the horn velocity before the collisions, there are multiple collisions between the horn and the impactor.

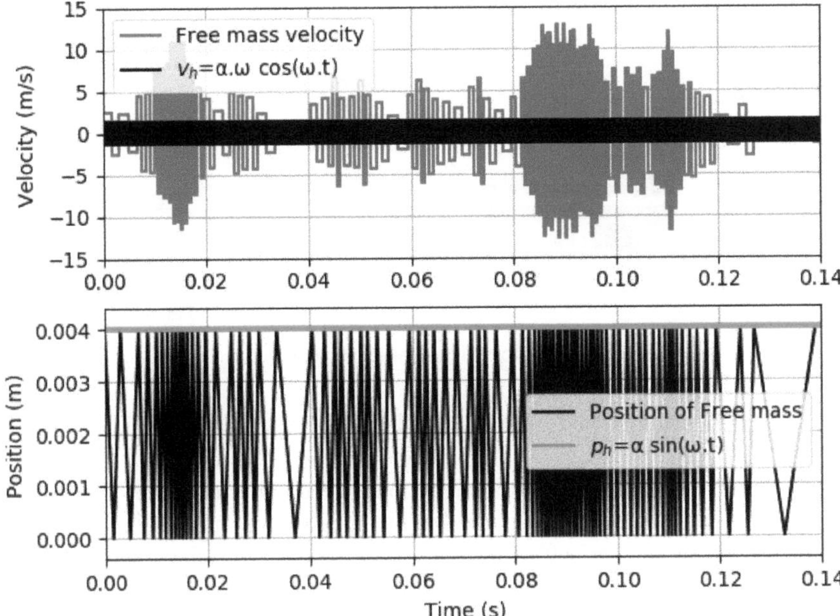

FIGURE 2.4 Velocity and position profile of free-mass and horn as a function of time assuming that there is no loss and gravity effects during collision (d = 4 mm, α = 10 μm, and f = 20 kHz).

2.3 THE FLEXURED STRIKER PIEZOELECTRIC ACTUATOR

Recently, a USDC configuration with flexured striker (constrained-mass) impactor has been developed to improve the controllability of the impacts as well as enhancing the capability of the rock breaking. The configuration of the free-mass compared to the constrained-mass are shown in Figure 2.5 left and right, respectively. Further details of an initial design of the constrained-mass configuration are shown in Figure 2.6.

In contrast to the USDC with a free-mass impactor, the constrained striker (impactor) is mounted on a diaphragm flexure and kept with a gap to the bit. The characteristics of the diaphragm flexure is an important property for the performance as it limits the range of the drill bit travels. The use of the diaphragm increases the compliance normal to the flexure plane, while constraining it in the other

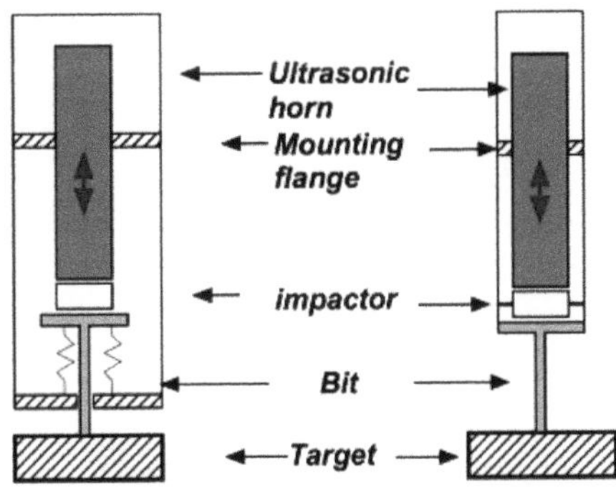

FIGURE 2.5 Schematic diagram of the piezoelectric actuator designs of the USDC with free-mass (left) and the flexured striker (right).

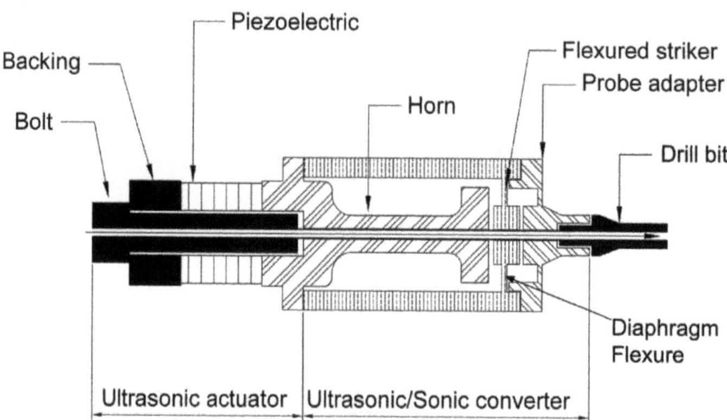

FIGURE 2.6 Cross-section view of the flexured striker piezoelectric actuator concept design.

directions and therefore reducing the parasitic motions of the impactor. Another distinct feature of this configuration is that the drill bit is attached to the housing through a higher stiffness structure, preventing the loss of impact energy from the transducer housing, and allowing the ultrasonic vibrations to be transmitted to the bit (dual frequency bit).

2.4 EQUIVALENT NETWORK CIRCUIT MODEL

For analytical modeling of piezoelectric transducers, methodologies developed by Mason and KLM are widely used as 1-dimension models (Mason, 1948, 1956; Oakley, 1997; Van Kervel and Thijssen, 1983). They present an equivalent circuit that separates the piezoelectric material into an electrical port and two acoustic ports using an ideal electromechanical transformer. Since the vibration properties of a piezoelectric transducer can be modeled as a function of the geometry and material data, it is a simple and fast method that is suitable for initial parametric studies. This section gives a brief summary of the information needed to implement a network model using KLM-based transfer matrix approach as an example.

The network representation of a piezoelectric transducer based on Krimholtz, Leedom, and Matthaei (Krimholtz et al., 1970) is shown in Figure 2.7. The electromechanical conversion is modeled as a perfect transformer with a winding ratio Φ:1, where Φ is the mechanical-electrical transformer ratio. Additional layers, such as backing and front mass, can be easily added using their thickness and acoustic impedance values represented as an acoustic transmission line.

The definitions of the circuit components are given:

$$
\begin{aligned}
C_0 &= \frac{\varepsilon_0 \varepsilon_r A}{t} \\
C_0' &= \frac{C_0}{k_t^2 \sin c\left[\dfrac{\omega}{\omega_a}\right]} \\
\Phi &= k_t \sqrt{\frac{\pi}{\omega_a C_0 Z_C}} \sin c\left[\frac{\omega}{2\omega_a}\right]
\end{aligned}
\tag{2.5}
$$

where ε_0 and ε_r are the dielectric permittivity of free space and the relative permitivity respectively, and A is the cross-sectional area of the piezoelectric layer. While k_t, ω_a, and Z_C are the electromechanical coupling factor, anti-resonance angular frequency, and the acoustic impedance of the piezoelectric transducer element, respectively.

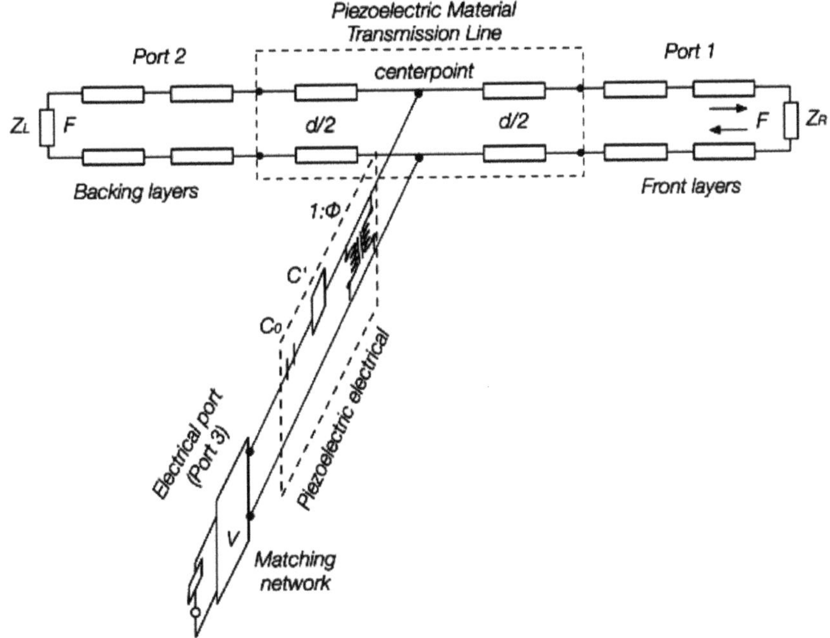

FIGURE 2.7 KLM-equivalent circuit model for a piezoelectric transducer. Voltage is applied on the electrical port and force is developed across the front acoustic port.

The mechanical delay line of the piezoelectric transducer, front, backing layer is represented by Eq. (2.6):

$$\begin{pmatrix} \cosh[k.l] & -Z_l \sinh[k.l] \\ -\sinh[k.l]/Z_l & \cosh[k.l] \end{pmatrix} \tag{2.6}$$

where Z_l, k, and l are the acoustic impedance, complex propagation constant ($k = \alpha + j\beta$, $\beta = 2\pi/\lambda$ and α is attenuation), and the thickness for piezoelectric and the front layers, respectively.

The relationships between the circuit quantities and a transfer matrix (ABCD) of the piezolectric are given by:

$$\begin{pmatrix} F_R \\ u_R \end{pmatrix} = \begin{pmatrix} A & B \\ C & D \end{pmatrix} \cdot \begin{pmatrix} V_g \\ I_g \end{pmatrix} \tag{2.7}$$

where V_g and I_g are the source voltage and current, respectively, and F_R and u_R are the mechanical force and velocity, respectively.

The overall transfer ABCD-matrix coefficients for a piezoelectric transducer are then obtained by the product of the individual matrices corresponding to the electrical matrix, electromechanical matrix, the piezoelectric element, and the front layers. The parameters listed in Eq. (2.9) are as follows:

$$\begin{aligned} V_g &= V_{i1} + V_{i2} \\ I_g &= (V_{i1} - V_{i2})/Z_T \\ u_R &= F_R/Z_R \end{aligned} \tag{2.8}$$

where the subscript $i1$ and $i2$ indicate a plane wave in the positive and negative direction, respectively. Z_R and Z_T are the radiation impedance and electrical impedance of the transducer, respectively.

The transmit transfer function and electrical impedance of the transducer can be obtained by substituting Equations (2.8) into (2.7),

$$tf(\omega) = \frac{F_R}{V_T} = \frac{2Z_R}{B - Z_T A - Z_R D + Z_T Z_R C}$$
$$Z_T = \frac{A_t Z_R + B_t}{C_t Z_R + D_t}, \tag{2.9}$$

where A_t, B_t, C_t, and D_t are matrix coefficients resulting from the inverse of the matrix defining the parameters for the transmit ABCD network.

Details of the KLM and Mason piezoelectric models can be found elsewhere (Krimholtz, Leedom, and Matthaei, 1970; Mason, 1948, 1956; Oakley, 1997; Van Kervel and Thijssen, 1983; Sherrit et al., 1999). It should be noted that 1-dimension models have been proven to be useful for various transducer applications. However, it is found to be difficult to accurately model horn-based piezoelectric transducers, such as Langevin and Tonpliz transducers, due to parallel acoustic elements like the stress bolt, the complex horn geometry, and the influence of lateral mode vibrations. More advanced modeling, such as finite element modeling, is necessary to accurately predict other features such as the mode shape and electrical impedance over wider frequency ranges.

2.5 FINITE ELEMENT MODELING OF PIEZOELECTRIC ACTUATOR

For the design of the horn-based piezoelectric actuator, a finite element model with ABAQUS solver was used (Abaqus Users Manual, 2013). The USDC or the piezoelectric actuator with a flexured striker, generally uses force from the piezoelectric actuator to bounce free-mass or constrained-mass striker. Therefore, the most important parameter in these mechanism is to achieve large vibration velocity as the power output is proportional to this vibration parameter. The vibration velocity (v_{rms}) of piezoelectric material for a given voltage is as follows:

$$v_{rms} = \frac{\omega L d_{33}^{*} Q_m}{\sqrt{2}\, h} V \tag{2.10}$$

where ω is angular frequency, d^*_{33} is the effective piezoelectric strain coefficient ($= n \times d_{33}$), Q_m is mechanical quality factor, h is the length along the poling direction, and L is the length along the the vibration direction, and applied voltage, V (Uchino, 2003). Equation (2.12) points out that materials with high piezoelectric strain coefficients and mechanical quality factors increase the vibration velocity of piezoelectric actuators.

Modal analysis of a piezoelectric horn can be used to identify the mode shapes and natural frequencies of a chosen piezoelectric actuator design under free-free boundary conditions. Figure 2.8 shows simplified geometry of the ultrasonic horn actuator that was used for the modeling. Note that owing to mechancial limitations of piezoelectric materials under high power drive conditions, the application of pre-stress is essential to maintaining the piezoelectric materials in compression during the electrical excitation. Excessive preload can depolarize the piezoelectric material and therefore care must be given during pre-stressing. In general, 20~30 MPa preload is the recommended value. Since the pre-stressing on piezoelectric materials affects the elastic and dielectric properties, the model should include the pre-stress effects in the analysis. In this model, the general static step is created prior to the frequency step in order to include pre-stress effects on the frequency response. The results of a general steady state after pre-stress using a bolt load module is shown in Figure 2.9,

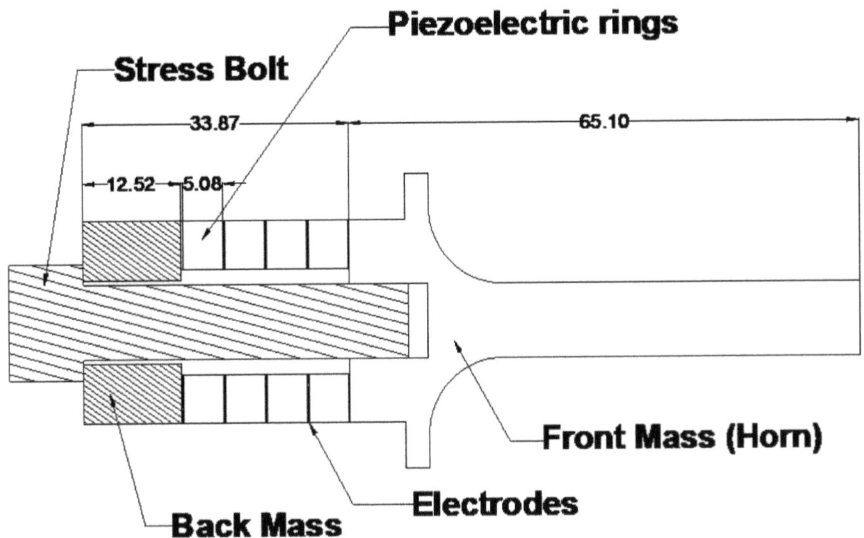

FIGURE 2.8 Simplified geometry used for finite element modeling for horn-based piezoelectric actuator.

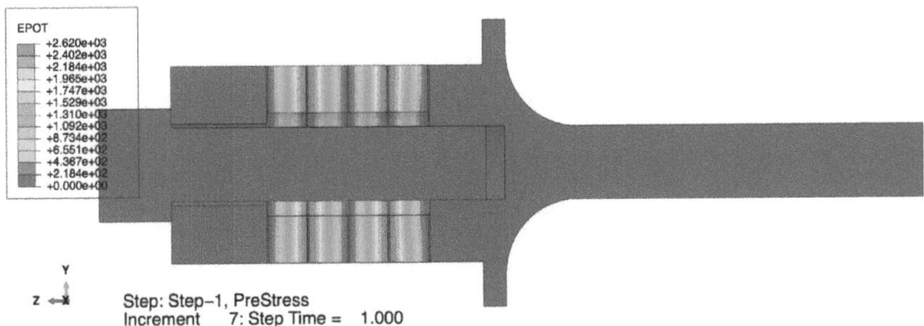

FIGURE 2.9 Open-circuit voltage (EPOT) output from piezoelectric rings after pre-stress step for horn-based piezoelectric actuator.

showing a voltage (2620 V) generated from piezoelectric elements due to pre-stress effects, which corresponds to 20 MPa of bolt preload, according to the piezoelectric constitutive equation:

$$E_3 = -g_{33}T_3 + \beta_{33}^T D_3 \tag{2.11}$$

where T_3 and D_3 are the stress on the element in the direction of polarization and dielectric displacement, E_3 is electric field $(= V/t)$, β_{33}^T is free dielectric impermeability constant. Under open-circuit condition ($D_3 = 0$), the equation reduces to:

$$V_{oc} = g_{33}tT_{33} \tag{2.12}$$

where V_{oc} is open-circuit voltage.

The modal analysis of piezoelectric horn is then performed in order to identify the mode shapes and natural frequencies of the chosen piezoelectric actuator design under free-free boundary conditions using ABAQUS. For the spectral analysis, resonance (f_r) and anti-resonance (f_a) frequency can be obtained by setting the voltage difference between the electrodes to be equal to zero (short-circuit). The resonance frequency represents the mechanical resonance vibrating under short-circuit conditions, while the anti-resonance frequency represents the mechanical resonance vibrating under an open-circuit condition. Thus, two steps (one for short-circuit and the other for open-circuit condition) are performed after the preload step to extract the resonance and anti-resonance frequency from the frequency analysis.

Based on the frequency extraction analysis, the resonance and anti-resonance frequencies are determined to be as 21.834 kHz and 22.774 kHz, respectively. After identifying the first-mode resonance frequency of the actuator, a steady-state harmonic analysis was performed in order to simulate the electromechanical responses of ultrasonic horn actuator as a function of frequency. This step provides the steady-state amplitude of the piezoelectric actuator as a function of frequency. Note that two most important parameters to determine piezoelectric transducer performance operating at resonance frequency are the electromechanical coupling factor, k, and the mechanical quality factor, Q_m. The square of the former is defined as the ratio of the stored mechanical energy to the input electrical energy, and the latter is defined as the ratio of the stored energy to the energy lost in the resonator. Higher Q values give higher displacements at the resonance frequency.

Using finite element modeling, the spectral responses of the electrical impedance of the ultrasonic horn were determined and are shown in Figure 2.10 for the actuator that is free at both ends and held at the nodal plane (i.e., mechanically free condition). The resonance and anti-resonance frequency, corresponding to the frequency of minimum impedance and the frequency of maximum impedance, respectively, are similar to the ones from the frequency extraction step, exhibiting 21,835 kHz and 22,780 kHz (Figure 2.11). The calculated electromechanical coupling and mechanical quality factor from the impedance curve are $k_{eff} = 0.31$, and $Q_m \sim 1000$, respectively, where these values can be calculated using the following equations:

$$K_{eff} = \sqrt{1 - \left(\frac{f_r}{f_a}\right)^{2}} \qquad Q_m = \frac{f_r}{f_1 - f_2} \tag{2.13}$$

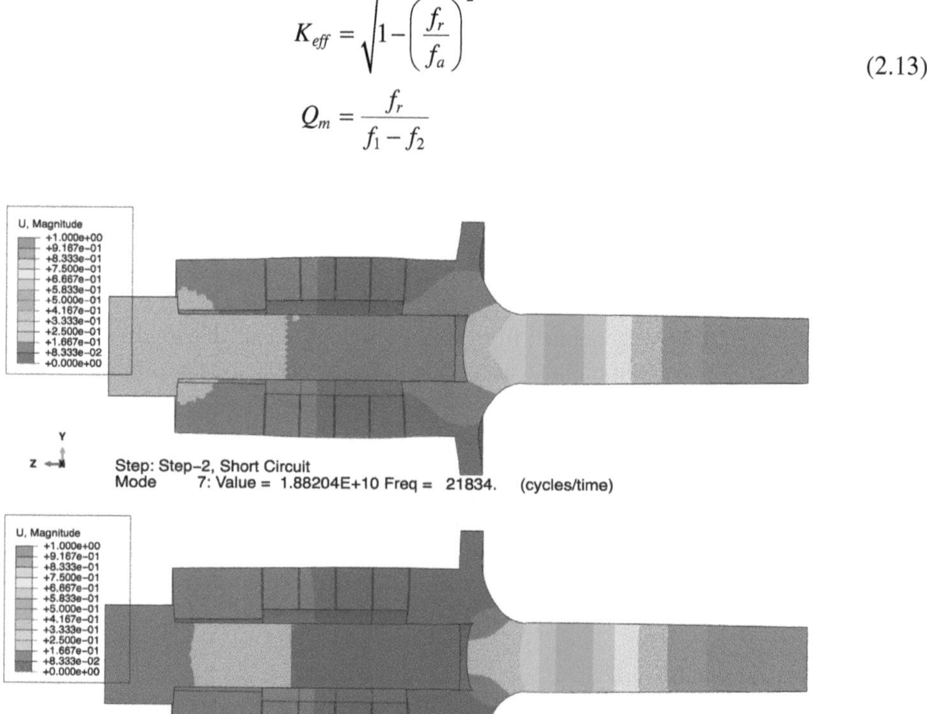

FIGURE 2.10 Modal analysis of piezoelectric actuators under short- (upper) circuit and open-circuit conditions (lower). The color represents displacement.

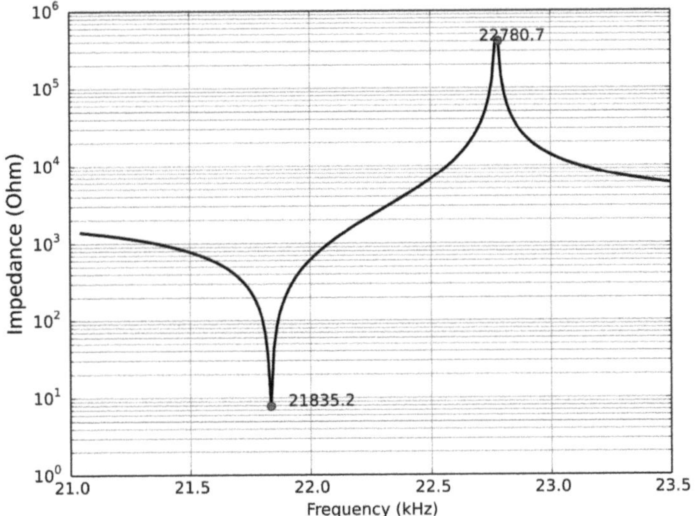

FIGURE 2.11 The results of modeling the electrical impedance of piezoelectric horn actuator as a function of frequency when the actuator is free at both ends.

where f_1 and f_2 are the frequencies at $\sqrt{2}$ the value of the maximum impedance of Z_m. The values for these parameters calculated from the impedance curve are $k_{eff} = 0.31$, and $Q_m = 1000$, respectively.

Note that the mechanical quality factor (damping) is highly dependent on the loading and drive conditions. The mechanical quality factor of this type of actuator with PZT8 material is generally ~1000 at small signal levels. However, the mechanical quality factor of a piezoelectric actuator will decrease with increasing drive level. To improve the model accuarcy for later impact model analysis, the electrical impedance and actual displacement amplitude of the fabricated piezoelectric actuator were measured, and the mechanical damping value was corrected accordingly for the impact analysis.

2.6 IMPACT ANALYSIS RESULTS USING FINITE ELEMENT MODEL

For the analysis and operation simulation of the bit with different types of flexure configurations, a simplified quarter finite element model was applied using ABAQUS/Explicit (see Figure 2.12). During the analysis, the software traces the translation movements of the piezoelectric actuator and the flextural striker as well as the vibration of the bit as a function of time. It calculates the time and location of the striker/horn and striker/bit collisions. The movements and vibration due to the impact are recorded along with the impact force and time. The software then proceeds to calculate the next impact.

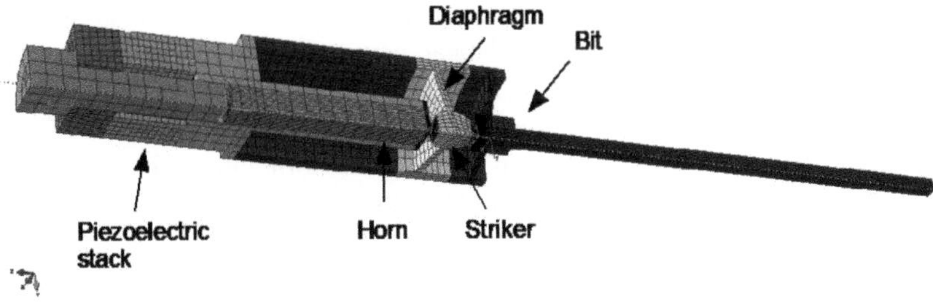

FIGURE 2.12 Model configuration of the actuator with flexured striker design.

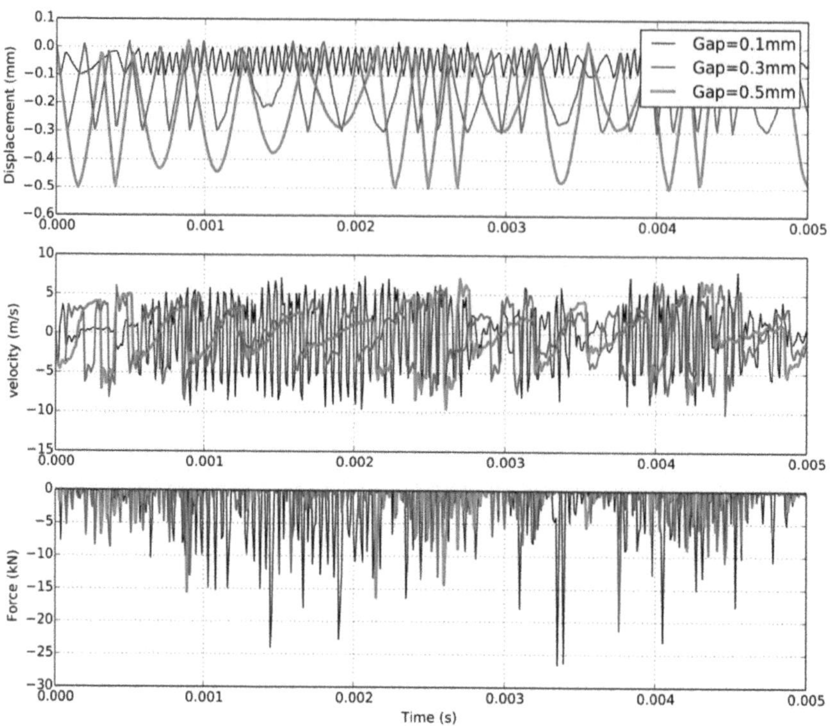

FIGURE 2.13 Time-dependent displacement (top), velocity (middle), and force (bottom) of striker for various values of the gap between striker and bit.

TABLE 2.1
The Properties of the Typical Kidney Stone. σ and ρ are Poisson's Ratio and Density, Respectively. YM, SM, and TS are Young's Modulus, Shear Modulus, and Tensile Strength, Respectively

Material	σ	ρ (kg/m³)	YM (GPa)	SM (GPa)	V (m/s)	TS (MPa)
BegoStone(15:3)	0.27	1995	27.4	10.78	4142	16.3(dry)/7.12(Wet)

Source: Esch et al. (2010).

The results of the finite element modeling of the flexured striker configuration are shown in Figure 2.13. This figure is showing the displacement, velocity, and force of the striker as a function of time when excited with a 100 V amplitude (or 20W of input power) at 20 kHz. The reaction force of the striker was calculated using the generated acceralation values of the striker multiplied by the mass of the striker (5 g). For the analysis, the gap between the impactor and the bit was varied from 0.1 to 0.5 mm. Note that when the horn vibrates at 20 kHz, the striker vibrates at reduced frequencies, 5–10 kHz, but with higher velocities >4m/s. The generated impact force from the striker is found to increase with a decrease in the gap between the horn and the bit, reaching >10 kN as shown in the bottom of Figure 2.13 on occasion when the gap is in the range of 0.1~0.5 mm (Vila and Malla, 2016).

To simulate the rock-breaking capability on a target material (representing kidney stones), we used BegoStone (power-to-water ratio = 15:3) model, whose properties are summarized in Table 2.1. This material has high compressive strength but low tensile strength, and is intended to simulate

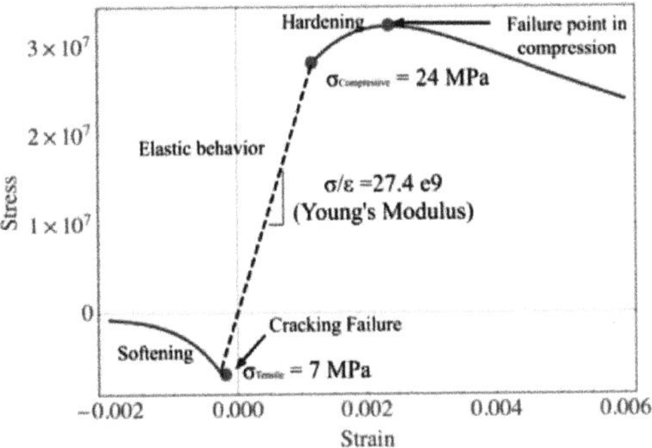

Young's modulus:	27.4 GPa
Poisson's ratio:	0.27
Density:	1995 kg/m3
Compressive initial yield stress:	24 MPa
Compressive ultimate stress:	32 MPa
Tensile failure stress:	7 MPa

FIGURE 2.14 Response of kidney stone-simulated material, soft BegoStone (15:3), to uniaxial loading in compression and tension. The main failure mechanisms are cracking in tension and crushing in compression.

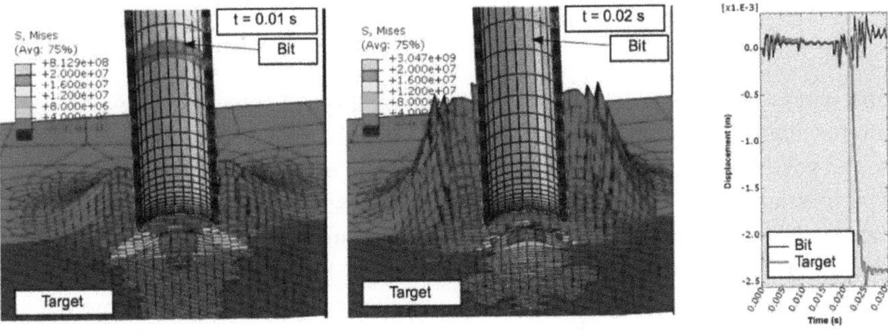

FIGURE 2.15 Equivalent stress distribution in the target material at 0.01 s (left) and 0.02 s (middle). Right figure shows displacement of bit and target material as a function of time.

materials that are damaged by brittle failure. The brittle failure model assumes that the material is elastic in compressive loading and a crack forms when the principal tensile stress exceeds the tensile strength of the brittle material according to Rankine failure criterion. Before failure, the material is linear elastic in tension, as shown in Figure 2.14. The details of brittle cracking model can be found in ABAQUS Theory Manual (Abaqus Users Manual, 2013).

An example of the time dependent dyanmic analysis results is shown in Figure 2.15, exhibiting the equivalent stress distribution on the drill bit and target (kidney stone-simulated material) at selected time (0.01 and 0.02 s). The displacement of the bit and target elements at the bottom of the bit as a function of time is also shown in Figure 2.15 (right). When the target material is subjected to high cyclic loading, cracks develop with time, and when the stress is exceeding the target material strength, it leads to the permanent breakage on the target.

Experiments were conducted to verify the performance of the flexure striker actuator. For this test, soft and hard BegoStones were used with 33–34 mm diameter and 9–10 mm thickness. The test results are shown in Figure 2.16.

2.7 CONCLUSION

In this chapter, we presented design configurations, modeling, and testing of piezoelectric transducers using, as an example, a flexured (constrained-mass) striker configuration. Equivalent circuit models, such as Mason or KLM models, were useful to determine initial electrical responses, as it requires low computing power and run time. However, it was found to be difficult to model the

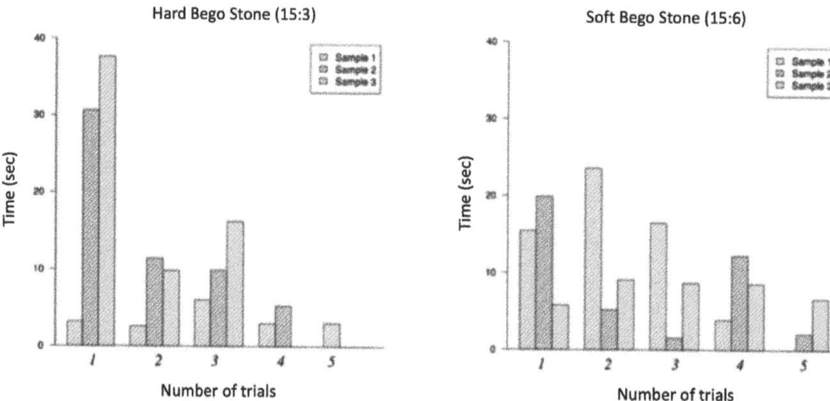

FIGURE 2.16 Time to break hard (15:3) (left) and soft (15:6) (right) BegoStone using flexured striker configurations. Dimensions of the tested samples are 33–34 mm in diameter and 9–10 mm thickness.

effects of pre-stress bolt and more complex geometry. A finite element model was used as it allows to model more complex geometries, pre-stress bolt and joint effects on the electrical and mechanical responses, and gave a good correlation between model and experimental results.

The experimental results of the recently developed constrained-mass striker configuration have shown advantages over the previously developed mechanism of free-flying mass USDC designs in terms of controllability. As expected, the dynamic impact force generated from the striker varies with the gap between the horn and bit, increasing with the decrease of the gap. Experimental investigations on BegoStone showed that a BegoStone with 9–10-mm thickness can be broken in less than 1 min using the developed prototype.

The recently developed piezoelectric stack and flexured striker was found to produce a hammering action which provided many benefits. The characteristics of operation under low power with small tip displacement and low axial preload can be beneficial to sample acquisition tasks in planetary exploratory missions that are constrained by very tight power and mass budgets at planets with low gravity.

ACKNOWLEDGMENTS

The research reported in this chapter was conducted at the Jet Propulsion Laboratory (JPL), California Institute of Technology under a contract with the National Aeronautics and Space Administration (NASA). The authors would like to thank Boleslaw Mellerowicz, Honeybee Robotics Ltd., Pasadena, California; Patrick Harkness, University of Glasgow, Scotland; and Malla, Ramesh, Department of Civil and Environmental Engineering, University of Connecticut, CT, for reviewing this chapter and providing valuable technical comments and suggestions.

REFERENCES

Abaqus Users Manual (2013) 'Version 6.13-2', Dassault Systémes Simulia Corp., Providence, Rhode Island.

Badescu, M.,Y. Bar-Cohen, S. Sherrit, X. Bao, H. J. Lee, S. P. Jackson, B. C. Metz, Z. C. Valles, K. Zacny, B. Mellerowicz, D. Kim, and G. Paulsen (2017) '*Auto-Gopher-II: an autonomous wireline rotary-hammer ultrasonic drill*', in *Industrial and Commercial Applications of Smart Structures Technologies 2017*, San Diego, California. doi: 10.1117/12.2260243.

Badescu, M., S. Stroescu, S. Sherrit, J. Aldrich, X. Bao, Y. Bar-Cohen, Z. Chang, W. Hernandez, and A. Ibrahim, (2008) 'Rotary hammer ultrasonic/sonic drill system', in *2008 IEEE International Conference on Robotics and Automation*, Pasadena, California. *IEEE*, pp. 602–607.

Bao, X. Y. Bar-Cohen, Z. Chang, B. P. Dolgin, S. Sherrit, D. S. Pal, S. Du, and T. Peterson, (2003) 'Modeling and computer simulation of ultrasonic/sonic driller/corer (USDC)', *IEEE Transactions on Ultrasonics, Ferroelectrics, and Frequency Control*, 50(9), 1147–1160. doi: 10.1109/TUFFC.2003.1235326.

Bar-Cohen, Y. and Zacny, K. (2009) *Drilling in Extreme Environments: Penetration and Sampling on Earth and Other Planets*. Hoboken, New Jersey: Wiley-VCH.

Bar-Cohen, Y. S. Sherrit, M. Badescu, and X. Bao (2012) 'Drilling, coring and sampling using piezoelectric actuated mechanisms: from the USDC to a piezo-rotary-hammer drill', in *Earth and Space 2012: Engineering, Science, Construction, and Operations in Challenging Environments*, Pasadena, California, 375–384.

Esch, E. et al. (2010) 'A simple method for fabricating artificial kidney stones of different physical properties', *Urological Research*, 38(4), pp. 315–319. doi: 10.1007/s00240-010-0298-x.

Krimholtz, R., Leedom, D. A., and Matthaei, G. L. (1970) 'New equivalent circuits for elementary piezoelectric transducers', *Electronics Letters*. 6(13), pp. 398–399. doi: 10.1049/el:19700280.

Mason, W. P. (1948) *Electromechanical Transducers and Wave Filters*. Princeton, New Jersey: Van Nostrand.

Mason, W. P. (1956) 'Physical acoustics and the properties of solids', *The Journal of the Acoustical Society of America ASA*, 28(6), pp. 1197–1206.

Oakley, C. G. (1997) 'Calculation of ultrasonic transducer signal-to-noise ratios using the KLM model', *IEEE Transactions on Ultrasonics, Ferroelectrics, and Frequency Control*, 44(5), 1018–1026.

Sherrit, S. S. P. L. and Y. Bar-Cohen (1999) 'Comparison of the Mason and KLM equivalent circuits for piezoelectric resonators in the thickness mode', in *1999 IEEE Ultrasonics Symposium. Proceedings. International Symposium (Cat. No. 99CH37027)*, Lake Tahoe, Nevada. IEEE, 921–926.

Sherrit, S. S. A. A., M. Gradziel, B. P. Dolgin, X. Bao, Z. Chang, and Y. Bar-Cohen, (2002) 'Novel horn designs for ultrasonic/sonic cleaning, welding, soldering, cutting, and drilling', in *Smart Structures and Materials 2002: Smart Structures and Integrated Systems*. International Society for Optics and Photonics, San Diego, California, 353–360.

Uchino, K. (2003) *Introduction to Piezoelectric Actuators and Transducers*. University Park, Pennsylvania State University

Van Kervel, S. J. H. and Thijssen, J. M. (1983) 'A calculation scheme for the optimum design of ultrasonic transducers', *Ultrasonics*. Elsevier, 21(3), pp. 134–140.

Vila, L. J. and Malla, R. B. (2016) 'Analytical model of the contact interaction between the components of a special percussive mechanism for planetary exploration', *Acta Astronautica*. doi: 10.1016/j.actaastro.2015.09.016.

Zacny, K. G. Paulsen, B. Mellerowicz, Y. Bar-Cohen, L. Beegle, S. Sherrit, M. Badescu, F. Corsetti, J. Craft, Y. Ibarra, and X. Bao, (2013) 'Wireline deep drill for exploration of Mars, Europa, and Enceladus', in *2013 IEEE Aerospace Conference, Big Sky, Montana*. IEEE, pp. 1–14.

3 Subtractive and Additive Manufacturing Applied to Drilling Systems

Javaid Butt, Vahaj Mohaghegh, Shabnam Sadeghi-Esfahlani, and Hassan Shirvani

Anglia Ruskin University, Cambridge, UK

CONTENTS

3.1 INTRODUCTION

Drilling for oil and natural gas is a multi-billion dollar industry that will continue to grow with every passing day. The reasons for its growth are endless but it mostly comes down to the profitability that is associated with this operation (Allred et al., 2015). In simplest terms, drilling is a process whereby a hole (mostly cylindrical) is made in earth either through soil or rock formations (in most cases both) to create a well for the extraction of oil and natural gas. There are different types of wells that serve various functions. They include exploration wells (drilled for exploration purposes based on recommendations from geologists/scientists), appraisal wells (drilled to assess the properties of an already known oil reserve), development wells (drilled in fields of proven economic and recoverable oil/gas reserves), relief wells (drilled to mitigate blowout from a reservoir), and injection wells (drilled to inject steam or carbon dioxide to maintain the pressure of the oil). There is a reason for these different types as the profits from oil and gas reserves are enormous (Misstear et al., 2017). As an example, it has been estimated that only 1% additional recovery rate from the North Sea (Norwegian sector) alone would generate an additional €36.6 billion (Nguyen et al., 2016).

According to the US Energy Information Administration, in 2017, more than 27 million barrels of oil were produced offshore across 50 countries. The world's largest offshore oil-producing countries are Saudi Arabia, Brazil, Mexico, Norway, and the United States (More, 2020). The long-term outlook for the oil industry, however, may be more promising. For example, most of the global top oil producing nations will have distinct higher production maxima in 2020 than in 2011. It is projected that global oil demand will increase until at least 2035. Daily global oil consumption is expected to grow from 89 million barrels in 2012 up to 109 million barrels in 2035. Transportation and industry will continue to be the sectors with the highest demand for oil (Sönnichsen, 2020). However, extraction of oil is not a simple task as a key issue is water and gas breakthrough, where the unwanted phase of water or gas displaces the required phase of oil. One of the most critical being the premature closure of production capacity caused by unwanted breakthroughs of water or gas in wells. There is a strong and growing need to maximize recovery from new and existing oil fields, and designing drill bits that can help in achieving this goal is getting more and more attention (Augustine et al., 2017; Smalley et al., 2018).

There are different types of drilling operations and each of them uses a specialized drill bit. The types of drilling include but are not limited to auger (uses a helical screw), percussion rotary air blast (uses a pneumatic reciprocating piston-driven hammer to drive the drill), air core (uses hardened steel or tungsten blades on the drills), cable tool (uses a drill string with a heavy carbide-tipped drill bit), reverse circulation (uses a pneumatic reciprocating piston-driven hammer to drive a tungsten-steel drill bit), diamond core (uses an annular diamond-impregnated drill bit), and vibratory (uses a sonic drill head that operates based on vibrations). These drilling types (and associated drill bits) have their place in the oil and gas industry where they are used according to the requirements and desired results. Regardless of type, drill bits must satisfy two primary design goals: maximize the rate of penetration (ROP) of the formation and provide a long service life. These two goals have resulted in extensive research over the years on the design of the drill bits (Cirimello et al., 2018; Dong and Chen, 2016; Gilles et al., 2019).

A few examples that have led companies in the USA in improving the performance of their drill bits by design changes include TerrAdapt adaptive drill bit (from Baker Hughes, Texas), AxeBlade (from Smith Bits—a Schlumberger company, Texas), Cruzer depth-of-cut rolling element drill bit (from Halliburton Corp, North Dakota), and ReedHycalog Tektonic drill bit (from National Oilwell Varco, Texas). TerrAdapt adaptive drill bit features self-adjusting depth-of-cut (DOC) elements. This polycrystalline diamond compact (PDC) bit automatically adjusts to different rock layers without surface intervention by engineers. AxeBlade ridged diamond element bit can improve rates of penetration across various formations and steering response in directional drilling. Cruzer technology makes use of rolling elements at the tip of the drill bit to help maintain DOC control while reducing equipment wear and tear. The cutter technology features a pointed cutting edge to increase drilling efficiency. The chainsaw configuration of the Tektonic drill bit delivers higher ROP than conventional bits (Oil & Gas Journal, 2017). All of these improvements have been made possible due to changes in design that have been investigated for optimal performance.

Making design changes for drill bits is one aspect but manufacturing them is an entirely different area of research. Traditionally, drill bits have been produced using a combination of joining processes and subtractive manufacturing (SM). The process of SM works by removing material from a large metal piece to achieve the final shape and includes a variety of processes including milling, cutting, shaping, turning, etc. The overlap between drilling and SM is owing to the fact that they are both very old fields of work. SM has been able to provide the design flexibility, material requirements, and tight tolerances that the drilling industry needed for decades but times have changed and new competitors have joined the market. Atop that list is additive manufacturing (AM) that produces parts by adding layers on top of each other. AM is an umbrella term and encompasses a variety of methods that include but are not limited to fused deposition modeling, direct metal laser sintering, electron beam melting, selective laser sintering, inkjet printing, laminated object manufacturing, and stereolithography. Just like SM, AM can also work with a multitude of different materials. AM has several advantages over SM as listed in Table 3.1 but perhaps the two most important ones are

TABLE 3.1
Comparison between AM and SM

Additive Manufacturing Processes	Subtractive Manufacturing Processes
Build parts by adding layers on top of each other.	Remove material to build a part.
Lower lead time for complex parts.	Higher lead time for complex parts because of tool changes.
Easy to use for intricate and hollow objects.	Milling undercuts and intricate shapes can be difficult.
Easy setup as no programming or tooling is required for new parts.	New parts require tooling setup and programming for new part geometries.
A wider range of materials can be used including polymers, photopolymers, resins, metals, ceramics, flour, etc.	Wide variety of materials available but can process more alloys compared to AM.
Less waste of raw material.	More raw material is scrapped.
Recycling of powder metal is a norm that saves costs.	Recycling of raw material is very difficult.
Build size has always been an issue with AM.	Bigger parts with longer lengths can be easily made by SM.
For metal parts, some form of post-processing is always required to improve surface finish.	Minimum post-processing is needed for metal parts.
Heat treatment is often utilized to relieve residual stress in metal parts and to improve their mechanical properties.	Heat treatment can be utilized, but the effects are not as noticeable as with metal AM parts.
Residual stress can lead to the production of defective metal parts.	Residual stresses are not as high as in the case of metal AM.
Operator skill is not a huge factor for the quality of the finished product. However, the operator needs to have an in-depth understanding of different process parameters e.g., part orientation, support material placement.	Operator skill can play a part in getting the required results which could lead to variability.
Porosity of parts is a major concern.	SM is not affected by porosity issues.

freedom of design and reduced lead times. With more complex designs for better performance of new drill bits, SM is struggling to deliver the specifics especially in terms of internal channels/passages. Most drill bits have internal passages to direct drilling fluid, conveyed by the drill pipe from surface pumps, through hydraulic nozzles directed at the bottom of the wellbore to produce high-velocity fluid jets that assist in cleaning the old cuttings off the bottom before the next tooth contacts the rock. Since AM can deposit layers on top of each other, complexity of a design and development of internal channels is not an issue but SM struggles with such designs as the tool is unable to reach undercuts and inside a part at certain angles. This chapter will shed some light on the usage of SM and AM for the manufacture of drill bits with the help of case studies. SM is still widely used for the manufacture of drill bits (for small holes in walls to large ones in earth) but AM has started to give strong competition to SM owing to its ability to manufacture any shape without the need for extra tooling.

3.2 DRILL BIT

This is a cutting tool used to remove material to create holes, almost always of circular cross sections. Drill bits come in many sizes and shapes and can create different kinds of holes in many different materials. In order to create holes drill bits are usually attached to a drill, which powers them to cut through the work piece, typically by rotation. A simple drill bit is shown in Figure 3.1.

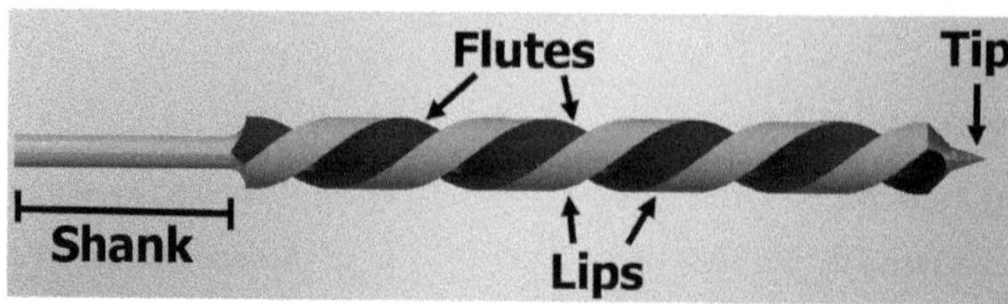

FIGURE 3.1 A simple drill bit.

The tip of the drill cuts through the material and the shank is the end of the drill bit that fits into the drill and secured by the chuck. A round shank helps to center a bit in the chuck more accurately, whereas the flat surfaces on a hex shank allows the chuck to grip the drill bit more securely. The chuck is the part of the drill that attaches the drill bit. The flute is a deep groove that typically twists around the bit, giving the waste material a path out of the hole whereas lips allow for accurate drilling. There is a wide variety of drill bits that are used for different applications. They include but are not limited to twist, brad point, auger, spade, forester, countersink, plug cutter, step, tile, spear point, masonry, etc.

Drill bits are used for a multitude of applications and are made from a variety of different materials ranging from steels to carbides. The applications also extend from small holes in the wall to hang a painting to large holes in the ground through rock formations to extract oil or natural gas. The oil and gas industry has been reliant on SM for decades due to its many advantages in terms of materials and dimensional tolerances. Drilling holes in the Earth's crust is a very lucrative business, especially in certain parts of the world where there is an abundant supply of naturally occurring oil and gas. The drills used for this purpose are extremely large in size and are designed in a specific manner to provide cooling channels to the drill bit as well as angled in ways that the debris can move away and not hinder the drilling operation. These drill bits can be broadly classified according to their primary cutting mechanism i.e., rolling cutter bits and fixed cutter bits. The former drill fractures or crushes the formation with tooth-shaped cutting elements on two or more cone-shaped elements whereas the latter employs a set of blades with very hard cutting elements (e.g., natural or synthetic diamond).

The rolling cutter bits come in different configurations including two, three, or four cone arrangements. They are further classified based on the manufacture of their cutting elements (teeth). The first are steel-tooth bits that have cones with wedge-shaped teeth milled directly in the cone steel itself. The second are tungsten carbide insert (TCI) bits that possess shaped teeth of sintered tungsten carbide pressed-fit into drilled holes in the cones. These bits have a complex design and take weeks to manufacture. On the other hand, the fixed cutter bits are simpler to manufacture compared to rolling cutter bits. The cutting elements do not move relative to the bit; there is no need for bearings or lubrication. The most common cutting element in use today is the polycrystalline diamond cutter (PDC) that has a sintered tungsten carbide cylinder with one flat surface coated with a synthetic diamond material (Wong et al., 2016) as shown in Figure 3.2. There is also a hybrid drill available in the market that combines both rolling cutter and fixed cutter elements for better performance. Regardless of which type of drill bit is being manufactured, traditionally various SM operations are carried out (e.g., cutting, milling, shaping, and turning) alongside other processes such as heat treatment, brazing, and welding. This trend has been impacted positively with other manufacturing operations, especially AM as it allows design freedom and less material waste. These two features have attracted several companies that have invested heavily in AM to improve their designs. This chapter will discuss the significance of both SM (Section 3.3) and AM (Section 3.4) with case studies that show how the manufacture of drill bits has improved with time and new technological advancements.

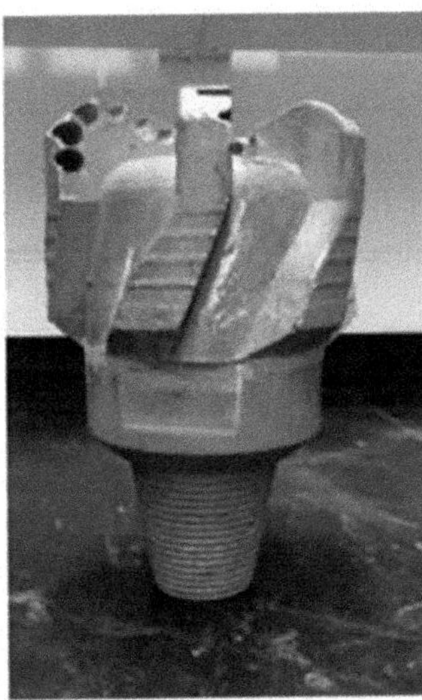

FIGURE 3.2 PDC drill bit. *Source:* Courtesy of Altaf et al. (2016).

3.3 SUBTRACTIVE MANUFACTURING

SM (CNC machining) is one of the oldest manufacturing processes and is an umbrella term that encompasses several processes as shown in Figure 3.3. The process involves "subtracting" the material from a large block to achieve the end product. Several new manufacturing processes have been introduced over the years including AM, but SM still remains one of the most commonly used methods due to its reliability and precision. Today's CNC machines comprise machine tools such as lathes, mills, routers, and grinders that are controlled through the use of numerical control technology—which is comprised of specialized G-code and CAD/CAM computer programs that instruct machine tools how to manufacture specific products. CNC machines can work both autonomously with the help of specialized programming as well as manually by an operator. It gives users the opportunity to design, shape, prototype, and manufacture in end-use materials.

CNC machines can be broadly classified into two categories i.e., conventional machining technologies and unconventional machining technologies. The former utilize wedge-shaped cutting tools and is well-suited for producing less complex large products with average surface finish. It uses tools such as lathes, milling machines, and drill presses. The latter removes extraneous material through methods involving mechanical, thermal, electrical, and chemical energy, or combinations of these types of energy. This is ideal for complex geometrical products and can work easily with extremely tough or brittle materials. Just like AM, there are options for desktop as well as large-sized machines from vendors all over the world with differing characteristics that make machining a more streamlined process. It is an appropriate choice for parts used for small and large volume production runs, to obtain specific finishes, or to obtain specific mechanical properties. The list of materials with which SM can work include ABS, Acetal Copolymer, acrylic, aluminum, brass, Delrin, HDPE, HMW, LDPE, Lexan, Lucite, Nylon, PEEK, PVC, Phenolic, plexiglass, polycarbonate, polypropylene, Rulon, Teflon, UHMW, Ultem, and wood (Newman et al., 2012). SM is capable of providing a very smooth surface finish to its products, which is something that AM struggles with at times. The finish can be functionally important if, for example, parts must slide, and it can be cosmetically important if the prototypes are to be used in market testing. However, because SM involves the removal of material, milling undercuts can sometimes be difficult. Machining also tends

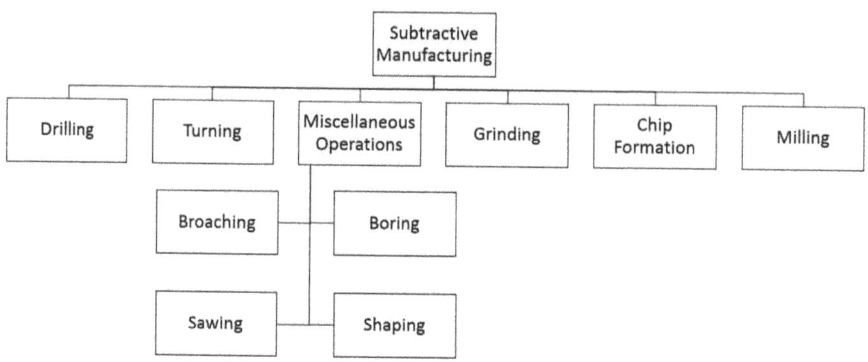

FIGURE 3.3 Classification of SM processes.

to be somewhat more expensive than AM because of high levels of scrap. Notwithstanding that SM generally has longer lead times compared to AM when dealing with complex geometries due to continuous tool change.

One of the biggest drawbacks of SM has been the programming needed to convert a CAD file into machine executable code. This problem has been rectified by software developments. A company called Proto Labs has found a way to automatically translate a 3D CAD model into instructions for high-speed CNC milling and uses a large cluster of computers to quickly generate code. Such enhancements have allowed SM machinery to be able to translate various file types as well including .STL, .DXF, .3DM, and .IGS/IGES. It is quite a breakthrough and some credit goes to AM for this as the open source platforms for AM have helped in the software development for SM as well. Another development in SM is the geometry it can now support as the new improvements in software help in the production of more intricate geometries. AM is still the choice for complex products, such as a sphere within a sphere or parts that interlock in unusual ways. The advantages of SM include a wide selection of end-use materials, good dimensional control and surface finish, and a high degree of repeatability suitable for end-use manufacture. The drawbacks are that there is more material waste compared to AM, and geometry limitations, especially when complex internal channels are needed in a product.

For decades, machining processes were considered the best option for a variety of products including drill bits. Drilling is used for a multitude of applications from small holes in walls of homes to large holes in the earth's crust through rock formations. Oil and gas industry has been making use of various drilling operations to extract, explore, and access natural reserves for decades. Drill bits are used for these drilling operations and they also come in a variety of different configurations depending upon their application. These drill bits include but are not limited to auger drill, PDC drill, down-the-hole drill, air core drill, and tungsten-steel drill. All of them make use of SM and its associated processes during their production. These processes include cutting, milling, trimming, grinding, polishing, turning, etc. For decades, SM has met the design and material requirements for drill bits and the next three sections provide case studies that explain different drill bits and the use of SM in their production.

3.3.1 Case Study 1: Auger Drill Bit

An auger is a drilling device that includes a rotating helical screw blade known as flighting (a spiral twist, or twists, through which the waste material escapes) to act as a screw conveyor to remove the drilled material. A simple auger bit is shown in Figure 3.4. Due to their construction and design, they offer several advantages. These bits do not require much downward pressure, which makes them very effective as they can create good quality holes with ease. They also allow for an efficient

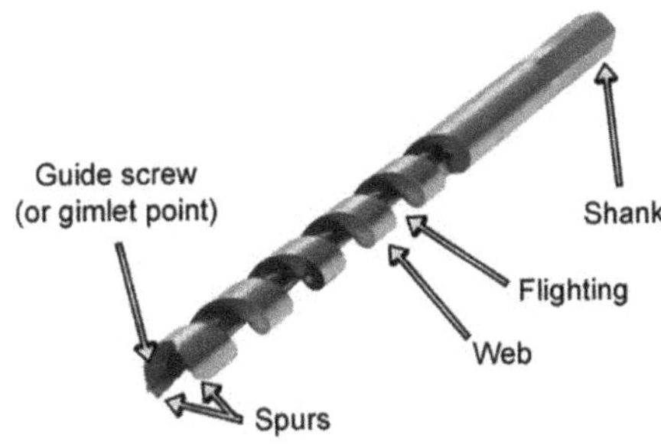

FIGURE 3.4 Simple auger drill bit. *Source:* Courtesy of Wonkee Donkee Tools (2014).

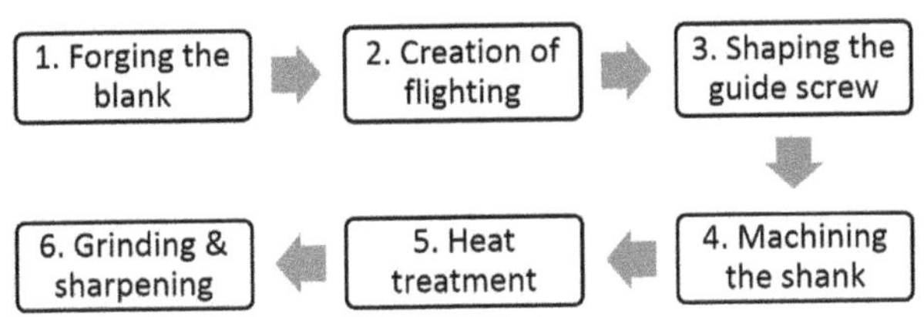

FIGURE 3.5 Production process for auger drill bit.

evacuation of shavings due to the wide and deep flighting. Single-twist auger bits are the most efficient at ejecting waste material, and can bore deep holes without having to be withdrawn at any stage. Even though the advantages make this drill a very attractive option, there are also disadvantages that should be considered. They work at a comparatively slower speed due to the angle and size of their cutting lips.

These drills are used for various applications ranging from making holes in wood to digging postholes (the auger in this case is known as an earth auger). These augers can be operated manually or powered through electric motors or internal engines if connected to a tractor in case of an earth auger. The production process of the auger drill bit is shown in Figure 3.5. From the design standpoint, flighting is very important. The auger's pitch is designed depending on the outside diameter, so that the feeding speed can guarantee the holes' uprightness. The common failure that the auger faces is chip choking, which can raise the torque and cause the drill to stick. Therefore, the design screw should be shaped appropriately according to the application (Griffin et al., 2018). Numerical analysis is very helpful in achieving optimal drilling results from auger bits and should be utilized for new applications (Qianhui, 2014).

As can be seen from Figure 3.5, steps 2, 3, 4, and 6 are operations linked to SM. After heating the blank in step 1, it is then reheated in step 2 and then clamped into a twisting machine, which creates the characteristic spiral shape of the auger. Flighting is created in step 2 and then in step 3, the tip is reheated and slammed under a drop hammer that forces the end point into a die. The waste material is removed from the end, leaving the basic guide screw shape that is then grounded to get the desired geometry. The shank is machined on a lathe in step 4. Heat treatment is used on step 5 to achieve desired mechanical properties for the auger drill bit. In the last step, the auger is subjected firstly to grinding to remove any deformities during the heat treatment and secondly the inside of the flighting is grounded to a smooth finish followed by sharpening of the lips and spurs of the auger. The operations of machining, shaping, grinding, and sharpening (steps 2, 3, 4, and 6) come under SM, hence showing its importance for the manufacture of auger drill bits.

3.3.2 Case Study 2: Polycrystalline Diamond Compact (PDC) Drill Bit

This drill bit is considered to be the workhorse for the oil and gas industry due to its capabilities that allow for better drilling operations. The PDC cutters on a PDC bit are used to shear rock with a continuous rotary motion. The shear strength of rock is usually lower than its compressive strength. Therefore, in soft formation drilling, PDC bits have a high rock breaking efficiency. These are fixed cutter bits which means that there are no moving parts that allow for a smoother operation. The design variables of PDC cutters are the core of the bit design and include cutter size, cutting angles, number of cutters, cutter distributions, and bit profile. Extensive research has been undertaken since 1976 (year of their first production) to improve the design parameters. Much like any other design, simulation tools have played a major role and the PDC drill bits have improved significantly as a result of such tools (Yahiaoui et al., 2013). Optimization strategies have been employed to improve the effectiveness and efficiency of the bit cutter design to reduce the lateral unbalanced force of the bit and to make the wear more uniform (Ali et al., 2018). This has resulted in an improved design with a longer life and a higher rate of penetration.

There are two specific designs for PDC drill bits, i.e., matrix body bits and steel body bits. The difference lies largely based on the materials that are used for their manufacture. The materials used for the matrix body are rather brittle composite material comprising tungsten carbide grains metallurgically bonded with a softer, tougher metallic binder. This limits the impact toughness of a matrix body drill bit, but its resistance to abrasion and erosion makes it a desirable option for specific applications. Being a composite makes this type a bit less predictable compared to a steel body drill bit. The composition of tungsten carbide particles is also a factor that needs to be taken into consideration for optimal performance (Bruton et al., 2014). On the other hand, the steel body is quite different from the matrix body in terms of material properties. They can withstand high impact loads but without appropriate coating would be damaged by abrasion or erosion. They are also much more predictable and can be simulated quite easily to predict the drill bit's behavior. Steel body bits are also relatively larger than the matrix body because steel is ductile and can withstand high-impact forces. Blade height is a feature that takes full advantage of steel and hence the reason for the larger size of the steel body drill bits. Due to comparatively higher erosion resistance, matrix body PDC bits are preferred in erosive drilling conditions. On the other hand, steel bodies are stronger than composite material bodies. Both types have their pros and cons; therefore, the decision should be made based on the application.

A molding process is utilized for the manufacture of matrix body bits but a steel body is easy to produce on a CNC machine as seen in Figure 3.6. This is a time-consuming process and is largely dependent on SM along with other inspection and heat-related processes. Steps 1, 2, 4, 5, 6, 8, and 14 come under the category of SM that goes to show its importance for the manufacture of PDC steel body drill bits that have been a popular choice for the oil and gas industry for decades. Although new design strategies have resulted in extreme modifications of such drill bits, SM is still a part and parcel for their manufacture. This is a testament to the reliability and reproducibility of its results. It is to be noted that in the case of steel body drill bits, it is quite common to use SM to rebuild damaged cutters and this is a massive plus point in low-cost drilling conditions (Jadoun, 2009).

3.3.3 Case Study 3: Diamond Core Drill Bit

Diamond core drilling is an accurate and faster method of drilling concrete, masonry, steel, and asphalt structures. This drilling process eliminates the chances of fractures and reduces spalling which makes it user friendly. It is vibration free, non-percussive and does not cause damage to the adjoining structure. These aspects make it a safe operation and the reason for its preference over other drilling methods. This drilling method is popular for both residential and commercial applications due to its versatility. The drill bit used for this method is known as a diamond core drill bit and very much like the PDC drill bit, this drill bit also makes use of diamonds for cutting/drilling

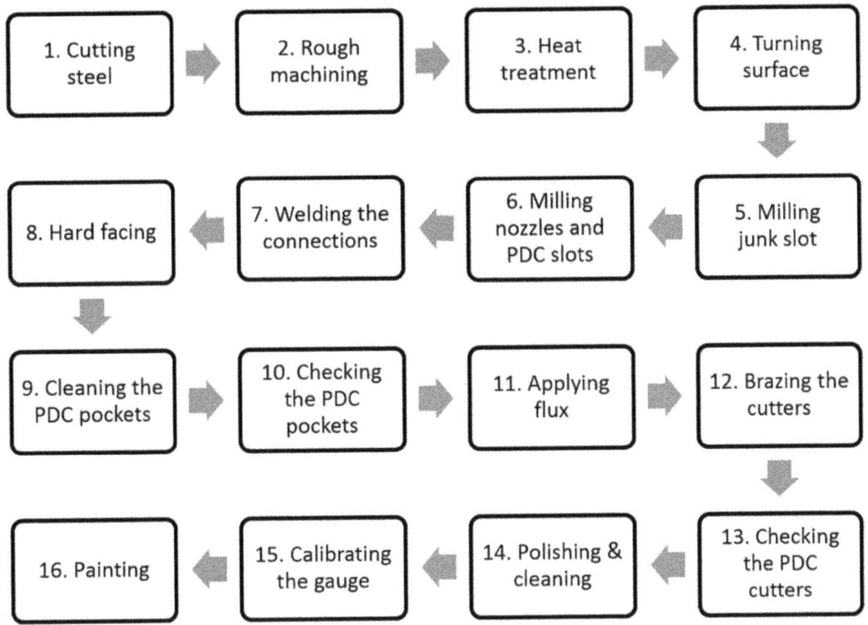

FIGURE 3.6 Production process for steel body PDC drill bit.

operations. Industrial-grade diamonds are used in the construction of this drill bit; therefore, it can easily cut through concrete, masonry, steel, and asphalt structures. These drill bits offer a plethora of advantages that include compact size, light weight, very little noise and vibrations, the ability to run on different power sources (electric, hydraulic, pneumatic), and the production of very little dust during operation. This drill can be used in both vertical and horizontal operations. They can make more accurate cuts and can be used to make holes of different sizes, which is a considerable advantage over other drilling mechanisms (Butler-Smith et al., 2015). They can also be used to create openings for routing cables, installation of the load-carrying devices, or to perform concrete sample analysis (exploration diamond drilling).

There are three different types of diamond drill bits, i.e., impregnated diamond bits, surface-set diamond bits and polycrystalline diamond bits (Flegner et al., 2016). The first type has diamonds impregnated in the crown matrix as shown in Figure 3.7. This helps in the cutting process as during grinding the matrix wears down and exposes the diamonds. The second type has diamonds imbedded in the crown's surface. The depth that these stones are set depends on the size of the stones and will normally be about 2/3 of the overall diameter of the stone. The third type is produced by sintering together multiple layers of diamond particles at very high temperatures and pressure (discussed in Section 3.3.2). Extensive research has been undertaken to improve the efficiency of these drill bits due to their inherent advantages (Wang et al., 2015; Yan et al., 2016; Zhang et al., 2015). There are a variety of variables that can affect the performance of a diamond drill bit. They include but are not limited to diamond content, diamond grit size, bond type, bond hardness, drill wall thickness, drill depth/diameter, and mounting type (Sun et al., 2015).

Regardless of the drill type, water is used as a lubricant and coolant as intense heat is generated during drilling operations. Like other drill bits, diamond core drill bit is hollow which allows water to flow inside and around the inner core to keep the drill bit cool and removing the debris. This is possible due to the internal channels that can be made using traditional SM. Much like other drill bits, SM is considered to be the most viable option for drill bit manufacture. The production process for impregnated diamond drill bit is shown in Figure 3.8. As can be seen, there are multiple steps

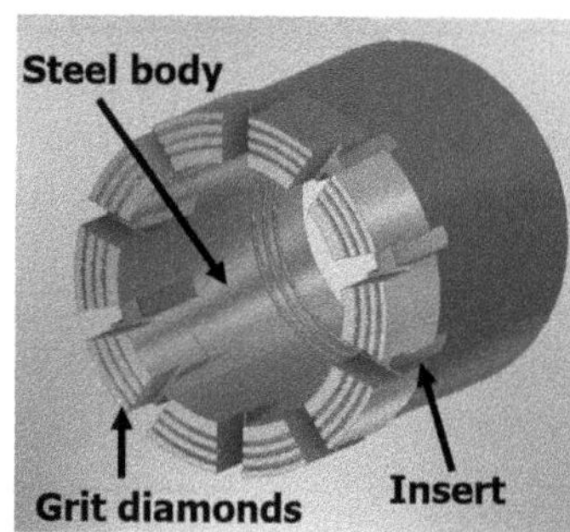

FIGURE 3.7 Impregnated diamond drill bit.

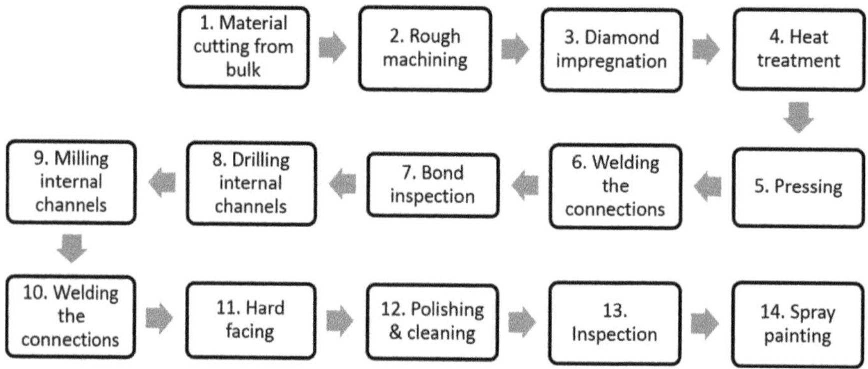

FIGURE 3.8 Production process for impregnated diamond drill bit.

(1, 2, 5, 8, 9, 11, and 12) that require traditional SM methods to achieve the final geometry of the drill bit. SM has been a trusted companion for drill bit manufacturing for decades and has provided the required results at a consistent level. However, there are limitations to this method and this is where AM has been able to give tough competition to SM, owing to its capability to provide freedom of design. The next section will discuss the impact of AM in the manufacturing industry and how it has helped shape the future of drill bits in particular.

3.4 IMPACT OF ADDITIVE MANUFACTURING

AM is an umbrella term that is referred to a group of techniques that can produce 3D objects with the use of 3D CAD data by adding layers on top of each other. The steps involved in the manufacturing operation are shown in Figure 3.9. There are several commercially available methods that are being used extensively for the manufacture of end products that use metal as build material. Table 3.2 shows four broad classifications for metal AM methods.

For a substantial time, AM methods were not considered precise enough to manufacture precision tools. This hindered their applications but this notion has been proved wrong due to the extensive research conducted in the past couple of decades (Wong and Hernandez, 2012). AM methods have emerged as market leaders for complex designs and cost-effectiveness. The focus from the industrial sector has been on metal AM methods as they can provide a ready-to-use product rather than a prototype. The product can be tested for its mechanical properties and performance rather than simply be used for visualization and aesthetics. It is no secret that the adoption/use of AM

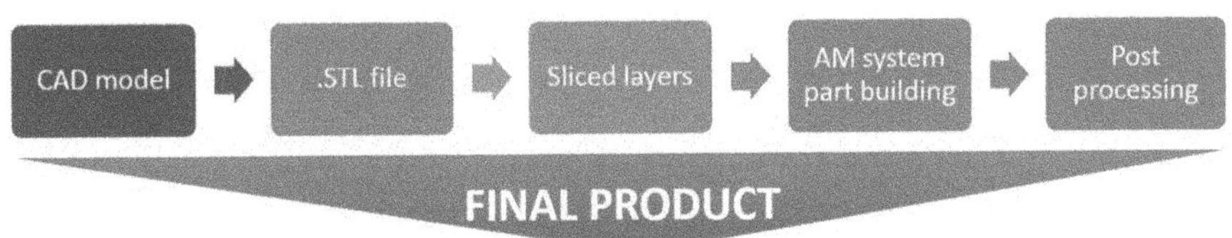

FIGURE 3.9 Steps in additive manufacturing.

TABLE 3.2
Classification of Metal AM Methods

Process	Materials
Powder bed	
Electron Beam Melting (EBM)	
Direct Metal Laser Sintering (DMLS)	Metal powder
Selective Laser Melting (SLM)	
Selective Laser Sintering (SLS)	
Powder feed	
Laser Engineered Net Shaping (LENS)	Metal powder
Laser Consolidation (LC)	
Wire feed	
Electron Beam Additive Manufacturing (EBAM)	Metal wire
Plasma transferred arc selected free form fabrication (PTAS FFF)	
Sheet laminates	
Laminated Object Manufacturing (LOM)	
Ultrasonic Consolidation (UC)	Metal laminate, metal foil
Composite Metal Foil Manufacturing (CMFM)	

offers significant advantages not just to the business but the product as well. Companies generally follow a set of factors important to them to decide on the strategic alignment of their business to new and untested manufacturing methods. AM on the other hand is backed up by decades of intensive research and real-world engineering applications that are growing on a daily basis (Butt et al., 2016; Gibson et al., 2014; Herzog et al., 2016).

AM differs drastically from the traditional SM in notable ways but at the same time there are some similarities that allow a link to be formed between the two principally different manufacturing operations. Both require the use of a CAD package (e.g., SolidWorks or Autodesk Inventor) and post-processing to enhance mechanical properties or improve product finishing/aesthetics. The post-processing operations for SM are more established but they are equally applicable to AM as well depending on the material of the product. For SM, the 3D CAD model needs to be converted to a technical drawing that is easier to translate to g-codes for the CNC machine. AM methods remove this issue completely as the 3D CAD file only needs to be converted to a machine-readable format (e.g., STL) and this job is done by the proprietary software packages of the AM systems.

Furthermore, CNC machining largely depends on the skills of the operator and the end product made by different operators will have variability. On the other hand, AM systems do not suffer from the operator's skills in the long run. In this case, setting the optimal parameters will provide reproducibility of results. However, it is crucial that an operator is experienced enough to prepare the build files to ensure optimal printing of a part. This is particularly challenging with metal AM systems because of parameters such as laser scan speed, laser spot diameter, laser power, hatch distance, layer thickness, etc., as these settings will differ from material to material. SM also requires the use of tooling and specialized fixtures which is a time-consuming and expensive process. AM has removed this aspect entirely due to the nature of its operation, i.e., adding layers on top of layers and using support material where needed. This capability allows AM methods to manufacture complex geometries with intricate features that would be impossible to manufacture otherwise or would take more time and resources with SM. This is the reason why the term "design freedom" is often associated with AM as it does not have the same constraints as SM and is capable of manufacturing complex and intricate geometries in a simpler and efficient manner.

AM has also helped in evolving the design methodologies. Traditionally there have been four focus areas for design, i.e., design for manufacturing (DFM), for assembly (DFA), for both (DFMA), and for disassembly (DFD). For conventional manufacturing, these approaches are linked and allow designers to develop efficient designs that can meet product specifications (DFMA) as well as strategies that consider the future need to disassemble a product for repair, or refurbishment (DFD). AM has made this aspect simpler by introducing design methodologies that are in line with its capabilities, i.e., direct part replacement, part consolidation and optimization based on design for additive manufacturing (DFAM). Direct part replacement allows parts to be made using AM methods quickly and with less waste of material compared to conventional methods. In this strategy, only the production process changes and not the geometry of the product. Part consolidation allows change of the product design to take advantage of AM's ability to make complex shapes. To streamline the final product, multiple parts can be combined into a single build by having feature-rich components, removing joints, and adding mechanisms. AM enables detailed features to be designed and built into components in a single operation (Gibson et al., 2015). Consolidation gives benefits such as reduced tooling costs, shorter overall processing time, shorter lead time, and replacement of complex assembly processes with a single, automated build. DFAM can optimize everything (Strano et al., 2013) for a product and make use of numerical analysis packages (e.g., ANSYS, LS-DYNA). This approach can be considered for different applications such as:

- **Hollow/porous parts**: This allows the manufacture of lattice structures to replace solids to reduce the mass while maintaining required mechanical properties.
- **Topological optimization**: Iterative design process can be utilized in ANSYS that will automatically reduce the weight while keeping the strength the same as before.
- **Improved aesthetics**: This gives designers the freedom to produce unusual and intricate shapes that can be easy to assemble and attractive to customers.
- **Increased surface area**: Complex and textured surfaces only possible with AM can be developed that will enable different interfaces with other systems.
- **Increased heat transfer**: Internal channels in a complex product are only possible through AM and they can allow for a greater heat flow per unit volume to increase the product's performance.
- **High-strength alloys**: AM allows the manufacture of fully dense and high strength alloys, including those that are not used due to poor machinability to increase product's reliability and performance.
- **Materials with anisotropic properties**: AM is capable of manufacturing metal foams that can exhibit different stiffness and thermal conductivity in different planes that can give a product bespoke properties.

- **Build the bill of materials**: AM systems can build an entire product (like a bike frame kit) in a single build which reduces lead time.
- **Mass customization**: Personalization is an important aspect in today's market; something that AM thrives at with the use of 3D scanning systems. A scan can be converted into a readable AM file to manufacture bespoke products for customers.

The aforementioned approaches not only help in the manufacture of better products but also reduce costs. Even though the capital investment in machinery for AM methods is high, technological advancements are continuously bringing these costs down. In terms of both equipment and materials, AM has seen a shift towards affordable 3D printers that are capable of doing more than before (Wang et al., 2017; Waran et al., 2014). This reduction has also encouraged a breakdown of multifunctional teams of designers, technicians, etc. that were traditionally needed for manufacturing operations to a more skilled labor approach, resulting in reduced labor costs. Waste reduction also saves costs as contrasted to SM that removes material which cannot be used. AM builds parts in layers with minimal waste. In addition to that, metal AM methods can recover a large portion of the metal powder that has not been melted and use it for future builds. Now because of the advent of Industry 4.0 (I4.0), AM has gained more popularity as it fits the mold of a digital manufacturing approach that can be used in conjunction with other pillars of I4.0 such as internet of things (IoT), system integration, simulation, etc. IoT is the extension of internet connectivity into physical devices and everyday objects. It involves electronics, internet, and associated hardware—these devices are programmed to communicate with one another through internet and can be remotely monitored and controlled (Dilberoglu et al., 2017).

Multi-national companies such as MAPAL (Aalen, Germany) and APS Technology (Connecticut, USA) have invested heavily in metal AM methods to produce drill bits. The reasoning behind such investments has hinged on two main factors, i.e., competitiveness and profitability (Gurova et al., 2013; Pathak and Saha, 2017; Rysava et al., 2016). System integration allows the bringing together of the component sub-systems into a single process that can allow for better functionality and data sharing with fewer delays and mistakes. Simulation for AM is a very useful tool. Because AM can manufacture products with bespoke properties, it is vital to have an idea of how to go about doing that. Numerical analysis packages, if used appropriately with the correct data, can provide useful results that can be validated through experimentation (Parteli and Pöschel, 2016; Schoinochoritis et al., 2017; Zinoviev et al., 2016). In case of metal AM, a software package called AdditiveLab is gaining popularity due to its dedicated protocols of powder metal AM methods that can help the user in ensuring that the optimal parameters for build are selected. In the same vein, there are some disadvantages associated with the use of AM, as commercially available metal AM systems are limited with regards to material selection and can require a bigger initial investment to set up the manufacturing operations (Cotteleer and Joyce, 2014; Zhai, et al., 2014). These issues also affect the manufacture of drill bits but in cases when the design is extremely complex, there is no other alternative but to use AM. The benefits of AM outweigh the risks and shortcomings quite easily, which is why it is gaining traction from academia and industries alike (Frazier, 2014; Murr et al., 2012).

To conclude, AM has had significant impact on the manufacturing industry and it will continue to make positive strides in the future as well. The discussions in this section form the basis for companies like MAPAL (Aalen, Germany) and APS Technology (Connecticut, USA) to heavily invest in metal AM methods to produce drill bits. The reasoning behind such investments has hinged on two main factors, i.e., competitiveness and profitability. These companies have realized the importance of metal AM, especially for design freedom to manufacture products that have improved performance at reduced costs and time. Three case studies are being presented here. Every case study is preceded by the AM process that is being used, followed by the description of the problem and the benefits that AM has provided.

3.4.1 Case Study 1 Process: LaserCUSING®

The patented LaserCUSING® process from Concept Laser GmbH (Lichtenfels, Germany) has patented the process of LaserCUSING® that is capable of manufacturing precise metal products. The term "LaserCUSING®," coined from the C in Concept Laser and the word FUSING, describes the process, i.e., fusion of metal powder. In this process, fine metal powder is melted locally by a high-energy fiber laser. It is important that the powder is of high quality, i.e., spherical grain structure so that precise parts can be manufactured. The powder material solidifies instantly and the next layer is generated. The contour of the component is created by redirecting the laser beam using a mirror redirection unit (scanner). The part is built up layer by layer (layer thickness generally vary from 15–500 μm) by lowering the build chamber, applying more powder and then melting again as shown in Figure 3.10. This patented process ensures a significant reduction in stress when manufacturing very large components. Much like other metal AM methods, it offers design freedom and near net shape (Becker et al., 2015; Sinirlioglu, 2009). It also allows for reuse of the metal powder that has not been melted during the build process, which makes it a green technology.

Concept Laser GmbH currently has a wide range of machines based on their patented LaserCUSING® process. These machines come in different sizes (build platform) fitted with a variety of laser powders to serve different industries. They all use the same Materialise Build Processor (Leuven, Belgium) for data preparation. Some technical details of these systems are given next:

- **Mlab cusing/Mlab cusing R**: For manufacturing parts with delicate structures.
  Build envelope: 50 × 50 × 80 mm^3 (x, y, z); laser system: fiber laser 100 W.
- **Mlab cusing 200R**: For high surface quality and the finest part structures.
  Build envelope: 100 × 100 × 100 mm^3 (x, y, z); laser system: fiber laser 200 W.
- **M1 cusing**: Suitable to produce small to medium-sized parts.
  Build envelope: 250 × 250 × 250 mm^3 (x, y, z); laser system: fiber laser 200 W (or 400 W).
- **M2 cusing/M2 cusing multilaser**: Designed in line with ATEX guidelines (ATEX directive consists of two EU directives describing what equipment and workspace is allowed in an environment with an explosive atmosphere. ATEX derives its name from the French title of the 94/9/EC directive: Appareils destinés à être utilisés en ATmosphères EXplosives) and thus make it possible to reactive materials.

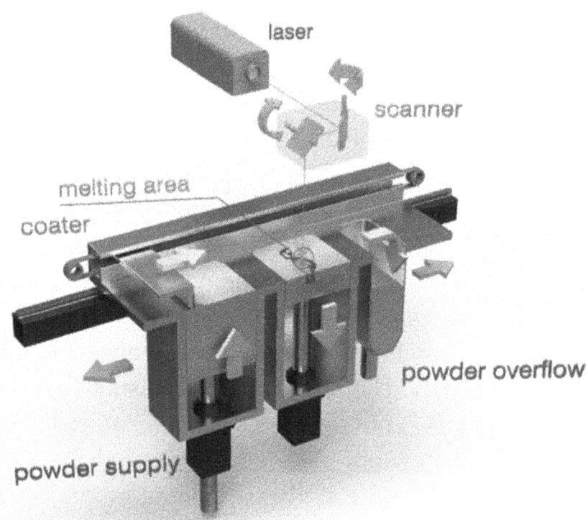

FIGURE 3.10 Patented LaserCUSING process from concept laser. *Source:* Courtesy of GE Additives Concept Laser (2017).

Build envelope: 250 × 250 × 350 mm^3 (x, y, z); M2 cusing laser system: fiber laser 400 W (or 200 W); M2 cusing multilaser laser system: fiber laser 2 × 400 W (or 200 W).

- **M LINE Factory**: Offers a new type of modular machine architecture which provides an unprecedented level of automation and innovation that allows economical series production on an industrial scale.

  Build envelope: 500 × 500 × up to 400 mm^3 (x, y, z); laser system: 3D-optics with maximum power of 4 × 1 kW.
- **X LINE 2000R**: World's largest metal melting machine for the toolless manufacture of large functional components and technical prototypes with series-identical (also reactive) material properties.

Building envelope: 800 × 400 × 500 mm^3 (x, y, z); laser system: fiber laser 2 × 1 kW.

3.4.1.1 Case Study 1: MAPAL Uses LaserCUSING®

MAPAL company from Aalen, Germany is a pioneer for high-tech drilling innovations in automotive, mechanical, tool, and plant engineering. According to them, high performance, long service lives, and rapid tool changes are the central requirements for modern tool design. They have created a new type of drill with the help of an AM process called LaserCUSING® (Section 3.4.1) and have termed it as the QTD insert drill. This drill has surpassed the company's expectations in terms of performance and longevity. Furthermore, AM has provided them with an opportunity to bring their complex design to life that in turn has enabled these QTD insert drills to have good chip deformation and reliable chip removal. The insert is held in a stable prism connection. These precision features make high cutting specifications and drill quality possible. Four types of QTD inserts are offered: for steel, stainless steel, cast iron, and aluminum. This insert drill was previously available in diameters of 13 mm and greater. It was done to ensure that a supply of coolant can be installed within the tool body to cool it down. The smaller the tool body, the greater the adverse effect of the standard central coolant supply on the tool's performance. This supply can significantly weaken the drill core that could result in its instability during operation, which is very dangerous.

Using LaserCUSING®, MAPAL has manufactured new QTD insert drills with a new steel tool body design that has spiral cooling channels not usually used for small diameters. The design uses hybrid manufactured parts—the tool shank is machined conventionally and the drill is additive laser melted—for more economical manufacturing. MAPAL has also introduced a new TTD-Tritan drill. MAPAL claims the TTD-Tritan is the industry's first triple-cutting-edged replaceable head drilling system. Interchangeable heads are the tools of choice for manufacturers aiming to streamline stock and reduce inventory, improve resource efficiency, and combat raw material prices (Industrial Laser Solutions, 2015). These incentives have led to huge investments from MAPAL into the AM research to develop interchangeable tooling systems. This has resulted in the development of a new triple-edged drill for replaceable head drilling applications. The TTD-Tritan was originally developed as a "universal variant" as it is suitable for machining cast iron, stainless steel, and carbon steels. The new complex design of the replaceable heads with an innovative cutting edge geometry will improve drilling performance whilst reducing machining costs and tool inventory (Production Engineering Solutions, 2016).

3.4.2 Case Study 2 Process: Direct Metal Laser Sintering (DMLS)

This process works with different metal alloys but has its limitations. The portfolio of materials that it can work with is growing. EOS GmbH started this process in 1994 that was capable of manufacturing metal parts without the use of binder or fluxing agent. The process is the same as for other AM technologies in that a 3D file in .stl format (stereolithography) is created and sent to the computer where it is sliced into layers to produce a three dimensional object. It works by fusing

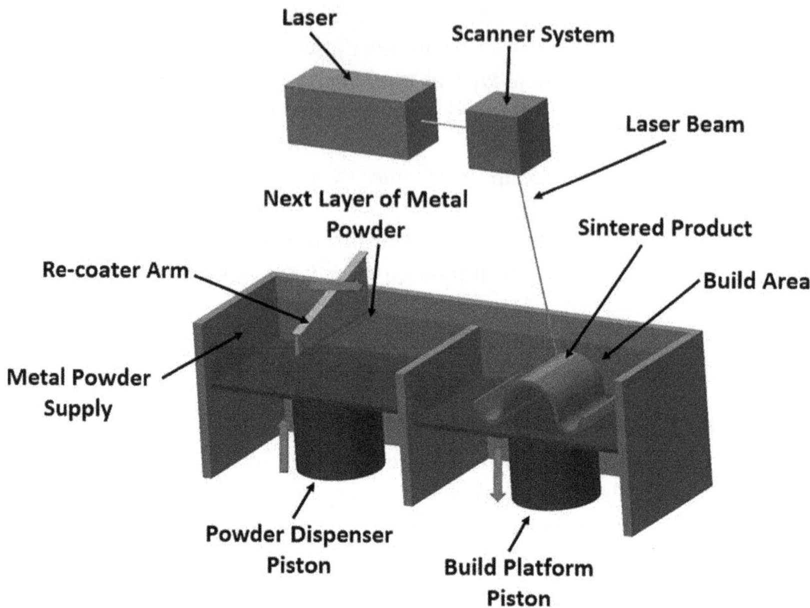

FIGURE 3.11 Direct metal laser sintering process.

together very fine layers of metal powder with the use of a focused laser beam as shown Figure 3.11. This technique requires the use of support structures as the excess powder is not sufficient to hold the object. Once a part is formed, it is essential that it undergoes some post-processing steps before it can be used for any other purpose. This technique can produce almost 99.99% dense metal components (Patterson et al., 2017). Upon its introduction in 1994, DMLS revolutionized the prototyping market in the sense that it used metal as base material; therefore, the prototype can now be formed of the same material as the production components. Thus, a clear and much better idea can be drawn from the testing of the prototypes. Speed is also one of the main advantages as there is no need for any extra tooling and the components can be made in a relatively short period of time compared to SM.

For the case study in Section 3.4.4, APS Technology has made use of EOSINT M280 that is based on the DMLS 3D technology developed by EOS. The technical specifications of the machine are:

Build envelope: 250 × 250 × 325 mm^3 (x, y, z); laser system: Yb (Ytterbium) fiber laser 200 W (or 400 W); diameter of laser beam at building area (variable): 100–500 μm; software: EOS RP Tools, EOSTATE Magics RP (Materialise).

3.4.2.1 Case Study 2: APS Technology Uses DMLS

APS Technology from Connecticut, USA specializes in the manufacture of oilfield drilling equipment. They have made use of DMLS process (Section 3.4.2) to solve a challenging problem related to the complexity of their drilling equipment. Drilling is a difficult endeavor, especially when the holes are several miles deep through various rock formations. This is made somewhat easier by adding complex hole geometries. This innovative technology uses a combination of Measurement While Drilling (MWD) and Rotary Steerable Systems (RSS), mounted immediately behind the drill bit to take real-time measurements of the borehole position and control the trajectory (Navarro-López and Suárez, 2004). Cutting rocks miles under the earth dangerous due to the build-up of heat; therefore, pressurized fluid is used to cool the drill head and remove debris. Since the pressure is very high, it causes damage to the drilling equipment, even destroying super tough Inconel and 17-4 stainless steel. To solve these issues, APS aims to help the well operators by offering them a variety of intelligent tools, including steerable drill motors, vibration dampers, modeling and analysis tools, and logging sensors, in addition to its MWD systems. This is where the AM process of DMLS comes into the picture. The system used by APS is the EOSINT M280 that can easily manufacture complex shapes in a relatively short span of time (EOS, 2015). The system uses a high-power Yb-fiber laser and precision scanning optics to trace tissue-thin slices of a CAD model onto a bed of fine metal powder. The company started off with a five-stage turbine used to power a steerable drilling head and its on-board MWD system. Each turbine contains several complex end housings and five sets of stators and rotors; of which several parts are printed using DMLS. The use of DMLS technology from EOS allowed APS to reduce its part count by making use of principles such as part consolidation and DFAM. The company has saved money by reducing the budget on machining and tooling with a new wave of ingenious designs being added to the company's portfolio.

3.4.3 Case Study 3 Process: ThermoJet™ or Inkjet Printing

This process is based on the 2D printer technique of using a jet to deposit tiny drops of ink onto a paper. In this case, the ink is replaced with thermoplastic and wax materials, which are held in a melted state. When printed, liquid drops of these materials instantly cool and solidify to form a layer of the part. This process has several advantages including excellent accuracy and surface finish. However, the limitations include slow build speeds, few material options, and fragile parts (Kim et al., 2006). Typical applications include jewelry, medical devices, and high-precision products (Calvert, 2001). Several manufacturers have developed different inkjet printing devices; e.g., 3D Systems has implemented their MultiJet Modeling (MJM) technology into their ThermoJet™ Modeler machines that utilize several hundred nozzles to enable faster build times.

The inkjet printing process begins with the build material (thermoplastic) and support material (wax) being held in a melted state inside two heated reservoirs as shown in Figure 3.12. These materials are each fed to an inkjet print head, which moves in the X-Y plane and shoots tiny droplets to the required locations to form one layer of the part. Owing to the speed of the process, both the build and support material instantly cool and solidify. After a layer has been completed, a milling head moves across the layer to smooth the surface. Milling leaves behind particles that are vacuumed away to preserve the integrity of the part being built. The elevator then lowers the build platform along with the part so that the next layer can be deposited. This process is repeated until the required height is achieved. After that, the part is removed and the wax support material is melted away.

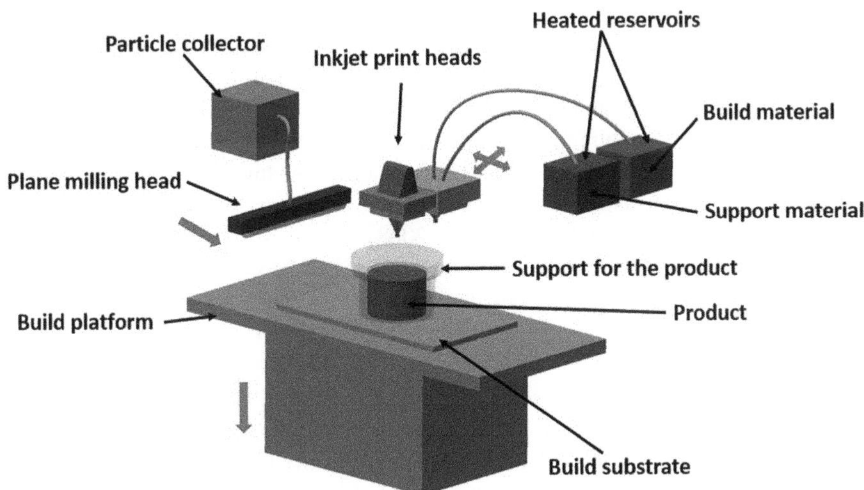

FIGURE 3.12 ThermoJet™/inkjet printing.

TABLE 3.3
Technical Specifications of ThermoJet™ Solid Object Printer

Parameter	X (Left/Right)	Y (In/Out)	Z (Up/Down)
Maximum print size (mm)	250	190	200
Resolution DPI (dots per inch)	400	300	600
Resolution (m)	64	85	42

The ThermoJet™ Solid Object Printer used in Section 3.4.5 is from 3D Systems. It effectively communicates complex designs to project teams, other departments, customers, partners, and vendors. There are several different variations of the machines made by 3D Systems based on this technology. The technical specifications of the ThermoJet™ Solid Object Printer are shown in Table 3.3.

3.4.3.1 Case Study 3: Polycrystalline Diamond Compact Drill Bit Body

This case study will discuss the manufacture of an improved design of Polycrystalline Diamond Compact (PDC) drill bit body by using ThermoJet™ three-dimensional printing (Section 3.4.3) and investment casting. Areas of interest include ease of manufacture, reduced lead time, cost-effectiveness, and precision of the PDC drill bit body. These drills are generally utilized for drilling a well aiming to find and developing new petroleum reserves. The success of the operation depends largely on the performance of the drill bit. This has led to significant research in areas such as cutter/formation connection, cutter performance, bit dynamics, and bottom hole assembly dynamics (Atici and Ersoy, 2009; Mensa-Wilmot et al., 2003; Motahhari et al., 2010). PDC drills are typically made from either carbon steel or stainless steel. SM has been the main manufacturing method for these drills where material is removed from the main part until the final shape has been achieved. The smaller and less complex part of the drill bit is subjected to this practice whereas the larger and complex section requires the production of a mold. Both methods are exposed to machining that leads to significant waste of material. Due to the size and complexity of the drill bit, multiple subtractive operations are needed that add to the time and cost of manufacture. Production time is counted in weeks and along with the associated costs, a full-sized PDC drill is not considered to be economically viable to manufacture using conventional means.

The basic sequence to produce a steel bit pattern is machining multiple bulk metal into a desired form by using CNC machines. The pattern can be two or multiple pieces of molds or dies. Molten steel is then poured in the combination of patterns. The shell mold is broken from casting at the end of the cooling period, and machining is done to achieve the final shape of the drill. Amongst the steps, casting only takes a small percentage of the time and a low investment for the fabrication process. Most of the time and investment are due to the pattern preparation since it includes expensive tooling and total cost of the material (Wells et al., 2008).

Several researchers have worked on the development of PDC drill bits but the focus here is on the work done by Altaf et al. (2016) where they developed a PDC drill bit by a combinational strategy of an AM technique (ThermoJet™/inkjet printing) and investment casting. Their approach included modifications in the original design to take advantage of AM capabilities. The empty space inside the wall was increased and changed in shape accordingly to actual drill bit flow line in as shown in Figure 3.13. A groove joint was inserted on the top side and bottom side model to ease merge after investment casting as shown in Figure 3.14. Furthermore, to save time and utilize less support material, the orientation of the parts was changed as shown in Figure 3.15. These parts will serve the role of sacrificial wax patterns and were 3D printed using the ThermoJet™ 3D printer.

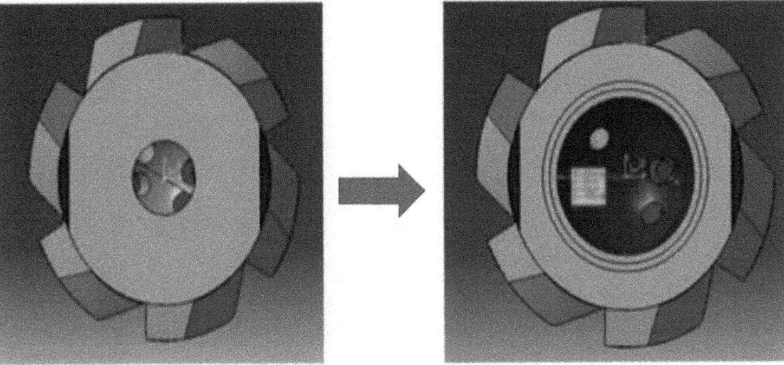

FIGURE 3.13 Original (left) and modified (right) top section of the PDC drill bit body. *Source:* Courtesy of Altaf et al. (2016).

FIGURE 3.14 Original (left) and modified (right) bottom section of the PDC drill bit body. *Source:* Courtesy of Altaf et al. (2016).

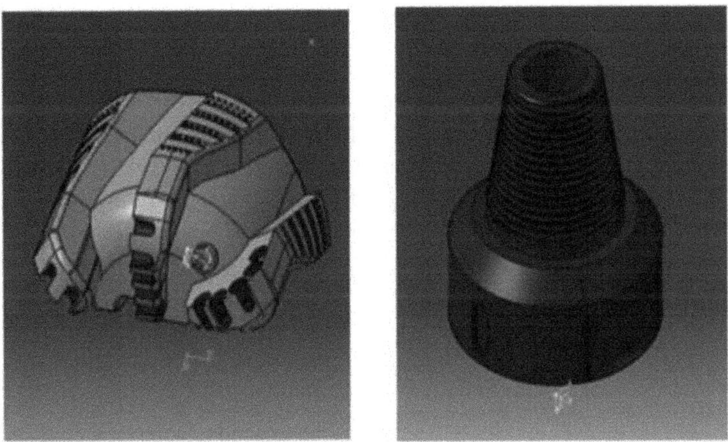

FIGURE 3.15 Orientation of parts for 3D printing. *Source:* Courtesy of Altaf et al. (2016).

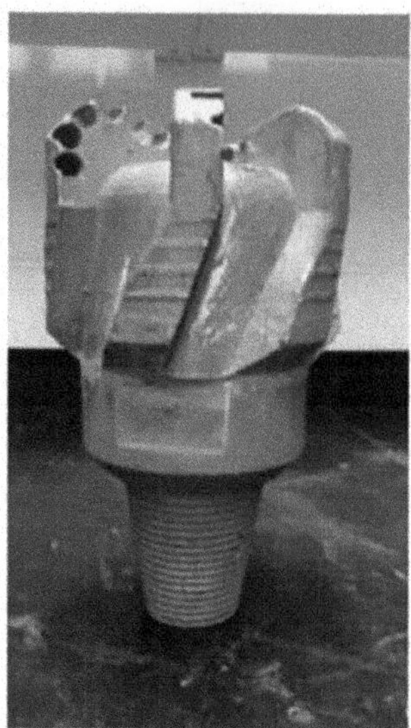

FIGURE 3.16 Investment casted PDC drill body before (left) and after (right) painting. *Source:* Courtesy of Altaf et al. (2016).

The wax patterns are converted into solid parts following a series of multi-step processes collectively known as investment casting. It allows the production of intricate geometries and features from a variety of metals with high precision and accuracy. Stainless steel was used to convert the wax patterns into solid parts as shown in Figure 3.16. Several issues were faced during this activity, especially the removal of support material after 3D printing from difficult-to-access sections as well as porosity of the final drill bit body (Pak et al., 2017). These two factors could potentially limit the use of this combinational strategy of ThermoJet™ three-dimensional printing and investment casting. However, this method has succeeded in developing a ready-to-use PDC drill in an economical and time-effective manner which goes to show that AM can work as an ally with other methods to achieve desired results. It is crucial to weigh the pros and cons of working with AM using a holistic approach that will benefit the product (e.g., complexity, materials) and the individual making it (in terms of cost and time savings).

3.5 SUMMARY

Oil and gas is a multi-billion dollar industry and deals with processes that involve the exploration and extraction of natural reserves among other global processes (e.g., refining, transporting, and marketing). Different drilling operations are used in these endeavors that makes use of specially designed drill bits. The manufacture of these drill bits is an important area and combines design optimization strategies such as design for manufacture/assembly/disassembly, etc., along with the manufacturing process itself. For decades, SM has met all the design and material requirements for the drill bits and has proved to be a reliable method. However, extensive research over decades to improve the performance of drill bits has led to more complex designs that SM is unable to create. This is where AM has joined the competition as it has the ability to iteratively add layers for the manufacture of a product and is especially well suited for internal channels/passages that are needed to allow for water circulation to cool down the drill bit. AM can work easily with SM as an ally to ensure that optimal results are obtained both in terms of design and cost-effectiveness. Both manufacturing methods have their pros/cons and should be used after considering all the pertinent factors.

ACKNOWLEDGMENTS

The authors would like to thank Sathish Nammi, Brunel University, London, UK; and John Paul C. Borgonia, Jet Propulsion Laboratory/California Institute of Technology, Pasadena, CA, for reviewing this chapter and providing valuable technical comments and suggestions.

REFERENCES

Ali, Z., Han, Y., Kuang, Y., Wang, Y. and Zhang, M., 2018. Optimization model for polycrystalline diamond compact bits based on reverse design. *Advances in Mechanical Engineering*, 10(6), https://doi.org/10.1177/1687814018781494.

Allred, B.W., Smith, W.K., Twidwell, D., Haggerty, J.H., Running, S.W., Naugle, D.E. and Fuhlendorf, S.D., 2015. Ecosystem services lost to oil and gas in North America. *Science*, 348(6233), pp. 401–402.

Altaf, K., Rani, A.M.A., Woldemichael, D.E., Lemma, T.A. and Zhi, C., 2016. Application of additive manufacturing/3D printing technologies and investment casting for prototype development of polycrystalline diamond compact (PDC) drill bit body. *ARPN Journal of Engineering and Applied Sciences*, 11(10), pp. 6514–6518.

Atici, U. and Ersoy, A., 2009. Correlation of specific energy of cutting saws and drilling bits with rock brittleness and destruction energy. *Journal of Materials Processing Technology*, 209(5), pp. 2602–2612.

Augustine, A., Junin, R., Jeffrey, G. and Onyenkonwu, M., 2017. Exploitation of thin oil rims with large associated gas cap. *International Journal of Petroleum Engineering*, 3(1), pp. 14–48.

Becker, T., Van Rooyen, M. and Dimitrov, D., 2015. Heat treatment of Ti-6Al-4V produced by lasercusing. *South African Journal of Industrial Engineering*, 26(2), pp. 93–103.

Bruton, G., Crockett, R., Taylor, M., DenBoer, D., Lund, J., Fleming, C., Ford, R., Garcia, G. and White, A., 2014. PDC bit technology for the 21st century. *Oilfield Review*, 26(2), pp. 48–57.

Butler-Smith, P.W., Axinte, D.A., Daine, M., Kennedy, A.R., Harper, L.T., Bucourt, J.F. and Ragueneau, R., 2015. A study of an improved cutting mechanism of composite materials using novel design of diamond micro-core drills. *International Journal of Machine Tools and Manufacture*, 88, pp.175–183.

Butt, J., Mebrahtu, H. and Shirvani, H., 2016. Microstructure and mechanical properties of dissimilar pure copper foil/1050 aluminium composites made with composite metal foil manufacturing. *Journal of Materials Processing Technology*, 238, pp. 96–107.

Calvert, P., 2001. Inkjet printing for materials and devices. *Chemistry of Materials*, 13(10), pp. 3299–3305.

Cirimello, P., Otegui, J.L., Sanchez, J.M. and Carfi, G., 2018. Oil well drill bit failure during pull out: Redesign to reduce its consequences. *Engineering Failure Analysis*, 83, pp. 75–87.

Cotteleer, M. and Joyce, J., 2014. 3D opportunity: Additive manufacturing paths to performance, innovation, and growth. *Deloitte Review*, 14, pp. 5–19.

Dilberoglu, U.M., Gharehpapagh, B., Yaman, U. and Dolen, M., 2017. The role of additive manufacturing in the era of industry 4.0. *Procedia Manufacturing*, 11, pp. 545–554.

Dong, G. and Chen, P., 2016. A review of the evaluation, control, and application technologies for drill string vibrations and shocks in oil and gas well. *Shock and Vibration*, 2016, 34, Article ID 7418635, https://doi.org/10.1155/2016/7418635.

Flegner, P., Kačur, J., Durdán, M., Laciak, M., Stehlíková, B. and Pástor, M., 2016. Significant damages of core diamond bits in the process of rocks drilling. *Engineering Failure Analysis*, 59, pp. 354–365.

Frazier, W.E., 2014. Metal additive manufacturing: a review. *Journal of Materials Engineering and Performance*, 23(6), pp. 1917–1928.

Gibson, I., Rosen, D.W. and Stucker, B., 2014. *Additive Manufacturing Technologies* (Vol. 17). New York: Springer.

Gibson, I., Rosen, D. and Stucker, B., 2015. Direct digital manufacturing. In *Additive Manufacturing Technologies* (pp. 375–397). Springer, New York.

Gilles, P., Olivier, S., Danny, T., Thomas, G., Bruno, C. and Julien, C., March 2019. *Modelling the 3D Bit-Rock Interaction Helps Designing Better PDC Bits.* In *SPE/IADC International Drilling Conference and Exhibition*, held at The Hague, The Netherlands. Society of Petroleum Engineers. https://www.onepetro.org/conference-paper/SPE-194134-MS.

Griffin, W.T., Phelps, T.J., Colwell, F.S. and Fredrickson, J.K., 2018. Methods for obtaining deep subsurface microbiological samples by drilling. *In Microbiology of the Terrestrial Deep Subsurface.*, Penny S. Amy (Ed) (pp. 23–44). Boca Raton, FL: CRC Press. ISBN 9781315895468

Gurova, M., Bonsall, C., Bradley, B. and Anastassova, E., 2013. Approaching prehistoric skills: experimental drilling in the context of bead manufacturing. *Bulgarian e-journal of Archaeology*, 3(2), pp. 201–221. https://be-ja.org/index.php/journal/article/view/be-ja-3-2-2013-201-221

Herzog, D., Seyda, V., Wycisk, E. and Emmelmann, C., 2016. Additive manufacturing of metals. *Acta Materialia*, 117, pp. 371–392.

Jadoun, R.S., 2009. Study on rock-drilling using PDC bits for the prediction of torque and rate of penetration. *International Journal of Manufacturing Technology and Management*, 17(4), pp. 408–418.

Kim, H.C., Lee, S. and Lee, S.H., 2006. Rapid tooling technology for producing functional prototypes using ceramic shell investment casting and patterns produced directly from ThermoJet 3D printer. *Journal of the Korean Society for Precision Engineering*, 23(8), pp. 203–210.

Mensa-Wilmot, G., Soza, R. and Hudson, K., January 2003. *Advanced cutting structure improves PDC bit performance in hard rock drilling environments.* In *SPE Annual Technical Conference and Exhibition*, Denver, CO.

Misstear, B., Banks, D. and Clark, L., 2017. *Water Wells and Boreholes*. New York: John Wiley & Sons.

Motahhari, H.R., Hareland, G. and James, J.A., 2010. Improved drilling efficiency technique using integrated PDM and PDC bit parameters. *Journal of Canadian Petroleum Technology*, 49(10), pp. 45–52.

Murr, L.E., Gaytan, S.M., Ramirez, D.A., Martinez, E., Hernandez, J., Amato, K.N., Shindo, P.W., Medina, F.R. and Wicker, R.B., 2012. Metal fabrication by additive manufacturing using laser and electron beam melting technologies. *Journal of Materials Science & Technology*, 28(1), pp. 1–14.

Navarro-López, E.M. and Suárez, R., September 2004. *Practical approach to modelling and controlling stick-slip oscillations in oilwell drillstrings.* In *Proceedings of the 2004 IEEE International Conference on Control Applications, 2004,* held in Taipei, Taiwan (Vol. 2, pp. 1454–1460). IEEE.

Newman, S.T., Nassehi, A., Imani-Asrai, R. and Dhokia, V., 2012. Energy efficient process planning for CNC machining. *CIRP Journal of Manufacturing Science and Technology*, 5(2), pp. 127–136.

Nguyen, T.V., Tock, L., Breuhaus, P., Maréchal, F. and Elmegaard, B., 2016. CO2-mitigation options for the offshore oil and gas sector. *Applied Energy*, 161, pp. 673–694.

Pak, M., Gumich, D., Zinnatullin, I., Mukkisa, S., Bits, S. and Gaynullin, I., March 2017. *Unique approach to bit design coupled with innovative rolling PDC cutter sets new performance benchmark drilling extremely abrasive sandstone formations, Usinsk Region Russia.* In *SPE Middle East Oil & Gas Show and Conference*, held in Mumbai, India, Society of Petroleum Engineers

Parteli, E.J. and Pöschel, T., 2016. Particle-based simulation of powder application in additive manufacturing. *Powder Technology*, 288, pp. 96–102.

Pathak, S. and Saha, G., 2017. Development of sustainable cold spray coatings and 3D additive manufacturing components for repair/manufacturing applications: a critical review. *Coatings*, 7(8), p. 122.

Patterson, A.E., Messimer, S.L. and Farrington, P.A., 2017. Overhanging features and the SLM/DMLS residual stresses problem: Review and future research need. *Technologies*, 5(2), p.15.

Qianhui, J., 2014. Research and application of auger-air drilling and sieve tube borehole protection in soft outburst-prone coal seams. *Procedia Engineering*, 73, pp. 283–288.

Rysava, Z., Bruschi, S., Carmignato, S., Medeossi, F., Savio, E. and Zanini, F., 2016. Micro-drilling and Threading of the Ti6Al4 V Titanium Alloy Produced through Additive Manufacturing. *Procedia CIRP*, 46, pp. 583–586.

Schoinochoritis, B., Chantzis, D. and Salonitis, K., 2017. Simulation of metallic powder bed additive manufacturing processes with the finite element method: A critical review. *Proceedings of the Institution of Mechanical Engineers, Part B: Journal of Engineering Manufacture*, 231(1), pp. 96–117.

Sinirlioglu, M.C., 2009. Rapid manufacturing of dental and medical parts via LASERCUSING® technology using titanium and CoCr powder materials. *RapidTech*, 2009(1), p. 718.

Smalley, P.C., Muggeridge, A.H., Dalland, M., Helvig, O.S., Høgnesen, E.J., Hetland, M. and Østhus, A., April 2018. *Screening for EOR and estimating potential incremental oil recovery on the Norwegian Continental Shelf.* In *SPE Improved Oil Recovery Conference*, held in Tulsa, Oklahoma. Society of Petroleum Engineers.

Strano, G., Hao, L., Everson, R.M. and Evans, K.E., 2013. A new approach to the design and optimisation of support structures in additive manufacturing. *The International Journal of Advanced Manufacturing Technology*, 66(9–12), pp. 1247–1254.

Sun, B., Bóna, A., Zhou, B. and van de Werken, M., 2015. A comparison of radiated energy from diamond-impregnated coring and reverse-circulation percussion drilling methods in hard-rock environments. *Geophysics*, 80(4), pp. K13–K23.

Wang, J., Cao, P., Liu, C. and Talalay, P.G., 2015. Comparison and analysis of subglacial bedrock core drilling technology in Polar Regions. *Polar Science*, 9(2), pp. 208–220.

Wang, X., Jiang, M., Zhou, Z., Gou, J. and Hui, D., 2017. 3D printing of polymer matrix composites: A review and prospective. *Composites Part B: Engineering*, 110, pp. 442–458.

Waran, V., Narayanan, V., Karuppiah, R., Owen, S.L. and Aziz, T., 2014. Utility of multimaterial 3D printers in creating models with pathological entities to enhance the training experience of neurosurgeons. *Journal of Neurosurgery*, 120(2), pp. 489–492.

Wells, M., Marvel, T. and Beuershausen, C., January 2008. *Bit balling mitigation in PDC bit design*. In *IADC/SPE Asia Pacific Drilling Technology Conference and Exhibition*, held at Jakarta, Indonesia. Society of Petroleum Engineers.

Wong, A., Bell, A., Williams, M., Isnor, S. and Herman, J.J., October 2016. New material technologies reduce PDC drill bit body and cutter erosion in heavy oil drilling environments. In *SPE Latin America and Caribbean Heavy and Extra Heavy Oil Conference*, held at Lima, Peru. Society of Petroleum Engineers.

Wong, K.V. and Hernandez, A., 2012. A review of additive manufacturing. *ISRN Mechanical Engineering*, 2012.

Yahiaoui, M., Gerbaud, L., Paris, J.Y., Denape, J. and Dourfaye, A., 2013. A study on PDC drill bits quality. *Wear*, 298, pp. 32–41.

Yan, G., Yue, W., Meng, D., Lin, F., Wu, Z. and Wang, C., 2016. Wear performances and mechanisms of ultrahard polycrystalline diamond composite material grinded against granite. *International Journal of Refractory Metals and Hard Materials*, 54, pp. 46–53.

Zhai, Y., Lados, D.A. and LaGoy, J.L., 2014. Additive manufacturing: making imagination the major limitation. *Jom*, 66(5), pp. 808–816.

Zhang, F.L., Liu, P., Nie, L.P., Zhou, Y.M., Huang, H.P., Wu, S.H. and Lin, H.T., 2015. A comparison on core drilling of silicon carbide and alumina engineering ceramics with mono-layer brazed diamond tool using surfactant as coolant. *Ceramics International*, 41(7), pp. 8861–8867.

Zinoviev, A., Zinovieva, O., Ploshikhin, V., Romanova, V. and Balokhonov, R., 2016. Evolution of grain structure during laser additive manufacturing. Simulation by a cellular automata method. *Materials & Design*, 106, pp. 321–329.

INTERNET LINKS

EOS. (2015, May 10). *APS Technology Oilfield Drilling Equipment Manufacturer.* https://www.eos.info/en/3d-printing-examples-applications/all-3d-printing-applications/aps-technology-industry-3d-printed-oilfield-drilling-equipment

GE Additives Concept Laser. (2017, June 22). *LaserCUSING*. https://www.concept-laser.de/en/technology.html

Industrial Laser Solutions. (2015, September 18). *Mapal relies on additive manufacturing to produce insert drills. https://www.industrial-lasers.com/articles/print/volume-30/issue-5/departments/update/mapal-relies-on-additive-manufacturing-to-produce-insert-drills.html*

More, A. (2020, August 18). *Offshore Drilling Rigs Market Size, Share 2020 Global Industry Trends, Sales Revenue, Industry Growth, Development Status, Top Leaders, Future Plans and Opportunity Assessment 2023*. MarketWatch. https://www.marketwatch.com/press-release/offshore-drilling-rigs-market-size-share-2020-global-industry-trends-sales-revenue-industry-growth-development-status-top-leaders-future-plans-and-opportunity-assessment-2023-2020-08-18

Oil & Gas Journal. (2017, November 6). *Smart bits advance drilling efficiency.* https://www.ogj.com/articles/print/volume-115/issue-11/special-report-drilling-technology-update/smart-bits-advance-drilling-efficiency.html

Production Engineering Solutions. (2016, September 13). *Triple benefits*. https://www.pesmedia.com/triple-benefits-mapal-cutting-tool/

Sönnichsen, N. (2020, June 22). *Global Oil Industry and Market—Statistics & Facts*. Statista. https://www.statista.com/topics/1783/global-oil-industry-and-market/

Wonkee Donkee Tools. (2014, August 4). *What are the parts of an auger bit? https://www.wonkeedonkeetools.co.uk/auger-bits/what-are-the-parts-of-an-auger-bit*

4 Onshore Drilling

Alfred Eustes, Nicole Bourdon, Deep Joshi, Kirtland McKenna, Ozan Uzun, Zachary Zody
Colorado School of Mines, Golden, CO
Saleh Alhaidari, and
Saudi Aramco, Dhahran, Saudi Arabia
Ahmed Amer
Newpark Drilling Fluids, Houston, TX

CONTENTS

4.1 SHALE REVOLUTION: WHAT MADE UNCONVENTIONAL COMMERCIALLY VIABLE?

In recent years, the development of unconventional oil and gas resources, such as shale gas, tight oil, oil sands, and gas coal seam gas has become economically viable and technically feasible. These resources play a crucial, ever-increasing role in the energy supply. The availability of unconventional oil and gas is orders of magnitude greater than the conventional deposits that were the primary target for drilling and production for the past two centuries. And unconventional reservoirs are becoming increasingly important around the world.

The revolution of the unconventional reservoir production has been historic. It was unpredicted, 10 years ago, that the United States would regain such position as the largest producer of unconventional resources in the world. Yet that it has happened is because of advances in drilling and completion technology. This represents a breakthrough on the scale of that achieved when the oil industry started offshore drilling shortly after WWII.

Production from unconventional reservoirs starts in 1821 where the gas was produced from shallow, naturally fractured shale formations in the Northeastern United States. Nonetheless, today's shale revolution did not certainly begin until 1985 when George Mitchel developed an economically viable exploitation model of the Barnett shale. It took almost 20 years and literally hundreds of wells to find the right formula, based on two transformative technologies: horizontal drilling and multistage hydraulic fracturing. Drilling and completion costs have been brought down by 50% and more.

This onshore shale revolution is due largely to the application of new technologies, efficient management, and utilizing every bit of data available, on every scale. These include new purpose-built drilling rigs designed for drilling wells in clusters (called pad drilling). These rigs have high pressure fluid systems, can walk on their own, can be easily moved, and have sensors and electronic data recording of drilling operational parameters.

New drilling fluid formulations have allowed for longer horizontal sections, reducing friction, and improving formation damage control.

Drill bits have evolved from tri-cone bits to polycrystalline diamond compact (PDC) bits. Today, the longevity and durability of PDC bits have allowed wells to be drilled in a single run, even in horizontal wells of 15,000 feet or more.

In the United States, the vast majority of oil and gas wells drilled today use directional drilling, especially horizontal drilling. This has required technology improvements in power, torque, and reliability in downhole hydraulic motors, called "mud motors." These are used because these systems do not allow for the rotation of the entire drilling string. To allow the entire string to rotate (which reduces friction), rotary steerable systems (RSS), where the entire directional control system rotates, have been developed. Downhole sensors to measure drilling and lithology parameters were created and built. Measurement-while-drilling (MWD) measures drilling parameters such as inclination, azimuth, bit orientation, temperature, pressures, forces, etc. Logging-while-drilling (LWD) measures lithological properties such as natural radioactivity, density, resistivity, etc. Both send signals to the surface through mud pulse telemetry, extra-long frequency radio, or physical wires.

The use of drilling information from the surface and subsurface in the form of drilling operational parameters such as time, weight on bit (WOB), rotational speed (both surface and downhole which are likely different), torque on bit, pump pressure and rates, rate of penetration, and mechanical specific energy to name a few parameters, have been used to optimize the drilling performance. In many ways, the use of this data has opened the eyes of many a drilling engineer about their drilling systems and insights into how to improve operations and especially performance. Non-productive time (NPT) has been the bane of many a drilling operation. By measuring and learning to optimize these parameters, drilling engineers and managers have significantly improved drilling times.

And the management of drilling operations has changed significantly. The days of drilling by the "seat-of-your-pants" are gone, at least if one is interested in cost and time savings as well as environmental protections and safety protocols for the protection of the public, crews, and wildlife. Management techniques developed over the last 20 years include the "Perfect Well," "Technical Limit," and "Drilling the Limit," which are all patented methods for improving drilling operations from before spud to after completion. These are methodical means for tearing down and rebuilding every aspect of drilling operations, determining bottlenecks, slowdowns, challenges, risks, logistical issues, and mitigation techniques.

Together, along with the hard work and efforts of operators large and small, service companies, drilling contractors, consultants, academia, and government laboratories, there has been a revolution in how wells are drilled today. This chapter covers some of these technology and management enablers in the onshore drilling arena.

Over the years, drilling rig performance, especially onshore, has proven to be the most important aspect of the drilling industry, especially in periods of low oil prices. The number of active drilling rigs is closely watched to estimate the health of the petroleum industry. Figure 4.1 shows the annual rig count and oil price. The close correlation between the oil prices and the rig count can be seen in this figure.

4.2 TECHNOLOGY ENABLERS

There have been significant improvements in drilling rig design, technology, and usage.

4.2.1 RIG TECHNOLOGY

All onshore drilling rigs contain five main subsystems: power, hoisting, rotary, well control, and circulation subsystems. Over the years, these main subsystems have remained the same, but the individual components have developed significantly, making the process of drilling a borehole increasingly efficient and safe. In this section we discuss the development of various rig subsystems over the last decade which fueled the shale revolution in a low oil price environment.

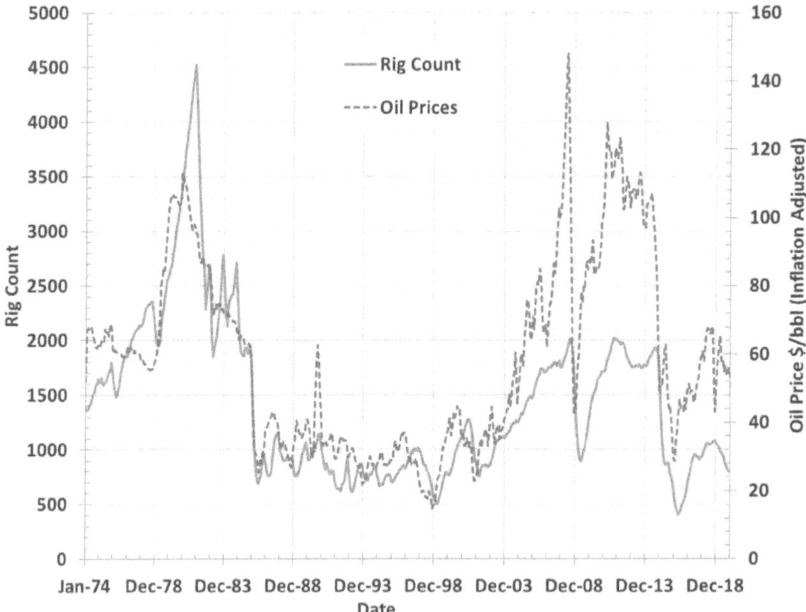

FIGURE 4.1 Rig count and oil price over time. *Source:* Courtesy of EIA data.

4.2.2 Super-Spec Rigs

The super-spec rig is defined as a 1500hp AC powered rig with at least 750,000 pounds-force hook-load, a "rig walking" system, and a 7500-psi circulating system (Patterson-UTI, 2018). Some additions such as a third mud pump and/or extra mud volumes can be made to the rig if needed for a well. These are the advanced onshore rigs that enable drilling of longer horizontal wells quickly, efficiently, and safely.

4.2.2.1 Rig Movement

Operationally, the rig movement operations are annotated as MIRU (Move In Rig Up) or RDMO (Rig Down Move Out). Figure 4.2 shows a self-propelled onshore drilling rig moving from a drilling pad on the North Slope of Alaska. The MIRU operation is conducted before the drilling activity begins. As the acronym suggests, the rig is brought to the well site on a truck or a trailer and raised upright. After reaching the wellsite, different rig components are connected to the electrical and hydraulic grids. The operation ends when the rig is deemed ready to begin drilling.

The RDMO operation is conducted after concluding drilling of a borehole and the rig is released from duty on that well. All the components of the drilling rig are disconnected and moved on trailers to the new drilling location. The drilling rig structure or the mast is then lowered on a trailer and carried to the new drilling location.

The MIRU and RDMO were designed for conventional boreholes, drilled in separate locations, making the trailer mounted move easier. Further system analysis review of the mobilization processes resulted in modifications resulting in an efficient rig mobilization system. The advent of onshore pad drilling—where multiple boreholes are drilled and completed from a single pad location—for unconventional reservoirs forced the companies to reconsider their mobilization strategies to become more efficient, faster, and safer in terms of rig breakdown and transportation to new site. Some unconventional wells have been drilled from single boreholes in what is termed multilaterals.

FIGURE 4.2 A rig self-transporting to a new drilling location. *Source:* Courtesy of Photo by A. Eustes.

FIGURE 4.3 Rig walking system. *Source:* Courtesy of Photo by A. Eustes.

To economically move the rig a short distance, drilling contractors developed the "walking" rig. In these systems, once a rig is transported to a location, all surface components are connected to the rig. After drilling the borehole, the hydraulic mechanisms slide (skid) or walk the rig slowly over to the new borehole, saving the time taken to MIRU/RDMO. Figure 4.3 shows an example of a rig walking system. The surface equipment such as the power and circulation subsystems are stationary and are connected to the rig substructure via umbilical hoses. These rigs can move laterally, and some can move perpendicular too. This allows a single rig to drill several wells in a row and potentially step over and drill another row. Since this process can be accomplished with the rig full of pipe standing in the mast, frequently wells are drilled in batches. That is, a series of wells may have their surface casing sections drilled sequentially down a row. Then, after changing out the equipment, the next set of sections are drilled, returning the rig to the first drilled hole. In this way, the equipment for the various sections of borehole only need to be changed out once instead of by each well, saving significant time.

4.2.2.2 Drill Control

The traditional driller's console has gone through minor changes over the years, but it still uses analog data visualization, rotary switches to control rotation, pump rates, and push buttons to activate or deactivate a system. Figure 4.4 on the left shows a traditional driller's console with all its panels. The increased drilling of unconventional boreholes in a low-cost environment has inspired the industry to move to a more streamlined console, resulting in much more informed operations. The consoles now contain a robust auto-driller software which automatically controls drilling rate by altering the weight on bit in real time. Figure 4.4 on the right shows an example of the advanced driller's console. Note the joystick controls as opposed to the traditional brake handle.

Electronic data recording (EDR) is also becoming more prevalent. The use of this data both in real time and historically has allowed drillers to fine tune their operational parameters and to determine time losses and NPT. This is discussed in detail later in this chapter.

4.2.2.3 Mud Pumps

Drilling fluid (called "mud") is essential in drilling a well efficiently. The mud has wide-ranging applications such as cooling and lubricating the drill bit, transporting the drill cuttings, and providing the necessary hydrostatic pressure to stabilize the borehole. Mud is pumped through one or more pumps that can either be single acting (fluid on one side of a piston) or double acting (fluid on both sides of a piston). Rig pumps typically have two or three stroke pistons, called duplex and triplex pumps respectively. Most rigs of any quality have triplex since they are more efficient in terms of fluid flow and energy utilization than duplex pumps.

In the last decade, the petroleum industry has focused on horizontal wells, drilling an average lateral (horizontal length) of 10,000 ft. A limiting factor in limiting the length of the lateral section is the hole cleaning efficiency—efficiency with the cuttings are circulated out of the borehole. Mud flow rate and pump pressure are essential in determining the lateral length as those parameters dictate the cuttings transport properties. Too little and the cuttings pile up in the horizontal borehole leading to plugging and sticking. It must be high enough to provide necessary hole cleaning and borehole hydraulics while remaining within safety limits of all equipment and subsurface pressure gradients (Husband et al., 2007). Traditional onshore rigs have 5000-psi mud pump systems but long laterals need significantly higher pump pressures. This has driven the rig manufacturers to upgrade the mud pumps to more than 7500-psi mud pumps by using a discharge manifold, increasing pump efficiency, using more durable seals, and by using more efficient bearings (Ardoin, 2014). These

FIGURE 4.4 Traditional drilling console (left) and advanced drilling console (right). *Source:* Courtesy of Photo by A. Eustes.

modifications have enabled the super-spec rigs to drill longer, stable horizontal boreholes efficiently and quickly, resulting in lower operating costs.

4.2.2.4 Hoisting System

The hoisting system on a rig facilitates the vertical movement of the pipe in and out of the borehole. The hoisting system includes draw-works, block and tackle, derrick/mast, and miscellaneous equipment.

Traditional rig power to the hoisting system has been the DC traction motors and Silicon Control Rectifiers (SCRs) coupled to mechanical transmission systems, but these are now rapidly becoming obsolete. New rigs have adopted a synchronous alternating-current drive system. The AC driven draw-works allow for much finer control of drilling operation, improved fuel efficiency, and reduced maintenance requirements. The AC draw-works can operate at a single speed, minimizing the need for transmission and clutch assemblies. All these benefits have exponentially increased application of AC drives in hoisting systems, resulting in lower tripping time, increased performance, and lower drilling costs (Drilling Contractor, 2002). There are also hydraulically powered rigs, but those have not seen as wide-ranging usage.

4.2.2.5 Rig Automation

From a cable tool rigs to the rotary systems, drilling rigs have been modified continuously to minimize the physical risk to workers on the rigs. Increasingly, the drilling rigs are moving their operations from mechanized to semi-automated or fully automated. Safety is the largest motivation to automate the operations on a rig. Reducing the number of crew in the vicinity of heavy machinery significantly improves crew safety. The other major driver for automation is efficiencies of operation and consistency of operating. Automation of the rig also makes the drilling operations financially and operationally more efficient (Eustes, 2007). A more automated rig also makes sense in harsh environments, taking the crew out of the elements. However, industry has not created the autonomous rig just yet.

Over the last decade, the drilling community has increasingly adopted rig automation. The complexity of the drilling process causes drilling automation to be a combination of numerous automated subsystems. The hierarchical construct of such an automated system relies on different levels of automation (LOA) for different subsystems. For an example, in the geosteering process, where the borehole is optimally placed in a reservoir based on real-time geological and log measurements, the downhole directional system requires a supervisory control whereas the surface systems require a shared control.

Drilling automation can be classified as downhole automation, surface automation, and remote automation. Downhole automation refers to automated control of directional profile, vibration detection, and drill string health monitoring. Surface automation refers to the automated control of the hoisting system, rotary system, and circulation system. And remote operation means just what it says, operations that control the rig that are some distance from the site.

The biggest impediments to the development of drilling automation are the slow adoption of drilling standards, the hesitation of different companies to collaborate, and a lack of knowledge about drilling system automation. The increased availability of advanced communication systems, and data processing capabilities, with technical drivers like increasing complexity and cost of wells, are driving the petroleum industry to rapidly develop and adopt the drilling automation systems (Macpherson et al., 2013).

4.3 DRILLING FLUIDS

Drilling to extract hydrocarbons is a commercial activity driven by cost controls and the never-ending need to improve performance, unlike drilling for purely scientific purposes. The cost of drilling is typically the most expensive stage in oil and gas extraction (Khodja et al., 2010). For conventional wells, drilling costs represent around 70% of the overall cost of the exploration wells and 50% of development wells expenditures (US EIA, 2016). For unconventional wells, such as those requiring hydraulic fracturing during completion to extract the oil or gas, the cost of drilling is approximately 1/3 the total well cost; most of the other 2/3 of the cost is completion related (US EIA, 2016).

On the average, drilling fluids cost is about 10% of the cost of drilling operations (West et al., 2006) but can reach up to 18% of the drilling operation cost in certain types of wells (Khodja et al., 2010) or when an event such as severe lost circulation leads to massive fluid replacement cost or the need for cementing or sidetracking. A well-designed drilling fluid based on the well objectives and conditions can mitigate the risk of unintended, and usually costly, consequences to the drilling operation. A study of the causes of non-productive time (NPT) and its impact on drill time shows a number of the fluids-related causes including lost circulation, stuck pipe, and wellbore instability (Figure 4.5).

The three factors critical to all fluids design (graphically shown in Figure 4.6) are:

1. Performance, including technical and operational issues.
2. Economic issues or cost minimization.
3. Health, safety, and environmental (HSE) aspects typically centered around regulatory compliance.

Technically, the drilling operation and drilling fluid plan(s) must adhere to the rules of physics, as well as the interaction between the chemistry of the fluid and the rock formation, which requires an understanding of other science domains such as the geology of the formation, fluid rheology, chemical reactions, and the mechanics of drilling.

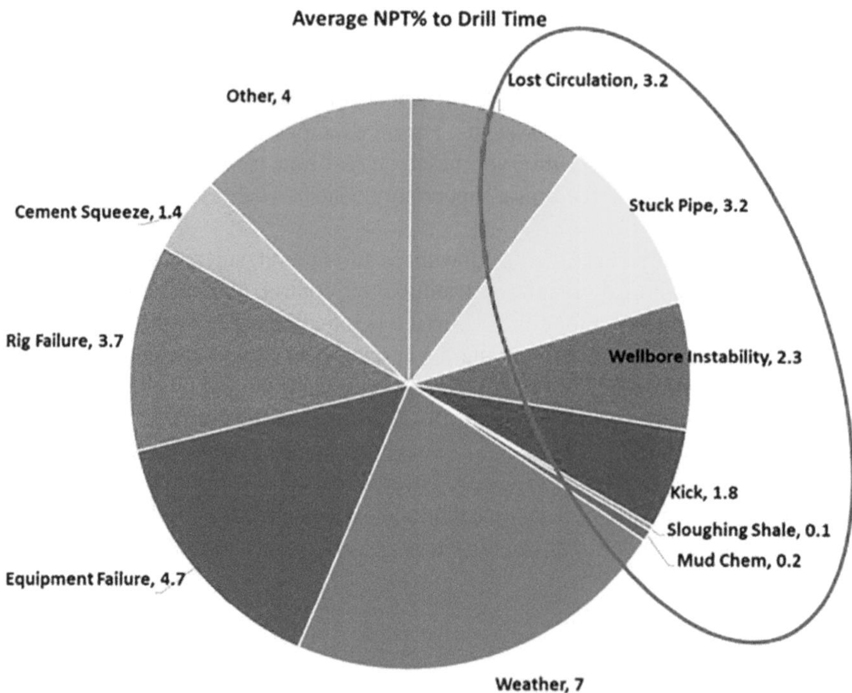

FIGURE 4.5 Average impact of non-productive time on drilling time highlighting the causes related to drilling fluids. *Source:* Courtesy of Amer (2019).

FIGURE 4.6 Critical aspects of drilling fluid design. *Source:* Courtesy of Amer (2019).

While the goal of all drilling operations is to achieve the best economics with lowest impact on the environment are clear to most, the expression of those goals can change quickly. Choosing a "low-cost" fluid may create additional costs in other parts of the field development phases with a worse-case scenario including damaging the producing formation, resulting in higher overall costs, lower production, and a requirement for expensive stimulation operations that may, or may not, work especially in unconventional reservoirs. Environmental perspectives may shift due to changing regulations or norms based on geographic location, objectives focused on the single well, or operator requirements. From a planning perspective, operational objectives and potential challenges must be taken into consideration to ensure flawless execution of the drilling program while achieving the best balance of technical performance, economic results, and environmental accountability.

The modern history of drilling fluids started in the early 1900s when clay was mixed into the drilling water to control fluid loss into the sand (Caenn, 2018). Over the next few years, drillers began to use "mud-laden fluid" as a means of controlling pressure (Caenn et al., 2011). By the 1920s, a major improvement was the addition of barite as a weight material to control borehole pressures (Stroud 1926a; Stroud 1926b).

The middle of the 20th century saw the incorporation of oil as the continuous phase in invert emulsion drilling fluids (Caenn et al., 2011). These invert emulsion drilling fluids are also known as oil-based drilling fluids and fall into the larger category of non-aqueous drilling fluids (NAF).

During the late 1960s, a new generation of polymeric viscosifiers was introduced to the industry allowing for a higher level of thixotropy (gel strength increases as shear rate decreases) and improved the overall functionality of the NAFs. NAFs made a massive transformation in the drilling industry due to their significant impact in improving stability and drilling performance.

Over the next few decades, HSE concerns pushed many operators from diesel to mineral oil with much lower benzene levels than diesel. By the 1990s, synthetic fluids were used as the continuous phase for a new NAF—synthetic-based drilling fluids (SBF)—which provided a lower environmental footprint for drilling fluids (Bleier et al., 1993). The HSE advantages and operational achievements of SBF opened the door for a new generation of SBF made from highly refined petroleum monomers without the benzene contamination present in less refined products.

The higher cost of the synthetics limited early use to offshore deepwater, which was at that time a drilling frontier posing multiple challenges. This category of SBF, under the standard classification Group III of the OGP (OGP, 2003), was further developed to comply with new environmental and waste management regulations banning the discharge of diesel-based NAF cuttings.

Regulations and cost are still drivers in the critical choice of NAF base fluid. For instance, SBFs and low-toxicity mineral oils are primarily used offshore, but mineral oil is not allowed in the Gulf of Mexico. Diesel is generally prohibited offshore but has become the dominant oil in unconventional land drilling.

Nowadays, drilling fluids are a mixture of advanced and complex, organic and inorganic compounds that are synthesized or processed for certain properties to allow drilling operations to succeed in specific environments like high temperature, high salinity, etc. While some would say that we have reached the limit with drilling fluids chemistry (with many of the more recent improvements being step changes rather than transformative leaps), the reality is that we have only scratched the surface and that there is much more optimization and innovation to be done.

The next wave of development will be energized by the availability, affordability, portability, and enhanced capability of current lab equipment combined with the promise of data analytics to optimize the selection of efficient products. This next wave of innovation will be driven to further enhance performance, optimize costs, and solve decades-old challenges. Still, more innovative research efforts are required to meet operational challenges and improve overall drilling efficiency.

4.3.1 Functions of Drilling Fluids

Drilling fluids have historically been viewed as providing key properties necessary for safe and successful drilling and, as such, are integral to the drilling process. Although all the functions are important, controlling subsurface pressures is the most significant by far from both an operational and a safety perspective. The functions in no specific order (beyond the control of subsurface pressure) are:

- Control subsurface pressure.
- Remove drilled cuttings from the hole.
- Suspend cuttings and weight materials while the fluid is circulating in the hole and release cuttings at the surface.
- Seal permeable formations.
- Promote borehole stability.
- Minimize reservoir damage.
- Cool, lubricate, and support the bit and drilling assembly.
- Transmit hydraulic energy to tools and bit.
- Assist in formation evaluation.
- Control corrosion of the tools and drill string.
- Minimize impact on the environment.

While these points are the legacy functions of drilling fluids, expansion of this list can be expected to include a new set of functions.

- Transmit data signals from downhole tools.
- Ability to be recycled from one well to another.
- Perform at higher levels of drilling speed or rate of penetration (ROP).
- Control subsurface losses.
- Allow for logging via better conductivity in the drilling fluid even in NAFs.
- Flexibility to be converted to suspension fluids, packer fluids, etc.
- Ability to replace cement using novel chemistries like geopolymers (Liu et al., 2016), epoxy resins, and other reactive polymers.

4.3.2 Drilling Fluids Classification

Drilling fluids, or "muds" as they are commonly known in drilling operations, can be classified into three categories (Figure 4.7) usually based on the continuous phase of the "fluid." The three categories are: aqueous fluids, non-aqueous fluids, and aerated fluids. Subcategories and generic system names are typically based on further analysis of the composition such as being inhibitive or not, salt

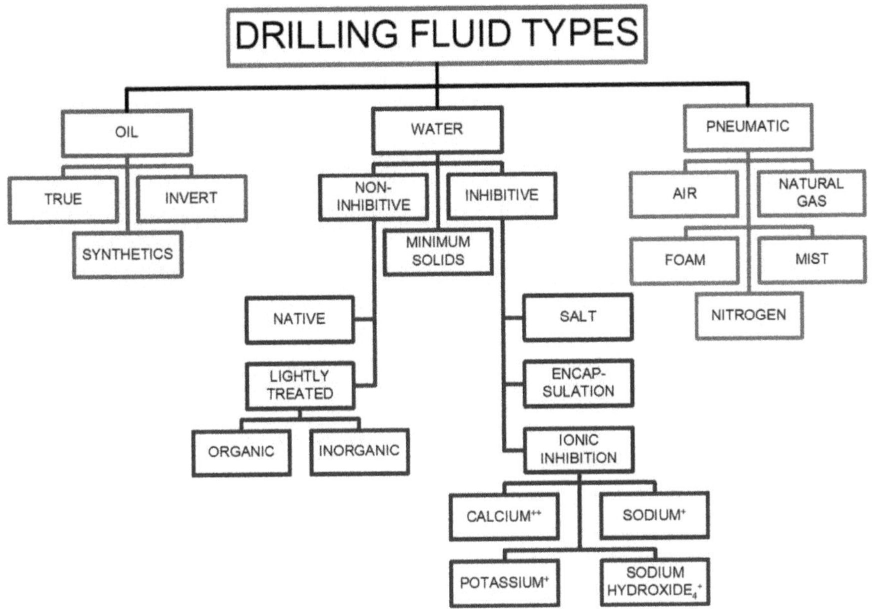

FIGURE 4.7 Classification of drilling fluid systems. *Source:* A. Eustes.

or freshwater for aqueous fluids, and both the internal and external phase types for NAFs as well as the ratio of one phase to another.

Aqueous Drilling Fluids, more commonly known as water-based drilling fluids (WBM), are very common drilling fluids systems. More than 90% of wells use this type of system in at least one interval of drilling. The first interval of drilling is traditionally drilled, or "spudded" with water or seawater, and incorporates the cuttings to form the "spud mud." WBMs used in deeper sections are much more sophisticated mixtures designed to provide the necessary high-tech performance to drill complex formations. WBM can be designed to provide a wide range of properties covering many drilling situations. Polymer and clays are the primary systems in WBM. Recent advances include high-performance water-based drilling fluids (HPWBM) for which the properties are carefully developed by high-tech additives to closely approach the performance of oil-based drilling fluids.

Aerated drilling fluids include gas or air, mist, and foam and have an inherently lower-than-water density. As such, these light-weight fluids are used for drilling low-pressured and highly depleted zones.

Non-aqueous fluids (NAF) are usually invert emulsions with the aqueous internal phase emulsified into a non-aqueous external or continuous phase. NAF fluids are frequently named for their base fluid which ranges in cost and sophistication from diesel to mineral oils, and at the high end of the spectrum, highly refined synthetic fluids.

An all-oil NAF has no aqueous-containing emulsion and is used mainly in the reservoir to avoid altering the wettability of the rock or for coring purposes to avoid an ion exchange between the drilling fluid and the rock of interest.

Fluids can be further categorized into invert or direct emulsions depending on which phase is continuous.

- **Invert emulsions systems**: have an aqueous (i.e., water or brine) internal phase emulsified into a non-aqueous (i.e., oil or synthetic fluid) external or continuous phase.
- **Direct emulsions**: are systems in which the oil is emulsified into the water or brine continuous phase and thus an adaptation to the chemistry of WBM. A direct emulsion system is preferred

in situations to lower the density of the fluid beyond the technical limit achievable with water, especially if salt formations will be drilled. Direct emulsions are also used for logging runs to measure geologic features since logging tools typically function better in a conductive media.

4.3.3 Testing of Drilling Fluid Systems and Additives

Testing of drilling fluids and component additives is a critical part of the process from R&D, to Quality Assurance and Quality Control (QA/QC) verification of the products, to field testing on the rig site to make sure the system and additives are functioning as expected and the well is behaving as predicted. The evaluation processes to identify the quality of these additives and systems include:

- A comprehensive study of the chemical compositions of each product plus concentration and activity is required.
- Evaluate product performance in the mud system.
- Compatibility analysis with other chemicals.
- Identify limitations imposed by in situ conditions such as temperature, pressure, and particle size.
- Calculate treatment cost per barrel.

Study other aspects such as formation damage, environmental regulations, and geological data which will impact selecting the best product at the lowest cost for the job.

To ensure the performance and quality of these additives, tests are performed to ensure the products meet the necessary criteria. Many of the significant additives and equipment used in testing drilling fluids are under the industry supervision of API Subcommittee 13 which deals with drilling fluids. Other properties are tested by using standard methods from ASTM or other accepted standards organizations.

QA/QC considerations also include vendor experience, vendor expertise, manufacturing capabilities, location, price, service, and quality. Both service companies and some chemical manufacturers supply drilling fluids additives. Regardless of the source, the additives must be quality checked to assure they will meet the demanding downhole requirements.

QA/QC is a program carried out to prevent any defective additives from getting to the field and disrupting rig operations. Quality Assurance are planned systematic preventive approaches and actions to ensure the chemicals are meeting the industrial standards. In a related fashion, Quality Control inspects and tests to ensure the chemicals conform to the standard. The combination of QA/QC ensures the drilling fluid operation will have consistent quality chemicals with predictable performance. This will help optimize the cost and performance of drilling and completion fluids.

A standard lab for drilling fluids testing requires specialized mud testing equipment and qualified personnel to conduct all standard, and non-standard, testing on drilling, drill-in, and completion fluids of all types. All testing equipment are subject to calibration and maintenance as needed or at regular intervals in accordance with documented procedures. Additional equipment might be required for a specific project or products not covered sufficiently by the general test equipment. Moreover, R&D generally has more expensive and precise instrumentation, some of which cannot survive the rig site environment which includes vibration detrimental to some equipment.

Table 4.1 identifies standard tests and the required equipment. The number on the left side of the table is supplied as an aid in referring to specific items within the chart only.

4.3.4 Learning and Development Skill Gaps

The drilling fluids domain is extremely broad. Diverse fluids are needed to meet the challenges posed by the wide variety of wells that include extended reach, deepwater, high-temperature/high pressure, shelf drilling, and tight shale to name a few drilling challenges. Even within this short

TABLE 4.1
Standard Tests for Drilling Fluids Chemicals

	Test Functionality	Description and Equipment Used
1	Mixing and blending to prepare drilling fluid samples	High-speed mixers—the API recommends a single-blade impeller for mixing either water-based or oil-based drilling fluids
2	Moisture content	Direct-reading moisture analysis meter or by oven-dry method
3	Density	Mud balance
4	pH measurement	pH meter. Measuring range: 0–14 pH/resolution 0, 01 and temperature range: −50 +150°C
5	To simulate downhole conditions, drilling fluids must be subjected to temperature and pressure under shearing	Aging/roller ovens
6	Rheology measurements	Viscometer (FANN-35 or equivalent) is used to measure viscosity as a function of shear rate at temperatures up to 150°F
7	Water, oil, and solids content of drilling mud	Retorts
8	Sand content	Sieves
9	Filtration behavior and wall cake-building characteristics	Filter press
10	Filtration behavior and wall cake-building characteristics of drilling muds at temperatures up to 400°F	HTHP filtration test
11	Coefficient of friction	Lubricity tester
12	Stuck pipe tendency and the effectiveness of spotting fluids	Differential sticking tester
13	Shale stability	
14	Cation exchange capacity of reactive clays	Methylene blue test
15	Stability of the water-in-oil emulsion	Electrical stability meter
16	Resistivity of fluids and slurries	Resistivity meter
17	Activity of emulsified water in oil-based drilling fluids	Water activity (hygrometer)
18	Viscosity as a function of shear rate at temperatures up to 500°F	High-temperature viscometer (FANN 77 or equivalent)
19	Permeability plugging	Permeability Plugging Apparatus (PPA) for permeability of 5–190 micron; up to 500°F (260°C) and overbalances up to 2500 psig
20	Quality of filtercake built on inner surface of a vertically oriented, cylindrical ceramic core	Dynamic filtration test
21	Rheology measured at low shear rates	Brookfield viscometer (or other low-shear-rate viscometer)
22	Particle size distribution (PSD)	Mechanical sieve analysis (percentage of different grain sizes within a sample) or by light scattering techniques
23	Concentration of polymer (lb/bbl) in mud sample	PHPA polymer concentration test

(Continued)

TABLE 4.1 (*Continued*)
Standard Tests for Drilling Fluids Chemicals

	Test Functionality	Description and Equipment Used
24	Total soluble carbonate and total soluble sulfides in the mud or the concentrations of CO_2 and H_2S	Garrett gas train
25	Cloud point and glycol content	Cloud point and glycol content determination test by chemical analysis
26	Flashpoint	Open-cup and closed-cup flash point by chemical analysis
27	Pour point	Pour point by chemical analysis
28	Bit balling, effect of surfactant dispersing agents, and borehole cleaning chemicals	Accretion tests
29	Residual damage and effectiveness of cake removal chemicals	Drill-in fluid return permeability
30	Crystallization temperature determination of high-density brines	Variable pressure crystallometer—indicating range of −45°C to 40°C (−50°F to 100°F) with a resolution of 0.05°C (0.1°F). (PCT, TCT, FCT, LCT)
31	Turbidity measurement of high-density brine	Nephelometric Turbidity Units (NTU). A sudden change in turbidity indicates contamination with organic or inorganic fine solids that can cause reservoir damage
32	Oxygen scavenger efficiency	Dissolved oxygen meter. Oxygen scavengers are commonly added to drilling fluids and completion brines to mitigate the adverse effects of dissolved oxygen in downhole
33	Aniline point for drilling fluids liquid additives and base oils	Measurement of aniline point
34	Total polycyclic aromatic hydrocarbon concentrations in drilling fluids additives and base oils	Measurement of total aromatic
35	Total sulfur content	Measurement by chemical analysis
36	Olefin content	Olefins test by supercritical-fluid chromatography for drilling fluids additives and base oils
37	Identification and detection of trace metals content in drilling fluids chemicals	ICP analysis
38	Definitive structural information and determination of crystalline compounds of drilling fluids chemicals	X-ray Diffraction (XRD)
39	Quantitative and qualitative elemental analysis of drilling fluids chemicals	X-ray Fluorescence (XRF)
40	Qualitative or semi-quantitative determination of chemical compositions, external morphology (texture), and crystalline structure of drilling fluids chemicals	Scanning Electronic Microsope (SEM)
41	Quantitative testing of electrolyte in drilling fluids chemicals	Ion chromatography
42	Identify unknown compounds and monitor chemical reactions in situ of drilling fluids chemicals	Fourier Transform InfraRed (FTIR) analysis (ID of organics)
43	Determination of mercury, lead, and cadmium at a minimum level of 0.5 parts per trillion in barite and other drilling fluids chemicals	Mercury, lead, and cadmium analysis
44	Drilling fluids chemicals characterization	Thermogravimetric Analysis (TGA)

(Continued)

TABLE 4.1 (*Continued*)
Standard Tests for Drilling Fluids Chemicals

	Test Functionality	Description and Equipment Used
45	Identify volatile and semi-volatile compounds in drilling fluids chemicals	Gas-Chromatography/Mass-Spectrometry (GC/MS) analysis
46	Measuring soluble ions such as Na, K, Ca, Zn, Fe, Pb, Cd, Br, Cl, etc. in brine completion fluids	Glass membrane ion selective electrodes
47	Identify the ionic nature (anionic, cationic or non-ionic)	Multiple methods/equipment employed for this

list, there are sublevels of categorization, each of which dictates a unique set of circumstances and competencies required to meet those challenges. Furthermore, the specific competencies range even further with the diversity in application of knowledge from research, laboratory work, development work, field application, technical support, and finally to capturing lesson learned, creating databases, and planning artificial intelligence guidance.

API 13L (2017) points out that the skills needed are gained by a unique combination of formal training and practical experience—and this is especially true for the advanced skills necessary to support for complex wells in the field covering specializations in engineering, chemistry, rheology, geology, and environmental and regulatory matters.

Drilling fluids, like the rest of the upstream industry, has experienced a major crew change in the last few years resulting in an enormous loss of field experience and expert knowledge. Some of that knowledge is intuitive knowledge that has not been yet captured by existing algorithms or "learned" by the new crew. This loss has the potential to create knowledge gaps where newer crew members with less engineering experience are less likely to recognize bad data, understand all the parameters for optimal decision making, or react quickly in emergency situations.

Some view the crew change as a chance to hire a more digital workforce better able to provide the data science skills required for the coming changes in jobs due to automation. However, Accenture (2017) reported that more than 40% of the oil and gas companies surveyed believe they are 3–5 years away from having a digital workforce ready which could result in a significant gap in trained personnel and skills over the next few years.

4.3.5 Role of Automation in Drilling Fluids

Automation in the drilling fluids industry, like other industries, involves many steps and iterations from basic mechanization to becoming entirely controlled by the machine with little interference from the human operator. The drilling fluid industry is on the path to more and higher-level automation—first equipment, then gradually decision making will be guided, even initiated, by data analysis and algorithms.

Automation is more apparent in areas like real-time and predictive fluid measurements. Some of the earliest predictive and visual software programs were VIRTUAL HYDRAULICS and PRESSPRO RT developed by M-I SWACO (Schlumberger 2018a; Schlumberger 2018b). These software programs monitored well data and very accurately predicted equivalent circulating density, equivalent static density, temperature, hole cleaning, and downhole pressure when MWD was not available due to tripping pipe or setting casing. These were among the forerunners of exciting software to visualize and better understand drilling fluids in the downhole environment and set the stage for the software needed for automating drilling fluids monitoring. Examples of real-time rheometers and mud monitoring equipment are ideas that have met recent acceptance in the industry.

Automating a process downhole requires the equipment to:

- Provide precise and accurate data.
- Be able to handle a wide temperature and pressure range without affecting the precision or accuracy of the data.
- Be able to correlate the downhole real-time measurement to measurements in the laboratory or other field instruments.
- Be reliably applied to a variety of situations.
- Be sturdy enough to survive downhole conditions for anticipated length of service.
- Be able to clean, calibrate, and reset the instrument remotely.
- Provide a cost-effective benefit to using the instrument.

Test equipment continues down the path to be faster, more accurate, more capable, more compact, and most important, safer oilfield. Data capturing and processing is both a result of automation and setting the stage for the next step further on the road to automation.

4.3.6 Future of Drilling Fluids

As mentioned, the three factors or processes critical to the fluids design framework are (1) performance, including technical or operational challenges; (2) economic or cost minimization, and (3) HSE aspects. The future of drilling fluids research reflects the ongoing balance of this trio. The goal for all R&D is optimizing value-added performance and minimizing overall reservoir cost with maximum reservoir recovery while achieving environmental regulatory compliance in a healthy and safe workplace. An added challenge is the complexity posed by unknowns and different variables as the frontiers of drilling are pushed further and further. These drilling challenges give rise to numerous R&D efforts to develop novel fluid systems, additives, techniques, and equipment.

Traditionally, the key drilling challenges and focus of upstream R&D dollars, including drilling fluids, are related to deepwater and extended reach drilling as the overall investment in these wells is significantly higher than land operations, thus driving the need for lower cost and higher performance options in this high-tech arena. Deepwater is responsible for much of the interest in automation, data analytics, directional drilling, and better logging capabilities including synthetic logs. Recent statistics from IEA (2018) (4) suggest that the upstream R&D money is moving to shale and tight gas while investment in offshore R&D is decreasing. Time will tell if this becomes a significant trend or R&D investment in offshore technology will continue to lead the way for the drilling fluids technology (Dimataris, 2019).

R&D focus is typically on avoiding problems such as borehole instability, enhancing reservoir conductivity, and reducing formation damage while complying with environmental regulations at a lower cost. The future of the R&D programs for drilling fluids must play a pivotal role in the development of new environment-friendly systems, non-toxic systems, and must find an alternative to high-cost fluids. One key challenge to the development of drilling fluids systems and additives is finding the right field trial and collecting meaningful data to assess the success, or failure, of the new solutions. In order to overcome that, the vendors and the operators need to agree on common objectives, the right well for the trial, and the metrics for success. Ideas and concepts may look good in the lab, but the proof is a successful field trial. Getting the right field trial is further complicated by the desire of every operator to be the second to use any new technology. This predicament comes from the operator managing risks involving large-value assets while reaping the proven value of the newest technology (Dimataris, 2019).

Intellectual property is also a challenge due to the broadness of applications and the diversity of the geographic locations where these additives and systems would be used. To protect novel inventions, companies must file in many different countries, under many diverse rules, to provide patent protection worldwide. The cost of filing, maintenance fees, and legal fees represent a considerable investment for companies.

Drilling fluids innovation and development has gone through cycles of the high level of innovation and times when very little development was done. The causes of this cyclic nature reflect both the push of market cycles and the impact of changing R&D focus. A half-century ago, R&D was deeply embedded in the operator companies who ran their own R&D departments. The research typically focused on drilling with the objective of understanding various phenomena, including innovation in fluids. One key advantage when operators encouraged internal R&D is that the cycle was typically faster as wells were always available to conduct field trials with minimal resistance. In due time the knowledge gained was typically shared with the industry through patents or conference papers. Under this system, the monetary cost and gain of new technology were solely held by the operator.

During the recessions and downturns that the industry underwent, operators cut their research budgets, shut down many of their drilling fluids-related R&D programs, and outsourced R&D to the service companies. Service companies, in turn, ramped up their own R&D programs. The R&D programs focused on a shorter cycle from development to field with less focus on fundamental research. The goal was clearly aimed at improving fluids and product performance and away from the overall optimization of the drilling process. The turn of the millennia saw some of the R&D money shifting from service companies to their vendors. This does not appear to be an ongoing trend but bears watching to see what happens in the next decades as the industry weathers yet another downturn.

Two important trends are impacting drilling fluids as they are other industries—ESG and data science.

The acronym ESG comes from Environmental, Sustainability, and Governance. ESG reflects a broader movement looking past the relatively narrow confines of the immediate drilling operation and focusing on climate, ethics, scare resource management, human rights, and global impact. Table 4.2 expounds on the components of ESG.

Data science and analytics are key words for the digital revolution coming to all aspects of our lives, not just drilling fluids. However, the implications for any industry, including drilling fluids, are profound. The old adage that "you can't compare apples and oranges" can be rewritten by data science. Data science will allow the industry to compare apples and oranges in huge data sets that reflect the overall impact on a much larger scale and identify trends and impacts not readily apparent. H. H. Alkinani et al. (2019) and Al-Hameedi et al. (4) have used predictive data mining to study the impact of treatments in mitigating lost circulation events.

Despite the downturns and reconfigurations, the future could not be brighter as the only limiting factors in fluids development are chemistry and equipment. The authors would make the case that we've only scratched the surface of chemistry, while some may argue that it has reached its limits. There are areas that are yet to be explored, discovered, or transferred from other industries such as nano-chemistry, material science, etc. As for equipment, the large, bulky, and costly analytical

TABLE 4.2
What is ESG?

Environmental and Technology	Sustainability—Social and Human Issues	Governance and Economic Issues
• Climate stability • Natural resource mgt • Innovation in products and services • Energy and GHSS • Waste management	• Employee health and safety • Inclusion and diversity • Organizational capability • Performance culture • Impact on communities	• Corporate governance • Risk management • Transparency • Business ethics • Cyber security and privacy • Political and social climate • Human rights

equipment is being replaced today by new innovative equipment and techniques that can analyze more samples on a continuous basis, at a higher accuracy, faster, in a much smaller footprint, and survive the rigors of rig-site usage. The use of data science to mine the existing data on products, systems, and performance under different conditions offers a new platform for selecting the "best" choice and customizing solutions. The growing number of joint research initiatives done by operators and service companies are a vital way to achieve more innovative and practical solutions to real-world challenges.

4.4 PDC BIT EVOLUTION

Polycrystalline diamond compact bits have revolutionized the way drilling is done around the world. Originally developed in 1972, the widespread adoption and use of them didn't take off until the 2000s (Scott, 2015). These bits are more durable, harder, and generally better in many aspects than the historical drill bits. They also have the advantage of being tailor-made for additive manufacturing and customizable design processes. Today PDC bits represent roughly 80% of the drill bit market share and drill 90% of the total footage around the world (Scott, 2015).

PDC bits were held back by development and market problems. It was known in the 1970s how valuable the technology might become, but significant research was needed to bring them to an acceptable technology readiness level. PDC bits were known to work well in certain clean applications and were shown to save money per run as early as 1983 (Offenbacher et al., 1983). The US market in particular had a narrow window for the PDC bits as they were priced for sale and the drilling environment were heterogeneous with some detrimental formation elements to the cutters. Roller cone bits were adequate and cost effective for most operators and applications. There were also multiple market downturns between their introduction in 1972 and 1999, though this arguably helped bit development at times as bit manufacturers became desperate for market share. Their superior cutting ability was apparent, but problems with vibrations and durability in harder formations was a challenge.

Researchers from academia and industry slowly worked through these issues. Strategies for mitigating bit wear were worked on in the 1980s and 1990s. One important discovery was that PDC bits tend to whirl backward downhole, which reduces the rate of penetration and bit life. Plus, whirling backwards tends to peel off the diamond table facing on the compacts. This led to the development of anti-whirl bits, which had significant performance improvements, as shown by Shell researchers (Langeveld, 1992). This backwards whirl is a problem in drilling operations because of the amount of severe vibrations it induces. Vibrations both waste input energy and create high frequency, large magnitude bending moment fluctuations that result in connection fatigue and failure.

One of the best ways to measure drilling efficiency is via mechanical specific energy (MSE). The theory behind MSE is that input energy greater than the compressive strength of the formation will break it. The input energy needed is usually much greater than the compressive strength, as there is significant energy loss throughout the string. Vibrations are a major contributor to this, particularly violent ones such as backward whirl. More about the use of MSE will be covered further in this chapter.

By 2000, PDC bits were accounting for 24% of drilled footage. Between 2000 and 2015, PDC usage increased market share, on average, by over 4% per year. This dramatic shift, along with other emerging technologies, has fundamentally altered drilling operations and greatly increased drilling efficiency. It's no coincidence that the rise of PDC bit adoption and technology coexisted with the rise of horizontal drilling operations. PDC bits are much better than roller cones at mating with rotary steerable systems and give operators greater control over the directional drilling trajectory (Greenberg, 2008). Plus, treated carefully, PDC bits can easily last as long as the horizontal section of a borehole which precludes having to waste time in replacing worn out bits. This also reduces the time the borehole is exposed to borehole fluids and mechanical stress, leading to better borehole integrity.

4.4.1 Types of Drilling Bits

Drill bits designed for rotary drilling can be broadly categorized into roller cone and fixed cutter bits. Roller cone bits have teeth or inserts on a cone that rotates around an axle. These bits crush the bottom of the hole and scoop up cuttings via upward rotation. They are commonly made of steel or tungsten carbide. These are not as used as they once were.

PDC bits belong to the broader category of fixed cutter or drag bits. These bits have fixed cutters as opposed to rollers on an axle and rotate as one fixed piece. There are three main kinds of fixed cutter drag bits: natural diamond bits, diamond impregnated bits, and polycrystalline diamond compact bits. The rotation transfers energy into the rock and causes it to fail by shearing, plowing, or grinding (Hycalog, 1994).

Natural diamond bits are manufactured by embedding diamonds on the working surface of the hard bit matrix. These bits cut the formation by plowing. Diamond bits aren't used as much today for a variety of reasons. For one, using natural diamonds is expensive, even though they sometimes can be salvaged. They tend to be difficult to use as they are sensitive to flow, heat, and force (i.e. you can break them fairly easily). Diamonds are prone to shock forces and need to be kept very cool with high fluid circulation. These bits also require a lot of skill to use, as there is a very small amount of clearance available for fluid passage. Only the diamond tips should be touching the formation when operating; otherwise, the bit matrix will be worn out by being run up against the formation causing the diamonds to pop out of the matrix.

Diamond impregnated bits have many small, synthetic diamond crystals embedded into a tungsten carbide matrix, not unlike a three-dimensional "sandpaper." This makes the point of contact between the formation and the bit much harder, increasing durability and cutting capability. These bits don't see the same usage PDC bits, but they are particularly effective in significantly hard formations. Bit wear typically seen in hard formations doesn't matter as much, because as the bit wears down the matrix is chipped away, leading to the exposure of new diamonds.

Polycrystalline diamond compact bits can be distinguished from other diamond bits by the way that the diamonds are embedded in the bit matrix. PDC bits have the diamond grown synthetically onto the faces of small cylinders of tungsten carbide (the cutters). These cylinders are then embedded into a bit matrix or fixed to it with rods.

PDC bits have gained their massive advantage over other bits because of their incredible versatility. These bits have a long life and are applicable over a wide range of formations and rate of penetration. They have some limitations in high heat (Glowka and Stone, 1986) and in harder formations (Langeveld, 1992), but, overall, these bits offer the best combination of cost, performance, and durability for most applications. An advantage of PDC bits comes from the way in which they are manufactured; the cutters are embedded in a mold, a matrix powder is poured into the mold, and the whole system is heated until the matrix material melts and flows into passages and around the cutters. Therefore, the bit can be economically designed for a given situation. This has implications on bit design and allows the design to be specialized by formation. A typical PDC bit can be seen in Figure 4.8.

4.4.2 Market Dominance

The current market dominance owned by PDC bits is from many different factors. The culture in the drilling industry does not facilitate change well; and, the mentality onsite, for many reasons, legitimate or not, is very often a "we know what's best attitude" (Mitchell, 2001). This makes the adoption of new technologies and strategies slow in practice (i.e. if it's not broke, don't fix it). That being said, in the case of PDC bits, the technology was compelling, and, once adoption hit a critical point, usage quickly saturated the market. Some of the economic and technological limitations have already been discussed, so this section will explore some of the technical reasons as to why this was able to occur so quickly.

FIGURE 4.8 A typical PDC bit. The diamond crystals are small and show up as a black mass on the cutters. *Source:* Photo by A. Eustes.

A major enabler for PDC bit design has been the advancement in the computer capabilities. 3D modeling software and product lifecycle management (PLM) systems have helped facilitate application-specific bit design. The physics governing bit function are very complicated and the ability to use computational fluid dynamics (CFD), finite element analysis (FEA), and other simulation models have significantly helped improve bit design.

Computing capabilities allow researchers to run iterative simulations of such massive systems of equations and optimize certain parameters. For example, strategies for the proper design of the bit crown shape have been shown to be more or less reasonable by an iterative FEA model (Ju et al., 2014). Similar optimization models using CFD software have been used to optimize hole cleaning in horizontal boreholes (Greenberg, 2008), which are notoriously difficult to clean and often have pack-off issues (Mitchell, 2001).

The rise of these capabilities, combined with a low-margin-for-error competition between bit manufacturers, combined with the need for greater string control in horizontal drilling, helped lead to an important shift in bit design philosophy. It became standard to accept that the bit design process "is not entirely the responsibility of the bit designer, but a collaborative effort between the bit company and the operator" (Greenberg, 2008). This philosophy has helped the automotive manufacturing industry achieve great success via a culture of what is now known as lean manufacturing. By making suppliers an integral part of the design process, better designs and efficiency have been achieved, as well as stronger relationships formed (DeWardt, 1994). In the case of bit design, it was really inverted with the suppliers getting the operators involved, but the value of the philosophy does not change.

4.4.3 Application of Specific Bit Designs

Studies have been ongoing in PDC bit design optimization. This optimization is typically based on the type of formation intending to be drilled. A variety of different parameters can be optimized: the rate of penetration, durability, stability, steerability, hydraulics, etc. As computational power has increased, these designs can be improved through more complex modeling. The bit design process is a results-based, iterative one and the market is thus very competitive. The introduction of a lean style supplier/operator relationship has helped spur on successful bit technology. It has also pressured bit manufacturers into developing formation specific bits for operators for specific situations and formations, a "designer" bit formulation. This has already proven to be powerful in the development of bit technology and will be even more so once additive manufacturing technologies become efficient and practical.

It is known that bit durability is affected by rock-breaking energy consumption (Ju et al., 2014). That is to say, the higher the amount of energy transferred back into the bit from rock failure, the more quickly the bit wears. PDC bits are being designed to reduce this for specific formations and areas. By utilizing large geomechanical data sets and modeling tools, bit designers are able to design the "perfect" bit for a given formation and application. This involves reducing wear and maximizing ROP. A brief glance at the drill bit section of all the major suppliers shows that this is a large emphasis for each of them. This kind of competitive market will continue to drive bit technology forward at a rate faster than many aspects of field operations.

PDC bits will remain at the pinnacle of bit technology for the foreseeable future. Application-specific designs will become more and more optimized. The versatility in operational applications and ease of manufacture make these bits the most adaptable for varying drilling operations. Currently, PDC bits have limitations in ultra-hard and hot formations; however, designs are coming out that will soon overcome these limitations. This has potential to have large implications within the geothermal industry, where wells can be five times less efficient than those drilled in oil and gas because of these and other issues (Visser et al., 2014).

4.5 DIRECTIONAL DRILLING CAPABILITIES

4.5.1 Directional Drilling

Directional drilling is a term that is defined as a controlled borehole trajectory to reach a predefined target underground. In other words, "any well that is intentionally nonvertical is a directional well" (Mishka 2011). Horizontal wells are a subset of directional wells. The techniques to drill these types of wells are common.

The first directional wells were drilled around 1920s to 1930s in Huntington Beach, California to cross the property boundaries. This led to developments to understand the position of the borehole in reference to the surface location of the well. In the 1930s, magnetic single shot surveys (meaning a single survey point to determine gravity inclination and magnetic azimuth) become ubiquitous. Later discoveries in offshore California, the Caspian Sea, and the Gulf of Mexico lead to expanded and controlled application of directional wells. There are eight reasons for drilling directional wells:

1. Sidetracking: Sidetracking is done to change the course of the borehole due to a problem while drilling. This problem can be related to drilling (e.g., fish), or unexpected behavior of geological formations (e.g., fluid loss). Another application of sidetracking is the exploration of producing zones in adjacent areas.
2. Inaccessible locations: If the surface location above the reservoir is inaccessible due to geological structures (rough mountain/hilly terrain, lakes) or manmade structures (cities), directional wells can be used to reach the target reservoir.
3. Salt dome drilling: Salt domes are challenging to drill as they are prone to creating drilling operation issues. So, directional drilling is used whenever it is possible.
4. Fault controlling: Drilling through an inclined fault plane is challenging. The abrupt change in the formation has a significant effect on drilling, which makes it hard to keep the borehole trajectory as per plan.
5. Relief wells: Relief wells are drilled to control a well blowout.
6. Single surface location: Several directional wells that are reaching to different portions of the reservoirs can be drilled from a single platform.

7. Horizontal drilling: Horizontal wells are drilled in order to penetrate more of the reservoir in greater length so that the production will be improved.
8. Multi-laterals: Production can be enhanced further by drilling several lateral wells from a single main well.

4.5.2 Methods

There are five methods to drill a directional well. The stabilized BHA, jetting, whip-stock, mud motors, and rotary steerable systems (RSS).

4.5.2.1 Stabilized BHA

A Bottom Hole Assembly (BHA) is the part of the drill string that is not the drill bit and not the drill pipe. The BHA can be used to control the borehole trajectory and loading the required weight on the bit to break and penetrate the rock. The BHA can be as simple as a stack of plain drill collars or as complex as a combination of Measurement While Drilling tools (MWD), Logging While Drilling tools (LWD), mud motors or RSSs, and stabilizers, jars, heavy weight drill pipes, and many cross-over subs to connect it all together.

Different combinations of stabilizers in a BHA can be used for directional drilling purposes. Since the 1950s, BHAs were designed for building, holding, or dropping the inclination of the well. A stabilizer that is placed close to the bit produces a fulcrum effect forcing the bit to push to the upper side of the hole which results in a building tendency. If the stabilizer is moved away from the bit, the so-called fulcrum effect decreases. Thus, the weight of the BHA below the stabilizer forces the bit to push toward the lower side of the hole which results in a dropping tendency in inclination. Multiple stabilizers can act to centralize a BHA and keep the borehole from deviating.

If directional drilling objectives cannot be achieved by only one stabilizer, two or three stabilizers can be placed in the BHA with varying distances. In the case of use of several stabilizers, two critical elements should be considered for the directional drilling: the gauge (outside diameter) and the position in the BHA of the stabilizer. The position of the stabilizer has a similar effect as noted earlier, the distance between two stabilizers determines the tendency of the BHA below it (build, drop, or hold). The ability to change the diameters of the stabilizers have made it possible to gain more control over build and drop rates without pulling out the BHA especially with the use of adjustable-diameter stabilizers. The directional tendency of the BHA can be controlled by adjusting the diameter of the second stabilizer in the BHA.

The issue is that this process is passive. You must run exactly what you need when you run into the borehole. Naturally, if conditions of the formations such as hardness variations, dip angles, lithology changes to name a few, the BHA may not perform as expected (and likely will not!). A trip out of the borehole would be necessary to change the BHA for the new conditions. Then if conditions change again (likely), then the process starts anew. All in all, an inefficient process for drilling a well.

4.5.2.2 Jetting

In soft formations, jet drilling is an easy and cost-effective way to kick off a well. A drill bit (typically a roller cone bit) is dressed with one large and two small nozzles. This allows the fluid to achieve high velocity exiting the bit. The drill bit is oriented to the required direction when touching the bottom of the hole. Fluid circulation is started to initiate the washing action. The high velocity of the fluid erodes the rock; and, without rotating of the drill string, the bit can advance from three to six feet. Then, the rotation is started with normal drilling parameters to increase the initial curvature established with the jetting for the next 20–25 ft. The curvature is confirmed with a survey; and, if the desired inclination and azimuth is not achieved, the process is repeated until desired inclination and azimuth is achieved. The hardness of the surrounding rocks is the main factor that determines the efficiency of jetting. Sometimes severe doglegs (abrupt changes in inclination and/or azimuth) can occur over short sections of the hole due to a change in surrounding rocks.

4.5.2.3 Whip-stock

A whip-stock is a wedge-shaped steel casting ramp that is used to guide the bit into the wall of a hole to start deviation. The two types of whip-stocks are fixed and removable. As the name suggests, the fixed whip-stock stays in the hole after the desired deviation is achieved. A removable whip-stock is pulled out of the hole with the drill string. Both types can be used in an open or cased hole.

The edge angle of whip-stock is selected according to the desired deviation. A particular bit that has a diameter small enough to fit in a hole in the whip-stock is anchored to the whip-stock for setting. After the whip-stock is oriented to the desired azimuthal direction (called a "tool-face"), weight is applied to the toe of the wedge to prevent movement of the whip-stock when rotation begins. After the whip-stock is set into the position, more weight applied to the drill string to shear the anchor pin between the whip-stock and the drill bit, freeing the bit for rotation and drilling operations. Drilling continues along the shape of the whip-stock until the whip-stock ends (typically one joint length). Then, the BHA and bit is changed according to the directional plans. A new bit is run that has the size of the desired borehole and enlarges the borehole to the given size.

These are not used very often in open boreholes. They can get stuck easily. However, whip-stocks are commonly used in cased holes for creating a window in the steel casing. A mill grinds the casing along one side opening up the pipe to open a hole on that side. This allows a bit and drill string to drill out laterally from that location into other parts of the reservoir.

4.5.2.4 Mud Motors

Mud motors, located just above a bit, provide rotational power to the bit for drilling. The whole drill string does not necessarily need to rotate for the bit to drill. The focused rotational power reduces the wear on the components of the drill string and less energy is required to penetrate the formation. Also, by not rotating the drill string, the bit can be oriented in a specific direction. Note that the bit is not aligned with the borehole axis, facilitating the borehole trajectory to head off into that desired direction.

When a motor BHA includes Measurement While Drilling (MWD), it is called steerable motor BHA. By interrogating the MWD system and retrieving the inclination, azimuth, and bit direction, the directional driller can reorient the bit to follow the predefined borehole trajectory to reach a target as they continue to drill. The directional driller will continue to monitor the trajectory, intervening by rotating and sliding to keep the borehole headed to the target.

There are two modes used when drilling with Steerable Motor BHAs: slide and rotate modes. In the slide mode, as this is called in directional drilling terminology, the drill string is not rotating. The required deviation is achieved by changing the bit direction (called tool-face orientation (TFO)—described in the next paragraph) by using the rotation subsystem and then locking it into place. However, the rate of penetration is significantly decreased in sliding mode. In the rotate mode, the drill string rotates which cancels the directional effect of the bit orientation and results in a slightly over-gauge hole. Since the drill string is rotated, the drill string and mud motor's rotations are linearly added. It also means that there is no preferred direction for the bit to point (since it too is rotating) and allows for a borehole to be drilled straight. This is how the borehole is drilled straight even though the bit is not pointed along the borehole axis.

The tool-face orientation (TFO) angle defines the direction of the drill bit relative to the direction of the borehole (think as if one is looking at the bottom of the hole from inside the borehole). The high side of the hole is highest point in the borehole relative to gravity whereas the low side is the lowest point. These are listed as 0° and 180°, respectively. 90° would be directly turning right and 270° would be turning left. Therefore, a TFO of 140° would mean the borehole trajectory should turn right and downwards. The tool face angle is updated continuously or at each survey point while drilling by the signal from the MWD.

The directional driller examines the directional tendency of the BHA in the drilled formation. The rotary drilling is disturbed by the directional driller by changing the mode to sliding or changing the drilling parameters (flow rate, WOB, RPM) whenever an active directional control is needed to keep the borehole in the trajectory and keep the dogleg severity (DLS—the borehole curvature) as per need for the borehole.

Mud motors consists of four main components:

1. The top sub that consists of flexed sub and dump-valve is used to bypass the fluid when the assembly is run into and out of the borehole.
2. A power section consists of a rotor and a stator. This section converts hydraulic energy from the drilling fluid flow and pressure drops into mechanical power to turn the bit.

   The stator is a steel tube with a rubber lining molded into the tube. The rotor and stator have similar spiral profiles, but the rotor has one less lobe than the stator. When circulation begins, since the rotor has one less lobe than the stator leaving a cavity, fluid flows into that cavity pushing the rotor out of the way making it rotate within the stator. The cavity progresses down the stator/rotor, hence the name "progressive cavity motor." For each rotation cycle, the rotor rotates the distance of one lobe width. Therefore, the rotor must rotate for each lobe in the stator to complete one revolution of the bit box. As the number of the rotor and stator lobes increase, the motor's torque output increases but it slows down in rotation. Note that the lower the lobe ratio (i.e. 1/2), the faster it rotates but the lower the torque output. The higher the lobe ratio (i.e. 9/10), the slower it rotates but the higher the torque output.
3. The transmission shaft is attached to the lower end of the rotor and transmits the rotational force created in the power section to the bearing and the drive shaft sections.
4. The bearing section and drive shaft supports the axial and radial loads and also transmits the torque from transmission shaft to the bit.

Motors can be used both for vertical and directional wells. The deviation needed for the directional wells is achieved by motors that have bent subs or bent housings within the motor and stabilizers. A bent sub is located above the motor which puts the bend some distance from the bit. This has a significant effect on achieving the required deviation in softer rocks as this leverage imparts a significant side force. A bent housing puts the bend closer to the bit. This negates the side force effect but allows for a significant bit angle relative to the borehole axis to be made. Bent housings were developed as bit angle tends to be the primary directional method for deviation in harder rocks. The angle of the bent housing (0°–3°/100 ft) can be adjusted according to the required deviation.

Recently, several sections of mud motors have been the focus of development for increasing drilling efficiency and decreasing the risk of motor failure. The power section is the most critical part of the motor and is also responsible for most of the downhole motor failures, frequently due to the operating limits (temperature, time, fluids).

The type of the elastomer in the motor determines the amount of torque that is generated and the life of the motor. The type of stator is defined by the thickness and the chemical properties of the elastomer. Long sections of drilling with the same motor can cause the elastomer to shift and strip (called "chunking" in oilfield terms). Small deformations can decrease the torque that is generated. In some cases, chunks of shattered elastomer might separate and pass through the motor, plugging the bit nozzles, which might cause more problems, especially when drilling in harsh environments where well control is an issue.

The thickness and chemical properties of the elastomer is of vital importance. Soft elastomers have a longer lifetime than hard elastomers, but they tend to fail more often. Service companies have investigated several ways to overcome this issue. Recent high-tech motors have a new type of enhanced hard elastomers which can be used for longer times and can deliver high torque in high temperature or harsh environments. Another approach that is used is to reduce the thickness of the elastomers and build the stators with a special steel alloy that backs the elastomer, eliminating the need to shape the elastomer in the helix shape. This generates less heat and creates a better seal between the stator and the rotor. Higher torque can be created for longer times, which will increase the ROP throughout a drilling run.

4.5.2.5 Rotary Steerable System

In mud motor drilling, in order to direct the bit into the appropriate trajectory, the drill string must not rotate. This sets up potential static friction issues, which in a horizontal (or near horizontal hole), limits the lateral length. The frictional drag causes the drill string to helically buckle, resulting in what's called "lock up," making further progress impossible. A rotary steerable system (RSS) allows for the drill stirring to continue rotating while allowing the bit to follow the appropriate trajectory. This means that dynamic friction, lower than static friction, causes the dominate drag force and since it is lower, longer distances can be drilled.

There are two types of RSS—a point-the-bit system and a push-the-bit system. As noted earlier, directional control is affected by side forces (push-the-bit) or bit angle (point-the-bit), more likely both effects.

A point-the-bit system uses one non-rotating sleeve which is a geostationary reference. The motor axis is deflected while rotating and the bit is displaced laterally from its central axis, pointing it in the desired direction. Point-the-bit systems are slower to react to required well-path changes and the achievable curvature is less. A push-the-bit system uses one non-rotating sleeve with retractable pads. When a pad is opened and is in contact with the wall of the borehole, it pushes the bit in the opposite direction. As the pad rotates 180°, it retracts back. The pads are opening and retracting in sync with the string rotation, pushing the bit in the desired direction.

Most of the current RSS tools have a sensor which transmits near-bit inclination and azimuth data. This information allows the directional drillers to adjust the course of the RSS on the fly through downlinking.

Recent developments in the industry lead to hybrid RSS tools that combine the properties of motors (higher curvatures) and both types of RSS (push-the-bit and point-the-bit) with high ROP and less tortuosity. In some models, a motor is added to the BHA above the RSS so that additional RPM is provided to the bit which keeps high ROP, especially while drilling hard formations. This RSS is also used for performance drilling in which decreasing the drilling time and increasing the ROP are the main effort.

When the RSSs were first deployed in BHAs, the static surveys from MWDs were the only option for trajectory monitoring and thus for making directional decisions (the same case as mud motors). Because, the MWD sensor is far above the RSS, the offset between the bit location and the static survey point are at least 30 ft or more depending on the type of RSS and connections used in between. One could drill out of a formation before the information reached the surface.

To remedy this situation, some RSSs are equipped with sensors similar to those in a MWD to provide continuous surveys. This addition helped directional drillers to respond to any deviation from the planned trajectory quickly. Recently, a new feature has been added to RSS survey equipment which allows the tool to recognize and drive the tool to a desired inclination and/or azimuth. This new feature decreases the human input on the drilling process. This automated steering feature makes frequent, small adjustments to the RSS parameters by comparing the RSS survey data with the target. This makes RSS tools more autonomous in their operations. They have become downhole drilling robots.

4.5.3 Measurement While Drilling (MWD)

Downhole trajectory surveys are taken throughout the drilling operation as part of the process. Downhole surveys are vital to making sure that the drilled well profile follows the planned borehole trajectory. The planned trajectory must be followed to make sure the actual path intersects the target with the correct inclination and correct azimuth.

Surveys are taken at regular intervals by the MWD tool. Each survey produces two measurements at the desired depth: inclination and azimuth. These measurements are used to calculate the True Vertical Depth (TVD) and vertical section of the drilled well path. The drilling operation must be stopped while the survey is taken which is why these surveys are called static surveys.

The data is transmitted to the surface by a signal that is generated by the MWD tool. The best way to transmit this data is through drilling mud pressure pulsations or extremely long frequency electromagnetic waves. The data is encoded in binary data and is sent to the surface where is decoded and processed by the surface sensors and surface computers. This transmitted signal also includes data from other downhole sensors (pressure, torque, flow, temperature) and Logging While Drilling (LWD). This data stream is relatively slow, on the order of 20–80 baud.

Drilling companies use a variety of types of signals to transmit data: negative, positive, or frequency modulated pressure pulses in the drilling fluid column and electromagnetic waves.

1. Negative pulses are generated by a slight reduction in standpipe pressure. The dump valve is used to achieve this reduction by diverting the mud inside the drill string to the annulus. The opening of the dump valve reduces the pressure and closing of the dump valve reverts the pressure to its original value.
2. Positive pulses are generated by short and partial blockage of the drilling fluid flow through the drill string in the mud column. The blockage of flow creates an increase in pressure, and the release of blockage reverts the pressure to its original value.
3. Frequency modulation are continuous waves similar to the positive pulses; however, in continuous waves, a rotary valve is used that is capable of continuously blocking and unblocking the flow to create the signal as binary data. The blockage of flow creates an increase in pressure, and the release of blockage reverts the pressure to its original value. The frequency of the rate can vary, speeding up for higher frequencies (a 1) and slowing down (a 0).
4. Electromagnetic waves can be used to transmit ELF radio waves to the surface. However, electromagnetic waves cannot penetrate long distances as they are not strong enough and can be affected by lithological conditions (i.e. high iron content). They are limited in baud rates.

Recently, MWD tools have been developed to transmit continuous deviation data which is vital for fine control in directional drilling, especially with RSS tools. This development has led to improvements in the frequency and quality of the surveys. Traditionally, surveys are taken at the end of each stand (roughly every 90 ft). However, faster survey frequency can be as short as 10 ft. This increases the resolution of the trajectory calculations making for a more accurate trajectory. In addition, since the points between survey stations are assumed to be on a circular arc, the effect of tortuosity in the well path is not well defined with low resolution whereas high resolution surveys allow for drillers to see this effect and manage it appropriately.

4.5.4 Logging While Drilling

The petrophysical properties of the formation around the borehole (e.g., hydrocarbon saturation, lithology) while drilling is captured by logging while drilling systems (LWD). The tools look similar to MWD tools; but these measure formation properties, not drilling operational parameters. Four common formation measurements are: natural gamma ray, resistivity, porosity, and bulk density. Similar to plots generated by a wireline formation measuring tool, the output of the LWD is a log, which is a graphical representation of the petrophysical properties of the formation.

LWD started with simple natural gamma-ray logs in the 1980s and developed to catch up with wireline logs. Today, most of the service companies can provide all type of logs and tests that can be done with wireline including reservoir fluid sampling, NMR, and acoustic logs.

Development of LWD tools allows the driller and geologist to utilize geosteering; steering the borehole trajectory from geological condition, not geometric constraints. This process allows the borehole to be steered to have maximum contact with the reservoir for better fluid management. For example, a BHA that consists of LWD and RSS can drill wells with maximum standoff from the water-oil contact quite close to the reservoir top without leaving the formation.

Recently, several service companies commercialized the new LWD tools that decrease the number of different LWD tools in a BHA by combining several sensors in one tool. This reduction in the number of different tools decreases the BHA length and increases the capability of achieving better directional targets. In addition, reducing the offset of the sensors from the bit improves the directional decisions especially in geosteering operations.

Another critical development in the industry is the capability of mapping the reservoir while drilling. This cylindrical map ranges up to 30–60 ft around the borehole. In a field where the hydrocarbons are located in separate "sweet" spots, the directional decisions are taken according to the position of these spots as determined by geologists. The conventional way of taking decisions was checking the drill cuttings and logs. However, logs have a low depth of investigation which varies from 1–10 in. With these tools, better decision making on trajectory orientation can be made for effective reservoir management.

4.6 DRILLING DATA ANALYTICS

Drilling is one of the most data-intensive processes in the petroleum industry. Over the last two decades, the data acquisition and analyses used while drilling have improved significantly. Because to the range of drilling performance data available, assuming it is recorded, more companies are using this information to improve their operations. But there still is some room for improving the data acquisition systems.

4.6.1 Surface Sensors

All operational parts of the rig have sensors that help the crew continuously monitor the drilling operations. It is important to note that sensors record a specific data set from which other data can be derived. This means that sometimes the derived data has compounded errors inherited from the sensor data origin. An example is Rate of Penetration (ROP) which is derived from the block height over time.

4.6.1.1 Depth Measurement

Drilling depth is a derived signal, not a measured signal. It is calculated based on the rotational movement of the draw-work drum or the crown block sheaves. Typically, these sensors are rotary encoders that track the rotational movement of the draw-works drum as the drilling line spools or unspools. Based on this movement, the distance traveled by a line is calculated. This calculation depends on the line size, the number of layers of line wound around the draw-works drum (Florence and Burks, 2012). A rotary encoder can also be on the crown block and works similarly to the draw-works approach.

Another technique used to estimate depth is by using a pressurized depth sensor, which estimates the drilling depth based on a change in the hydrostatic pressure of drilling fluid. Being a non-contact measurement, the drilling depth is relatively more reliable than other drilling measurements but has relatively low resolution plus uncertainty regarding the density distribution of the drilling fluid. The low-resolution can lead to significant troubles, especially in longer drill strings.

The bit depth is calculated by adding all individual components in the borehole and stretch experienced by these components. These components are usually noted in a "pipe tally" along with their lengths. Initially, the pipe tally was updated manually by the drilling crew; but this led to extensive inaccuracies in estimating bit depth, and hole depth. These inaccuracies in the depth estimation can be minimized by using algorithms to detect the "rig state" or the type of operations being performed based on drilling data (Dunlop et al., 2004).

4.6.1.2 Flow Measurement

Since the drilling fluid can be a multiphase fluid, measuring the accurate flow rate can be quite challenging. These flow measurements are essential in estimating drilling fluid hydraulics, well control, and cutting lag (Dowell et al., 2006). Accurate tracking of flow-in and flow-out rates can help the crew predict potential downhole problems like loss circulation or kicks.

The most common type of flow-out sensors used on the drilling rigs are spring return paddles in the return line. Figure 4.9 shows a schematic of the flow-paddle sensor. It measures the flow rate coming out of the borehole using a strain gauge analog transducer. The changes in the resistance values are used to estimate reduction or increase in the mud rate. This only provides a relative estimate of return flow; no absolute flow rate value can be made. Newer data acquisition systems use a Coriolis mass detector to calculate flow. Because of their inability to handle the multiphase flow as well as expense and location limitations, they have not universally used (Florence and Burks, 2012).

Proximity and/or whisker switches are used on oil rigs to measure flow-in. The proximity switch acts as a relay switch activated by an electromagnet or a magnet whereas the whisker switch is activated when an external rod raises a ball bearing to initiate contact against a piston. The flow rate can also be measured with a flow meter, but it can have severely inaccurate measurements because of the presence of high-density solids in the mud which can sag at low flow rates (Cayeux et al., 2013). Flow meters that can handle large percentages of solids and multiphase is a major area of research. Several companies have come out with such flow meters. The jury is still out but results appear to be encouraging.

4.6.1.3 Pressure Measurement

All drilling rigs contain numerous pressure sensors and transmitters. The pressure on the top of the drill string, frequently called "standpipe pressure," can directly be used to estimate downhole pressures which can indicate downhole problems such as kicks or loss of circulation. The pressure sensors used on drilling rigs are usually a high-pressure diaphragm located on the standpipe or pump manifold. The first type of sensor measures the pressure exerted by the mud on a rubber diaphragm.

FIGURE 4.9 A "FlowShow" with the flow paddle sensor. *Source:* Photo by A. Eustes.

In the second type of sensor, the transducer face is in direct contact with the mud (Dowell et al., 2006).

The bottom hole pressures and the annular pressures are calculated based on standpipe pressures, rheological measurements, and flow rates. Some newer tools have a capability of measuring internal pipe pressures and fluid pressure in the formations at the bit both of which are essential during operations (Florence and Burks, 2012).

4.6.1.4 Force Measurement

Continuously understanding the weight of the drill string is crucial for any drilling operation. This is usually done by tracking the hook load. The hook load or the total load exerted by the drill string on the hook is measured using the same pressure transducers used to measure standpipe pressure. The hook load can be measured at the deadline, in the crown block, or at the top drive. Figure 4.10 shows a deadline anchor where the tension in the deadline (the anchored line in the hoisting system) is measured. The calculated hook load takes that tension and multiples by the number of lines on the traveling block (the mechanical advantage). The strain experienced by the deadline applies load on the hydraulic fluid which then is translated to a force measurement (Dowell et al., 2006).

By pushing the drill string to the bottom of the borehole, a portion of the weight of the drill string is transferred to the bottom of the hole. This exerted force is termed weight on bit (WOB). WOB is primarily generated by the drill collars that attempt to keep the drill pipe always in tension with the heavy wall pipe, acting as a transitional string between the drill collars and drill pipe. The transition from tension to compression must always occur in the drill collars. WOB is calculated as the difference of hook load on bottom and off-bottom. This calculation assumes a frictionless system leading to erroneous estimation of WOB. In recent years, downhole WOB measuring systems have been deployed by some contractors. In these sensors, a special pipe connected just behind the drill bit detects shear or compressional stress in the pipe wall (Florence and Burks, 2012).

4.6.1.5 Torque and RPM Measurement

Two of the most important parameters that affect drilling operations are RPM and torque. Both top drives and rotary mechanism on the surface and downhole motors contribute to the total torque experienced during drilling. In a Kelly drive, the rotary torque is estimated from the motor control system. For a DC system, the torque is estimated using the current through the motor using a toroid,

FIGURE 4.10 Weight indicator measurement off of a deadline anchor. *Source:* Photo by A. Eustes.

FIGURE 4.11 RPM sensor on rotary drive (left) and pump stroke counters for flow rate (the hoops going into the pistons (right)). *Source:* Photo by A. Eustes.

and for an AC system, the torque is obtained from the Variable Frequency Drive (VFD) (Florence and Iversen, 2010).

For modern top-drive systems (TDS), the torque sensor is a clamp that sits around the power cables to the TDS. It monitors the deformation of the Hall-effect chips by the magnetic field produced around the cable through current being drawn (Dowell et al., 2006). Figure 4.11 shows an example of the RPM and pump stroke sensors use on drilling rigs. The rotary speed in RPM is measured using a proximity switch or a limit switch. It is shaped differently to work for circular geometries but follows the same principle (Dowell et al., 2006). The pump stroke counter gives the flow rate when multiplied by the pump factor, the volume pumped per three strokes of the pump. Also, by "counting" strokes, one can determine the volume pumped as these pumps are positive displacement pumps.

4.6.1.6 Downhole Sensors

In addition to the MWD and LWD tools as discussed earlier, the modern drilling rigs employ a wide range of sensors in the MWD/LWD tools that improve a driller's understanding of the bottom hole conditions. Sensors that measure shock, downhole pressure, the torque on bit, and downhole weight on bit can be used to estimate downhole vibrations, the health of the bottom hole assembly (BHA), and formation pressures (Dowell et al., 2006). In this section, shock sensors and pressure while drilling sensors will be discussed. We will also discuss data transmission and sensor limitations.

4.6.1.7 Downhole Shock Sensors

Downhole vibration is the major cause of failure of MWD/LWD tools and mud motors. The health of these tools becomes essential in reducing the cost of drilling long laterals for unconventional reservoirs. The shocks can be characterized into lateral, axial, and torsional shocks. Several analytical models to estimate the severity of vibration have been used in the industry but their accuracy is questionable. The shock sensor included in the MWD tool consists of a single axis accelerometer. The sensors count the number of shock events within a certain frequency range over a time period. The sensors use pre-determined criteria to classify the drilling conditions as excellent, good, poor, and severe (Dowell et al., 2006).

Downhole shock levels can be correlated with the design specification of the MWD tool. If the tool is operated above design thresholds for a period, the likelihood of tool failure increases proportionally (Dowell et al., 2006). Application of the shock measurement can prevent "unscheduled events" or unplanned interruption to the drilling that costs substantial rig time. The inclusion of the shock sensors in the BHA and their application can prevent events like BHA failure, minimizing non-productive time.

4.6.1.8 Pressure While Drilling Sensors

The success of drilling operations in high angle wells and extended reach wells depends on the accurate tracking of both static mud density and equivalent circulating density (ECD). Maintaining the fluid density within the safe margins (above any pore pressure but below any fracture pressure anywhere in the open hole) is vital in ensuring safe operations. Inability to stay within the safety margins can lead to loss-circulation events, blowouts, or differential sticking (Ward and Andreassen, 1998). Accurate pressure management becomes important, especially while drilling High Pressure/High-Temperature (HTHP) wells where the safety margin is often less than 1 ppg (Charlez et al., 1998). Low margin wells in Deepwater GOM have been the primary driver of the development PWD and the development rheological models that predict downhole pressures both static, dynamic, and transitional. Sometimes there are only a few psi differences. Pressure while drilling (PWD) sensors are used to estimate downhole pressures and downhole mud density in real time, enabling efficient control of downhole mud density. With these sensors, the annular fluid is sampled through a drilling collar to a pressure gauge connected to the MWD tools (Ward and Andreassen, 1998). The PWD measurement can also be used to identify problems such as swab/surge pressures, loss circulation, barite sag in real time (barite particles settling out of the drilling fluid leaving the lighter weight base fluid), all of which improve drilling practices.

4.6.2 Drilling Data Visualization and Communication

The extensive data recorded on modern drilling rigs require complex data management systems which contain data processing, visualization, communication, and storage capabilities. The data recorded by the service companies and rig contractors through the Electronic Drilling Recorder (EDR) system on the rig is communicated to the offices through cloud-based servers in real time to ensure continuous monitoring of the rig (Petrowiki, 2015). These EDRs are a network of sensors, software, and displays that accumulates all measurements made on the drilling rig and displays them across the rig site. It includes both calculated and measured data traces. This drilling data can either be transmitted through a wired or a wireless network. Wellsite Information Transfer Standard Markup Language (WITSML) and OPC Unified (OPC-UA) architecture are used to transmit the data between different rig sites or between rig sites and offices. Figure 4.12 shows an example of an EDR output as shown on a screen in the drilling control center of a rig.

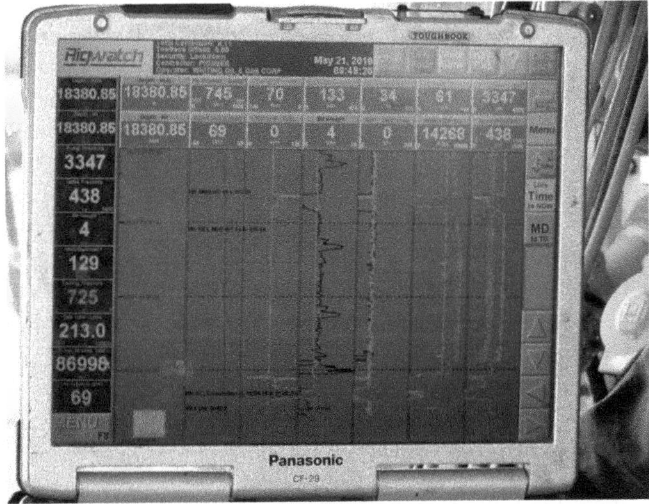

FIGURE 4.12 Snapshot of an electronic drilling recorder. *Source:* Photo by A. Eustes.

4.6.3 Wellsite Information Transfer Standard Markup Language

WITSML was defined to seamlessly transfer the drilling data between systems between different service companies and contractors to increase the efficiency of the decision-making process. It was developed by a special interest group, which comprises more than 50 companies (Pickering et al., 2009). WITSML enables interoperability from different servers and clients. One of the main advantages of standardizing WITSML was that it allows for timely access to the real-time data, helps execute efficient operations, and helps conduct safe operations. WITSML is updated continuously based on inputs from clients, suppliers, and the special interest group.

4.7 APPLICATION OF MSE IN DRILLING DATA ANALYTICS

4.7.1 Mechanical Specific Energy Evaluation

For years, various researchers in the petroleum industry have proposed different indices or factors that quantify the efficiency of the drilling process. Some of these factors are also used to evaluate different drilling problems. The most widely used factor among these is the Mechanical Specific Energy or MSE proposed by Teale1 for rock excavation process. Simply put, MSE is the energy needed to excavate a unit volume of rock. Teale defined it as the sum of work done by thrust and the torque. The Teale equation (shown here) can be used to calculate the MSE for any excavation operation (Teale, 1965).

$$MSE = \frac{WOB}{A_{bit}} + \frac{2\pi N\mathrm{T}}{ROP} \tag{4.1}$$

where

MSE = mechanical specific energy
WOB = weight on bit
A_{bit} = bit cross sectional area
N = rotational speed (RPM)
T = torque
ROP = rate of penetration

MSE has been used extensively in the drilling industry for both drilling optimization and mitigating drilling problems. The main reason for its wide application is the versatility of the factor and its ability to capture a multitude of information in a single parameter. The main application of MSE has been to increase the rate of penetration by monitoring drill bit performance. Several researchers have also applied MSE to detect bit dysfunctions, when energy supplied to the bit is wasted, such as drilling vibrations, bit balling, slip-stick, and whirling. In addition, attempts have been made to monitor bit wear and change in formations. Note that the numerical value of the MSE isn't as important as its trend over a depth range. This is because the parameters on which MSE depends have significant variability due to the complex and dynamic nature of the drilling process. Dupriest et al. (2005) show an example of this trendline analysis. Some studies have used the numerical value of MSE, but that application usually assumes homogenous subsurface properties.

4.7.1.1 ROP Optimization

The rate of penetration or the drilling rate is one of the most important drilling parameters as it directly affects the drilling cost. Maximization of ROP is one of the main drivers for any drilling company. As apparent from the Teale Equation, as MSE decreases, ROP increases as they are inversely related. In the field, drilling companies often conduct the drill off test to identify founder point or point above which ROP will change nonlinearly with respect to the WOB. MSE at this point will be selected as baseline MSE for a formation. If subsequent MSE values are close to the baseline MSE, the drilling is assumed efficient. The same process can be repeated with RPM variability, keeping WOB constant.

4.7.1.2 Bit Balling

Bit balling is a condition in which the drilling cuttings get accumulated between cutters, minimizing transfer of energy from the bit to the formation. A sustained rise in MSE in absence of decrease of WOB and/or RPM can indicate bit balling. The drilling crew can easily identify the problem and increase the hydraulic horsepower per square inch (HSI) to improve cutting cleaning. Dupriest et al. (2005) show an example of this. The crew observed an increase in MSE and suspected bit balling. They pulled the bit out and replaced the nozzles to increase HSI; this improved the cutting removal process and significantly reduced MSE resulting in an increase in ROP.

Being an extremely dynamic process, drill systems often experiences vibrations. If drilling vibrations aren't mitigated immediately, they can lead to catastrophic failures such as bit failure, mud motor ruin, and MWD/LWD destruction (and the associated costs which are extremely high). The vibrations are subdivided into three types: axial vibrations, lateral vibrations or whirl, torsional vibrations or stick-slip.

Comparing MSE with other measurements can confirm downhole drilling vibrations. They can be controlled by altering WOB and/or RPM. Dupriest et al. (2005) show a system where bit/BHA whirl significantly increased the MSE. The vibrations were mitigated by increasing the WOB which increased the ROP from 35 to 50 ft/h. The same paper shows a situation where the crew was trying to increase the WOB, but they experienced stick-slip resulting in higher MSE.

4.7.1.3 MSE Data Management

Operational standard practices and procedures use real-time MSE data analysis to continually improve drilling efficiency and performance. The ideal MSE is steady throughout the drilling process. Identifying sudden or significant changes in MSE can pinpoint the time at which drilling becomes prohibitively inefficient. Simultaneous display of drilling parameters including ROP, WOB, logging data, and MSE allows for incremental changes in procedure to take place during the drilling process. When trained to identify key trends in MSE data, drilling personnel in the field can effectively manage the drilling process by comparing the active drilling process to historical well records. The combination of real-time data and formation data from offset wells is key to identifying the point at which the drilling process should be significantly altered through a change of downhole equipment. The predictive capacity of MSE data can prevent costly mistakes such as abnormal borehole geometry, "fishing" a lost item out of the borehole, and other events that would cause a substantial increase in the safety risk and cost of the well. The recent advances made in MSE data capture and analysis must be coupled with a comprehensive management style to maximize the potential efficiency gains of MSE data analytics.

4.7.2 Non-productive Time Drivers

One of the largest improvements in drilling operations in the past 20 years has been the recognition and reduction of non-productive time (NPT). This is time spent working towards a goal that is not actually helping achieve that goal. In the case of drilling operations, anything that slows down or outright stops the construction of the borehole that was unscheduled can be considered non-productive time. Running a planned casing string is not NPT. Waiting for a part for the hoisting system for lifting the casing would be NPT.

An NPT driver can be defined as an event that interferes with the critical path of borehole construction and thus increases borehole construction time. A study in the early 90s by Transocean Offshore identified items on the critical path in drilling. Their conclusion? Anything going through the rotary table on the rig is on the critical path!

Pressure to reduce non-productive time is very real, as drilling expenses can be up to 31% of the cost of a given unconventional well. Many times, this leads to laziness or apathy about needed operational stoppages or borehole construction stages. Irrational decision making driven by a pressure to drill a hole quickly can lead to disastrous results such as explosive blowouts.

Human psychology plays an important role in causing non-productive time. The results aren't always disastrous, but much non-productive time is caused by confirmation bias of their decisions to ignore weak signals (Thorogood et al., 2014). Very experienced operators often ignore weak signals because they "know" what's going on downhole. This can lead to total drilling stoppage for no reason. A human factors perspective is almost never taken during incident investigation despite the encouragement from some in the industry to do so (Thorogood et al., 2014). This industrial culture makes data analytics a powerful decision-making process in regard to reducing non-productive time. It is difficult to argue with what data says. Intelligent monitoring of NPT drivers can significantly increase drill time without introducing unnecessary safety risks by anticipating and intercepting human factors before they enter the equation.

Different types of inefficiencies and time stoppages plague every kind of industry. Many different techniques have been developed over the years by various fields in order to prevent work stoppages or identify and eliminate process inefficiencies. Modern big data and computer science technologies play a large role in capabilities for combating these issues. A good example of a revolution in this sector is the adoption of lean manufacturing techniques by the American automotive industry. Core lean concepts focus on continuous improvement and reducing downtime. A shift in focus on "lean drilling" in terms of borehole construction has been encouraged (De Wardt, 1994). Leveraging data analytics to NPT drivers can help achieve this goal. The borehole construction process will never be as efficient and tightly controlled as the process to manufacture a car due to the uncertainties downhole, but lean techniques are effective when properly utilized outside of manufacturing.

Real-time operating centers (RTOCs) are an effective way of leveraging the power of data to help mitigate errors in decision making on site and thus reduce NPT time. Real-time operating centers are remote locations where subject matter experts can watch drilling operations and data streaming in real time. It was such a RTOC data stream that allowed investigators to piece together the events leading up to the Macondo Blowout in the Gulf of Mexico in 2010. RTOCs are successfully employed in a variety of other industries: space flight, aviation, and Formula 1 racing being appropriate examples (De Wardt, 2014). RTOCs give drillers support during the borehole construction process. Big data is important, but it isn't helpful if all it's doing is overwhelming operators in real time. High-frequency data collection and processing at RTOCs allow drillers to be sent back only pertinent data for informed decision making. It also opens up a line of communication between those on the rig site and those in the office, which, traditionally, has not been open. The idea of processing data and monitoring multiple rig sites at one centralized location is not new, but the slow adoption of better sensors and data practices in contracting is unlocking the power of this strategy.

In drilling, NPT is most commonly identified at a macro level by examining days versus depth plots (Figure 4.13). These plots show the total depth of a hole drilled as a function of time. Note that in drilling operations, it is common to set the independent variable (depth in this case) on the Y-axis with the dependent variable on the X-axis (days). That means a horizontal line is a zero slope. NPT drivers cause the slope of these charts to trend towards zero or total stoppage; in many cases, the slope is zero while operations are affected by NPT events. In essence, within the context of drilling, an NPT driver reduces the average rate of penetration or outright stops the drilling process.

4.7.2.1 Types of NPT in Drilling

Non-productive time can be driven by sources in many different areas. In this case, they will broadly be described as coming from three sources: the borehole, the surface, and logistics.

Issues stemming from the borehole include things such as lost circulation events, borehole stability problems, stuck pipes, and well control problems. Downhole events can broadly be described

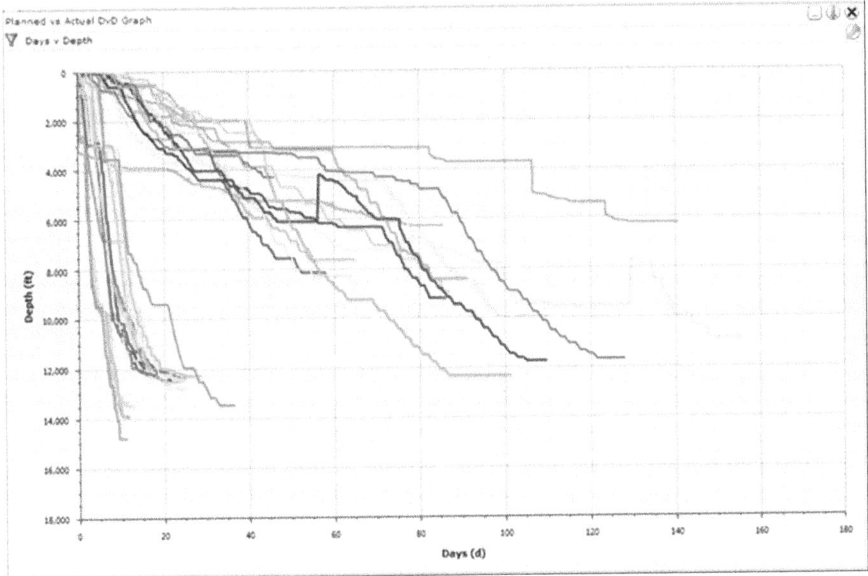

FIGURE 4.13 Typical days versus depth for oil and gas (left) and geothermal (right) drilling time curves. NPT is frequently, but not always, shown in the "flat times." *Source:* Courtesy of Visser et al. (2014).

as the most significant sources of NPT. Well control is the most important of these, as well control issues, such as kicks and blowouts, are a safety concern and can result in environmental damage, the loss of a well and rig, and death. It seems like every Hollywood movie that has a drilling rig in it has a well control incident gone horribly wrong. Recently, enough such events in succession (Montara, Bardolino, and Macondo) have helped completely change the way the industry views human factors within the context of NPT (Thorogood et al., 2014).

Problems originating from the surface are operational related and include things such as rig breakdown, tripping pipe, and time spent evaluating decisions. Events that can be broadly categorized as equipment failure can be dealt with via the standardization of rigs, sites, and equipment. Many of the NPT drivers originating at the surface are also attributable to human factors. As discussed previously, a combination of culture change and offsite support can help mitigate these.

Logistics issues manifest themselves when communication problems occur or when the time is spent waiting on equipment and transport. In the context of NPT logistics, issues are less on the rig site and more the responsibility of other entities. Supply chain issues, permitting hang-ups, and unpredictable weather can all drive NPT in this area. Data analytics and good operations research certainly help reduce the likelihood of logistics problems.

Many non-productive time issues at the surface stem from a natural conflict between different operators and contractors. The borehole construction process involves many different entities that prefer not to share more information than is necessary with one another. This is a problem for the efficiency of the process as a whole, as there's a myriad of ways that shared information could be beneficial to all.

4.7.2.2 The Big Three NPT Drivers

Three of the most important drivers of drilling non-productive time are all related to the borehole. Well control, lost circulation, and borehole stability are sometimes considered the "big three" NPT drivers. Fear of encountering these three issues also implicitly induces other NPT events meant to prevent the three from happening such as evaluation and tripping.

Well control issues involve kicks and blowouts. A kick occurs when formation fluids flow uncontrolled into the borehole during drilling. This happens when the pressure exerted by the formation fluids is greater than the pressure of the hydrostatic column in the borehole. Kicks can be induced by drilling operations or simply be a result of inaccurate subsurface data interpretation. A blowout is a kick that leaves a borehole uncontrolled such as one that breached the surface. Blowouts are a worst-case scenario for drilling.

Lost circulation events are the opposite of a kick. In this case, drilling fluids penetrate the formation due to being a higher pressure than the pore pressure or in some cases, greater than the rock itself can hold, and thus exits the circulation system uncontrolled. Lost circulation can actually lead to well control issues because of the drop in column pressure due to system volumetric loss. Lost circulation generally occurs in four different types of formations: natural caverns, natural fractures, induced fractures, and highly permeable formations. This can be very difficult to handle, as it's hard to tell exactly what's causing the event and how long an event will last. Many aspects of drilling mud engineering focus on preventing lost circulation in the borehole.

Setting casing points and creating a borehole construction plan depends on balancing the risk of taking kicks and experiencing lost circulation. In order to do this, the driller needs to keep the column in the fluid between the pore pressure and the rock fracture gradient. This margin can be very small and sometimes changes drastically with depth when lithology changes. These drastic changes in formation properties can be challenging to predict; but, a combination of quality downhole data and historical analysis can be used to anticipate these changes.

Borehole instability is usually in reference to sloughing shales or borehole collapse. The amount of unconventional plays in or through shale being drilled today makes this a constant battle. A shale sloughs when the wellbore stress is not counterbalanced by the stress within the wellbore created by the mud density gradient. The chemistry of the mud can impact this process by either reducing the strength of the rock by adsorbing water from the mud (either aqueous or non-aqueous invert) or increase the local stress field by absorbing to the clays and swelling, thus increasing the localized stress state. This chemical mechanism is primarily because shales contain clays that can absorb water and expand or be dispersive in water. Since shale layers are so tightly compacted to begin with, additional expansion between layers causes fracture. When this occurs at the borehole boundary because of formation fluid, the shale sloughs or breaks and falls into the borehole. Shale sloughing is one of the most prescient forms of borehole instability, but other instances of instability can manifest for a variety of different reasons. In situ stress failure, fluid circulation erosion, and chemical interactions between borehole and formation fluid can change the borehole geometry in a way that negatively impacts the operations. This often leads to enlarged boreholes.

Borehole issues also concatenate on each other and can result in total loss of a well if left unchecked. For example, a lost circulation event leads to a loss of hydrostatic pressure in the fluid column which can induce a formation kick. The sensitive balances in pressure need to always be kept in check. Since these three also exist within the hole itself, they cannot be physically seen. Interpretation of downhole data and feedback on surface equipment and the drill string are thus the only ways to predict the occurrence of these events. Data analysis then becomes integral in the diagnosis and prevention process. As it stands, a lot has been learned from the available data. The implementation of high-frequency sensors now and in the future will reveal a lot of new information about what is happening in a borehole at any given time.

4.7.2.3 Examples of Data Analytics for NPT Drivers

The ways in which data analytics are applied to reducing non-productive time are virtually limitless. Large data sets can be dissected from every imaginable angle. It has already been discussed how important good data practices can disrupt human factors and minimize rig site culture from leading to unnecessary non-productive time. There are also ways to directly mitigate major NPT drivers downhole by using real-time data streams or historical data analysis. This section focuses on a few different methods already being done to give the reader tangible examples. Examples for each of the major NPT drivers are shown.

The interplay between fluids in a borehole can be complicated. It can be difficult to determine whether or not the behavior is representative of a kick or simply borehole breathing in real time. Borehole breathing is a phenomenon that occurs when drilling mud is overpressured. While circulating, the mud induces microfractures and penetrates the formation. Once pumps are shut off, the lost mud forces its way back into the borehole. At the surface, this looks like a kick because the volume of mud in the mud pit rises. This often leads to unnecessary work stoppages to control the "kick." Deterministic models based on fundamental physics have been available to use for a while, but like any large-scale fluid dynamics modeling, these simulations are very computationally intensive and not practical for real-time decision making. Stochastic models have been developed based on historical data from wells in similar formations. These models can show a likely outcome but lack precision. This can be useful for large, complex systems that don't necessarily require precision. Work has been done in this arena for real-time kick detection with positive results (Ojinnaka et al., 2018).

4.7.2.4 NPT Based on Human Factors

Machine learning is an effective tool for creating predictive models. Machine learning is a type of artificial intelligence that uses statistical techniques to help a program "learn" over time. It is an iterative process that uses regression in a step-by-step manner to repeatedly generate successively more accurate predictions based on the input. These methods have effectively been used to estimate lost circulation in real time. Lost circulation events are difficult for drillers to handle and often the approach used in particularly tricky formations changes from operator to operator. Historical data in a given area on many different lost circulation materials used and their effectiveness can be used as a training set for an algorithm. This same algorithm can then be used to identify sets of best practices and recommend appropriate lost circulation material for a given borehole in future operations in those areas (Al-Hameedi et al., 2018a).

Some formations present a multitude of borehole stability issues. This causes problems such as stuck pipe, borehole fill, bridging, and cementing challenges. Real-time analysis of drilling geomechanics helps operators make informed decisions while drilling. Studies have been done that combine data from gas information, leak-off tests, formation integrity tests, and LWD resistivity and gamma-ray data in order to build temperature-dependent pore pressure models before drilling additional wells in the area. By taking data during the operation and constantly updating expected conditions downhole the severity of borehole instability issues can be mitigated. More useful information available to decision-makers helps (Schlumberger, 2010).

All of these examples show how different types of data analysis can be combined to combat non-productive time drivers. Potential and current applications are limited only by the quality of data available, the power of computing available, and insight of the engineers attempting to mitigate NPT problems.

4.8 DRILLING MANAGEMENT

Technical advances made in recording and transmitting downhole drilling parameters have extended the ability to visualize and predict subsurface features that determine site-specific design plans and construction. Management and operational practices have advanced along with recent technological discoveries to consistently improve worldwide drilling performance. A number of tools have been developed to aid in the management and planning phase of drilling and borehole construction. The "Perfect Well" and the "Technical Limit," to name a few of these techniques, are all trademarked names of the processes for drilling management developed and implemented over the last 20 years.

4.8.1 The Perfect Well®

A perfect well is a theoretical tool used to quantify the cost and time scale of an ideal well construction process. The perfect well represents the minimum time it would take to drill a well by only considering the physical limitations of the rig and the energy required to break apart rocks in the area. The parameters that determine the rate and cost of a drilling operation vary with geological area and the known lithology of an area. A benchmark standard that is analogous to a 100% efficient heat engine is needed to enable comparison of operational procedures in different areas. J. Ford Brett, the inventor, explains the special assumptions of the perfect well as

> one bit per hole section, with penetration rate limited by the horsepower available from the rig, the rock strength, perfect trip times, perfect casing and cementing operations, etc., to closely estimate the minimum time physics would let a particular well be drilled.
>
> (2006)

These physical parameters are used to calculate the "perfect time" and subsequently the perfect well ratio, defined at the actual time taken to drill a well divided by the expected "perfect time."

The perfect well and the associated perfect well-time ratio are starting points to develop incrementally better operations practices without factoring in human expectations into the goal. This objective analysis can be used to evaluate technical, operational, and economic areas of improvement before, during, and after the well construction process. As the process becomes established within an organization, the addition of new geologic and offset well data is essential to maintaining an accurate perfect well time calculation.

A mechanical earth model (MEM) is a three-dimensional model representation of rock properties and earth stresses that are of interest to drilling engineers. The parameters include Poisson's ratio, Young's modulus, unconfined compressive strength, friction angle, pore pressure, minimum and maximum horizontal stress, along with the direction of horizontal stress axes and vertical stress. The recording of subsurface temperature, ambient stress, and pressure data is then stored along with time and location markers in an accessible application that engineers of all disciplines can use in the planning, development, and abandonment phases of drilling and well construction (Bérard and Prioul, 2016). A comprehensive mechanical earth model enables drilling engineers to predict high pressure or troublesome areas before drilling into them (Plumb 2000).

The MEM is a modeling starting point that provides drilling engineers with the necessary data to design the perfect well for the region in question. The recording of temperature, ambient stress, and pressure data is then stored along with time and location markers in an accessible application that engineers of all disciplines can use in the planning, development, and abandonment phases of drilling and well construction (Bérard and Prioul, 2016). In high-risk areas such as tectonically active basins and the Gulf of Mexico, exploration and development activities require predictive capabilities beyond traditional offset geotechnical data that only computer models can offer.

The ability to refine the mechanical earth model and thus the perfect well while drilling has been advanced by logging while drilling (LWD) and measurement while drilling (MWD) technologies that provide engineers with real-time data from the borehole. Refining the assumptions made during pre-drill analysis to fit the modern real-time data collection ability would enable engineers to proactively adjust key parameters and result in safer and more efficient wells (Staveley and Thow 2010). Evaluation of drilling efficiency occurs by calculating the mechanical specific energy (MSE) of a "perfect well" and using this benchmark to pinpoint inefficiencies. Considering input from LWD/MWD downhole technologies is essential to compare the perfect well estimation with reality. The real-time data input can also pinpoint geologic formations of concern that necessitate a significant change in specialized downhole equipment including bits and bottom hole assemblies or even drilling fluid type. A perfect well model provides a benchmark that can be used during the drilling process to pinpoint when a change in procedure should take place. The latency in LWD/MWD technologies must be taken into account when making step changes in the drilling process. The creation of a "perfect well" can be further tailored to a specific geologic area using a mechanical earth model system resulting in even greater predictive capacities (Staveley and Thow 2010).

4.8.2 The Technical Limit

The technical limit, while similar to the concept of the perfect well, is different because it includes human expectations in regard to safety and operational performance. Data capture and analysis is central for evaluating drilling performance in real time and identifying the technical performance limit. Once this limit has been identified, proper training of all involved is needed to fully actualize the potential of the technology used in the project. Finding the limit also suggests ways for future drilling operations to overcome and stretch that limit for faster and safer operations.

The technical limit is a term used to describe an ideal standard of drilling performance given site-specific parameters that assume perfect technical teams, drilling conditions, and tools. The technical limit differs from the perfect well in that the technical limit is prescriptive in nature and aims to consider potential failures and safety concerns in addition to maximizing the rate of penetration (Akgun, 2007). The technical limit evolves over time as more technological advancements are made in drilling and new standards emerge in regard to safety and environmental protection. While the technical limit is an important component of process analysis, it should be critically analyzed to make sure human bias is not implicit in setting the technical limit (J. Ford Brett 2006).

Targeting the achievable ideal in the pre-construction planning phase allows operational inefficiencies to be identified. The technical limit has been targeted and performance has improved with pre-construction planning, site evaluation, and personnel training. This management technique has reduced operational time and construction cost while delivering higher than average production rates in complex and high-risk drilling projects (Bond et al., 1996). Setting objectives based on the technical limit offers a more objective approach to performance improvement than comparing the target well to a previous well.

Once performance targets have been formed via the technical limit, the field personnel and team members are often placed in an intensive training workshop to properly prepare them for the job (Bond et al., 1996; Dupriest et al., 2005). Thorough training and preparation before drilling begin to mitigate risks incurred by a lack of preparedness during abnormal drilling events that jeopardize health and human safety. Field personnel must be properly trained to identify and mitigate drilling issues during active drilling operations. Beyond being able to operate rig equipment, drillers need to understand the core concepts of drilling to identify and correct situations that lead to lower rates of penetration. Proper management and cultivation of a cooperative workplace environment have been shown to further advance the efficiency and the ability of modern drilling technology when dealing in geologically difficult areas. Combining new technologies with the improved management practices outlined previously is effective when dealing with troublesome formations, as seen in the application of vertical borehole drilling tools in tectonically active locations in South America (Barnes et al., 2001). Pre-well optimization analysis, daily drilling reports, and post-well evaluation reports have been shown to reduce well construction time by 15% (Devine et al., 2002).

Continuously updating the model made during the pre-drill analysis with real-time logging and drilling data is essential to continuously improve efficiency gains in project implementation. The mechanical earth model can be updated with logging while drilling (LWD) data, geological parameters from mud logging, vibration data from downhole accelerometers, and rig measurements of mudflow and rate of penetration. Simultaneously displaying drilling parameters, mechanical specific curves, and vibrational curves allow trained field personnel and remote engineering teams to modify the drilling process as soon as a significant increase in MSE is seen (Dupriest et al., 2005). Implementing active comparisons of real-time and historical data was shown to reduce well construction time and cost while advancing the knowledge-based modeling system required for accurate well planning.

Using the perfect well model with the technical limit can assist drilling organizations in investigating potential drilling and completions methods while assisting in the consistently accurate economic evaluation of a complex project (Bond et al., 1996). The increased time and attention required to pursue the technical limit have been shown to have economic returns due to decreased rig time and cost (Devine et al., 2002) in addition to an increased ability to manage risks in the form of early kick detection. In the case of slim-hole design, early kick detection has been proven to work when coupled with measured mudflow and standpipe pressure data and a predictive model driven by real-time rig data (Swanson et al., 1997). In this model, a dynamic twin borehole model was used to identify data trends that signify abnormal pressure conditions that can lead to a kick.

4.8.3 Management Challenges

Drilling process analysis should occur at various stages throughout the management process to identify areas of concern before they propagate inefficiency throughout the lifetime of the project. The pre-construction well analysis is composed of mud logging, geological and offset well data that is used to fine tune the broader well construction strategy. During drilling, daily reports and real-time drilling parameters and geologic data are collected from the rig. The post-well analysis compares the achieved performance to a benchmark standard determined by the perfect well and technical limit.

Large-scale or global operations require structured yet nimble operations that are enabled by data acquisition and analysis. Improved communication techniques now enable global operations to communicate and make engineering decisions based on a wealth of historical and project specific data sets. Specialized consultants can be contacted to provide unique risk analysis and mitigation techniques that further enhance the planning process (Bond et al., 1996).

However, the range of data sources and types used in drilling operations introduce uncertainty and make comprehensive pre-construction analysis difficult. Paper copies of well records, test results, and geological data is tedious and can be difficult to digitize. Relying on outdated data to plan drilling operations increases the economic and hazard risk of the operation. However, controlling for quality of historical data can result in valuable trends being identified within the historical data. Importing, categorizing, and displaying a variety of data sets, from geological data to offset well data, can greatly improve the predictive ability of pre-drilling situational analysis (Staveley and Thow 2010).

Knowledge management is an area that optimizes knowledge acquisition and transfers to address the chronic inefficiencies in drilling management and operations. The utilization of relevant current and historical data depends on the accessibility of that data and the ability to simultaneously visualize and correlate a number of data sets in one comprehensive view (Staveley and Thow 2010). The active drilling process requires real-time visualizations and updates to the modeling software to better inform the driller and engineering team. To harness these technological developments properly, a multidisciplinary team was employed during the planning phase of development.

4.9 SUMMARY

This technology of penetrating the subsurface on earth has been around for 3,000+ years. Yet over the last 20 years, a phenomenal increase in drilling operational efficiency and significant cost reductions in onshore oil and gas drilling has taken place. Part of this was forced upon the drillers from low commodities prices. The drillers looked at everything from the rigs, fluids, downhole equipment and bits, control systems, all the way to how the drilling operations were managed. The advent of more capable computer processors and storage has created new modeling and analytical approaches. In addition, today's massive data collection from capable sensors, the historical analysis of said data, as well as predictive analytics has led the industry into unexpected insights into their operations, leading to elimination of wasteful practices and far better equipment usage and personnel efficiency. The results have led to lighter weight drills that use lower power and yet perform significantly faster and safer.

Coming in the future is the automation of all aspects of drilling. This is driven by safety, manpower reductions, and a desire for consistency in operations. Given the uncertainties in Mother Nature, this will be the industry's next greatest challenge. It will be overcome.

ACKNOWLEDGMENTS

The authors of this chapter are greatly appreciative of the time and efforts of all the reviewers. Their comments and concerns have been addressed. The chapter editor also acknowledges the phenomenal contribution by the various chapter authors, undergraduate and graduate students at the Colorado School of Mines, including Ahmed Amer of Newpark Drilling Fluids and recent PhD graduate, Dr. Saleh Alhaidari with Saudi Aramco. The chapter editor also acknowledges Nicole Bourdon for her work in finalizing this chapter. They all deserve enormous credit for the intense work that was accomplished. The authors would like to thank Dodie Ezzat and Mary Dimataris for their input and help with this chapter. The authors would like also to thank Christopher Yahnker, JPL/Caltech, Pasadena, CA; Fred Bruce Growcock, Occidental Petroleum Corporation, Houston, Texas; and Arthur Hale, Aramco Services Company, Houston, Texas, for reviewing this chapter and providing valuable technical comments and suggestions. Finally, the team appreciates the patience and efforts of the primary editors of this book.

REFERENCES

Abimbola, Majeed, Khan, Faisal, Khakzad, Nima, and Butt, Stephen. n.d. "Safety and Risk Analysis of Managed Pressure Drilling Operation Using Bayesian Network." *Safety Science* 76: 133–144.

Accenture. 2017. "Drill Deeper into Digital—2017 Upstream Oil and Gas Digital Trends Survey." Accenture. https://www.accenture.com/us-en/insight-2017-upstream-oil-gas-digital-trends-survey

Akgun, Ferda. 2007. "Drilling Rate at the Technical Limit." 1(1): 99–118.

Al Dushaishi, Mohammed F., Nygaard, Runar, and Stutts, Daniel S. n.d. "Effect of Drilling Fluid Hydraulics on Drill Stem Vibrations." *Journal of Natural Gas Science and Engineering* 35: 1059–1069.

Al-Hameedi, Abo Taleb T., Alkinani, Husam H., Dunn-Norman, Shari, Flori, Ralph E., and Hilgedick, Steven A. 2018a. "Real-Time Lost Circulation Estimation and Mitigation." *Egyptian Journal of Petroleum* 27: 1227–1234.

Al-Hameedi, A.T.T., Alkinani, H.H., Dunn-Norman, S., Flori, R.E., Hilgedick, S.A., Aklhamis, M.M., Alshawi, Y.Q., Al-Maliki, M.A., and Alsaba, M.T. 2018b. "Predictive Data Mining Techniques for Mud Losses Mitigation." *SPE Kingdom of Saudi Arabia Annual Technical Conference*, Dammam, Saudi Arabia, April 23–26, 2018. SPE-192182-MS. https://doi.org/10.2118/192182-MS

Alkinani, H.H., Al-Hameedi, A.T.T., Dunn-Norman, S., Flori, R.E., Alsaba, M.T., Amer, A.S., and Hilgedick, S.A. 2019. "Using Data Mining to Stop or Mitigate Lost Circulation." *Journal of Petroleum Science and Engineering*, 173: 1097–1108. Available from https://www.sciencedirect.com/science/article/abs/pii/S0920410518309483

Amer, A.S. 2019. "Lost Circulation, an Old Challenge, in Need of New Solutions." SPE Distinguished Lecturer 2020–2021.

API RP 13L. 2017. Recommended Practice for Training and Qualification of Drilling Fluid Technologists, 2nd ed. Available from www.techstreet.com.

Ardoin, R. 2014. "High-Pressure Mud Pump Systems Boost Well Drilling Efficiency & Depth Range." Upstream Pumping accessed November 7, 2019. http://www.upstreampumping.com/article/drilling/2014/high-pressure-mud-pump-systems-boost-well-drilling-efficiency-depth-range.

Azar, J. J. and Samuel, R. 2007. *Drilling Engineering*. Tulsa, OK: PennWell Corporation.

Barnes, Michael, Vargas, C., Rueda, F., Pan American Energy, Garoby, J., Pacione, M., Hughes, B., Huppertz, A., and Inteq, B.H. 2001. "SPE/IADC 67695 Combination of Straight Hole Drilling Device, Team Philosophy and Novel Commercial Arrangement Improves Drilling Performance in Tectonically Active Region."

Bérard, T. and Prioul, R. 2016. "Mechanical Earth Model". E&P Defining Series, Oilfield Review. Schlumberger.

Bleier, R., Leuterman, A.J.J. and Stark, C. January 1993. "Drilling Fluids: Making Peace with the Environment." *Journal of Petroleum Technology*, 45(1): 6–10. SPE-24553-PA. https://doi.org/10.2118/24553-PA

Bond, D.F., Scott, P.W., Page, P.E., and Windham, M. 1996 "Step Change Improvement and High Rate Learning Are Delivered by Targeting Technical Limits on Sub-Sea Wells" IADC/SPE 35077. Presented at the *IADC/SPE Drilling Conference and Exhibition*, New Orleans, LA, March 12–15, 1996.

Bourgoyne, A.T., Young, F.S.T., Chenevert, M.E., and Millhelm, K.K. 1991. Directional Drilling and Deviation Control. In *Applied Drilling Engineering Chapt. 8*. Richardson, TX: Textbook Series, SPE, 351–473.

Brett, J. Ford, Oil and Gas Consultants International OGCI. 2006. *The Perfect Well Ratio: Defining and Using the Theoretically Minimum Well Duration to Improve Drilling Performance*," AADE-06-DF-HO-13 Presented at the *AADE 2006 Drilling Fluids Conference*, Houston, TX, April 11–12, 2006, 1–7.

Caenn, R. 2018. "The Recognition of Mud-Laden Fluid." *Drilling Fluids History*. Last Modified March 26, 2018. http://drillingfluidshistory.org/index.php/date-categories/1901-1929/mud-laden-water

Caenn, R., Darley, H.C.H., and Gray, G.R. 2011. *Composition and Properties of Drilling and Completion Fluids*, 6th Ed. Waltham, MA: Gulf Professional Publishing, Elsevier, 42.

Cayeux, E., Daireaux, B., Dvergsnes, E.W., and Florence, F. 2013. "Toward Drilling Automation: On the Necessity of Using Sensors That Relate to Physical Models." *SPE/IADC Drilling Conference and Exhibition*, Amsterdam, the Netherlands.

Charlez, Ph. A., Easton, M., Morrice, G., and Tardy, P. 1998. "Validation of Advanced Hydraulic Modeling Using PWD Data." *Offshore Technology Conference*, Houston, TX.

Contractor, Drilling. September 2002. "Contractors Building State-of-the-Art Land Rigs." Accessed November 1, 2019. https://www.drillingcontractor.org/dcpi/2002/dc-sepoct02/Sep2-landrig.pdf.

Cunha, J.C. and Kastor, R. Introduction to Rotary Drilling. In Miska, S.Z. and Mitchell, R.F. (Eds.), *Fundamentals of Drilling Engineering*. Society of Petroleum Engineers, Richardson, TX, pp. 1–54.

De Wardt, J.P. January 1, 1994. "Lean Drilling—Introducing the Application of Automotive Lean Manufacturing Techniques to Well Construction." Society of Petroleum Engineers. IADC/SPE 27476, Presented at the *IADC/SPE Drilling Conference and Exhibition*, Dallas, Texas, February 15–18, 1994.

De Wardt, J.P. March 1, 2014. "Industry Analogies for Successful Implementation of Drilling-Systems Automation and Real-Time Operating Centers." IADC/SPE 163412, Presented at the *IADC/SPE Drilling Conference and Exhibition*, Amsterdam, Netherlands, March 5–7, 2013.

Devine, L., Ellins, M., Scotchman, A., May, N., and Oasis, B.H. 2002. "IADC/SPE 74522 Systematic Team Approach to Drilling Optimization Reduces Well Construction Time by 15%, Ghadames Basin, Algeria." Presented at the *IADC/SPE Drilling Conference*, Dallas, Texas, February 26–28, 2002.

Dimataris, M. 2019. "Production Optimization: What's Next for Digital in Drilling & Production." Prepared for Upstream Intelligence in conjunction with DDDP 2019. Available from www.upstreamintel.com

Dowell, I., Mills, A., Ridgway, M., and Lora, M. 2006. "Drilling Data Acquisition." In *Drilling Engineering*, edited by R. F. Mitchell. Richardson, TX: Society of Petroleum Engineers.

Dunlop, J., Lesso, W., Aldred, W., Meehan, R., Orton, M., and Fitzgerald, W. 2004. World Intellectual Property Organization Patent No. WO 2004/059123 A1. System and Method for Rig State Detection. Edited by Schlumberger Technology Corp.

Dupriest, Fred, and Koederitz, W. 2005. "Maximizing Drill Rates with Real-Time Surveillance of Mechanical Specific Energy." *SPE/IADC Drilling Conference*, Amsterdam, The Netherlands.

Dupriest, F.E., Witt, J.W., and Remmert, S.M. 2005. "IPTC 10607 Maximizing ROP with Real-Time Analysis of Digital Data and MSE," no. 1: 1–8.

Eustes, A.W. 2007. "The Evolution of Automation in Drilling." *2007 SPE Annual Technical Conference and Exhibition*, Anaheim, California.

Florence, F. and Iversen, F. 2010. "Real-Time Models for Drilling Process Automation: Equations and Applications." *SPE/IADC Drilling Conference*, New Orleans, Louisiana, USA.

Florence, F.R. and Burks, J. 2012. "New Surface and Down-hole Sensors Needed for Oil and Gas Drilling." *IEEE International Instrumentation and Measurement Technology Conference*, Graz, Austria.

GeoForce Even Layer Thickness Motors—Halliburton. 2018. Accessed December 10, 2018. https://www.halliburton.com/en-US/ps/sperry/drilling/directional-drilling/downhole-drilling-motors/geoforce-endure-motors.html.

Geo-Pilot® Rotary Steerable Systems—Halliburton. 2018. Accessed December 10, 2018. https://www.halliburton.com/en-US/ps/sperry/drilling/directional-drilling/geo-pilot-and-geo-pilot-xl-rotary-steerable-systems.html.

GeoSphere Reservoir Mapping-While-Drilling Service. 2018. "GeoSphere Reservoir Mapping-While-Drilling Service" Brochure—Schlumberger." Accessed December 10, 2018. https://www.slb.com/services/drilling/mwd_lwd/mapping_while_drilling/deep_directional_resistivity_lwd.aspx?t=3&libtab=5.

Glowka, D.A. and Stone, C.M. 1986. Effects of Thermal and Mechanical Loading on PDC Bit Life". *SPE Drilling Engineering*, 1(3), 201–214

Greenberg, Jerry. 2008. "Improved Design Processes Allow Bits to Fit Specific Applications, Set New Drilling Records." Drilling Contractor, July/August 2008, 80–84.

Honeywell. 2017. "Sensors and Switches in Oil Rig Applications." Accessed February 14, 2019. https://sensing.honeywell.com/honeywell-sensors-switches-oil-rig-application-note-000756-4-en.pdf.

Hycalog. 1994. PDC Bit Technology Manual, Hycalog Industries, Columbia. "Rig Walking System." Accessed November 5, 2019. https://www.columbiacorp.com/oil-gas/rig-walking-systems/.

International Energy Agency. 2018. Insights Series 2018—State of Play of Upstream Investment. https://webstore.iea.org/insights-series-2018-state-of-play-of-upstream-investment

Ju, P., Wang, Z., Zhai, Y., Su, D., Zhang, Y., and Cao, Z. 2014. "Numerical Simulation Study on the Optimization Design of the Crown Shape of PDC Drill Bit." *Journal of Petroleum Exploration and Production Technology* 4(4): 343–350.

Khodja, M., Khodja-Saber, M., Canselier, J.P., Cohaut, N. and Bergaya, F. 2010. "Drilling Fluid Technology: Performances and Environmental Considerations." *Chapter 13 in: Products and Services; from R&D to Final Solutions*, edited by Igor Fuerstner. InTechOpen, Rijeka, Croatia. Last Accessed November 29, 2019. https://www.intechopen.com/books/products-and-services--from-r-d-to-final-solutions/drilling-fluid-technology-performances-and-environmental-considerations

Langeveld, C.J. 1992. "*PDC Bit Dynamics*." Paper 23867 presented at the *1992 IADC/SPE Drilling Conference, New Orleans, February 18–21*. Society of Petroleum Engineers. doi:10.2118/23867-MS

Macpherson, J.D., de Wardt, J.P., Florence, F., Chapman, C., Zamora, M., Laing, M., and Iversen, F. 2013. "Drilling-Systems Automation: Current State, Initiative, and Potential Impact. SPE-166263-PA." *SPE Drilling & Completion* 28(04): 298–308.

Mishka, S.Z. 2011. "Directional Drilling." In *Fundamentals of Drilling Engineering Chapt. 8*, Richardson, TX: Textbook Series, SPE, 449–585.

Mitchell, John. 2001. *Trouble-Free Drilling*. Woodlands, TX: Drilbert Engineering.

Offenbacher, L.A., McDermaid, J.D., and Patterson, C.R. 1983. "PDC Bits Find Applications in Oklahoma Drilling," paper SPE *11389 presented at the 1983 IADC/SPE Drilling Conference*, New Orleans, February 20–23.

OGP. 2003. "Environmental Aspects of the Use and Disposal of Non Aqueous Drilling Fluids Associated with Offshore Oil & Gas Operations." Report #342. Available from www.iogp.org

Ojinnaka, M.A., and Beaman, J.J. 2018. "Full-Course Drilling Model for Well Monitoring and Stochastic Estimation of Kick." *Journal of Petroleum Science and Engineering* 166: 33–43.

Patterson-UTI. 2018. Accessed January 2020. http://q4live.s22.clientfiles.s3-website-us-east-1.amazonaws.com/516069926/files/doc_presentations/2018/2018-09-04-PTEN-Presentation-Barclays-v4.pdf.

Petrowiki. 2015. "Rigsite Data Systems." Society of Petroleum Engineers. Accessed November 10, 2019. https://petrowiki.org/Rigsite_data_systems.

Pickering, J.G., Whiteley, N., Rochford, J., Sheil, K., and Lowe, J.P. 2009. *WITSML Real Time Inter-Operability Testing*. Houston, TX: SPE Digital Energy Conference and Exhibition.

Plumb, R., Edwards, S., Pidcock, G., Lee, D., Stacey, B., 2000. "The Mechanical Earth Model Concept and Its Application to High-Risk Well Construction Projects." IADC/SPE 59128, Presented at the IADC/SPE Drilling Conference and Exhibition, New Orleans, LA, February 23–25, 2000.

PowerPak Steerable Motors. 2018. PowerPak Product Sheet—Schlumberger. Accessed December 10, 2018. www.slb.com/services/drilling/drilling_services_systems/directional_drilling/drilling_mud_motors/powerpak.aspx?t=3&libtab=4.

Schaaf, S., Pafitis, D., and Guichemerre, E. January 1, 2000. Application of a Point the Bit Rotary Steerable System in Directional Drilling Prototype Well-Bore Profiles. Society of Petroleum Engineers. doi:10.2118/62519-MS

Schlumberger. 2010. Real-Time Drilling Geomechanics Reduces NPT. Case Study by Schlumberger.

Schlumberger. 2018a. PRESSPRO RT Real-Time Downhole Performance Measurement Software. https://www.slb.com/services/drilling/drilling_fluid/drilling_software/presspro_rt.aspx

Schlumberger. 2018b. VIRTUAL HYDRAULICS ECD and ESD Management Software. https://www.slb.com/services/drilling/drilling_fluid/drilling_software/virtual_hydraulics.aspx

Scott, Dan. 2015. "A Bit of History: Overcoming Early Setbacks, PDC Bits now Drill 90%-Plus of Worldwide Footage." Drilling Contractor Anthology Series—DC Drill Bits. International Association of Drilling Contractors IADC.

Staveley, C. and Thow, P. 2010. "Increasing Drilling Efficiencies Through Improved Collaboration and Analysis of Real-Time and Historical Drilling Data." SPE 128722, Presented at the SPE Intelligent Energy Conference and Exhibition, Utrecht, Netherlands, March 23–25, 2010.

Stroud, Ben K. 1926a. "Application of Mud-Laden Fluids to Oil or Gas Wells." US Patent 1,575,944 (March 9, 1926). Available from www.uspto.gov

Stroud, Ben K. 1926b. "Application of Mud-Laden Fluids to Oil or Gas Wells." US Patent 1,575,945 (March 9, 1926). Available from www.uspto.gov

Swanson, B.W., Gardner, A.G., Brown, N.P., Murray, P.J. et al. 1997. "Slimhole Early Kick Detection by Real-Time Drilling Analysis." *SPE Drilling and Completion* 12(1): 27–32. https://doi.org/10.2118/25708-PA.

Teale, R. 1965. "The Concept of Specific Energy in Rock Drilling." *International Journal of Rock Mining Science* 2:57–73.

Technology, Tianjin Elegant. 2019. "Fast Moving Land Rigs." Accessed November 10, 2019. http://www.sovonex.com/drilling-equipment/api-fast-moving-land-rigs/.

Thorogood, J.L., Crichton, M.T., and Bahamondes, A. 2014. "Case Study of Weak Signals and Confirmation Bias in Drilling Operations." IADC/SPE 168047, Presented at the IADC/SPE Drilling Conference and Exhibition, Fort Worth, TX, March 4-6, 2014.

US Energy Information Administration (EIA). 2016. "Trends in U.S. Oil and Natural Gas Upstream Costs." US Department of Energy, Washington, D.C., March 23, 2016.

Visser, C. et al. 2014. "Geothermal Drilling and Completions: Petroleum Practices and Technology Transfer Final Report," NREL/CSM FY AOP 3.2.1.1

Ward, C. and E. Andreassen. 1998. "Pressure-While-Drilling Data Improve Reservoir Drilling Performance." SPE-37588-PA. *SPE Drilling & Completion* 13(01): 19–24.

West, G., Hall, J. and Seaton, S. 2006. "Drilling Fluids." *Petroleum Engineering Handbook*, edited by L.W. Lake. Richardson, TX: SPE. Last Modified April 26, 2017. https://petrowiki.org/PEH:Drilling_Fluids

Wilson, A. 2015. "Drilling-Systems-Automation Roadmap: The Means to Accelerate Adoption. Society of Petroleum Engineers". *Journal of Petroleum Technology* 67(9): 137–138.

XBAT Azimuthal Sonic and Ultrasonic LWD Service—Halliburton. 2018. Accessed December 10, 2018. https://www.halliburton.com/content/dam/ps/public/ss/contents/Data_Sheets/web/H010024.pdf.

5 Offshore Deepwater Drilling

Marcio Yamamoto,
National Maritime Research Institute, Mitaka, Japan
José Ricardo Pelaquim Mendes, and
University of Campinas, Brazil
Kazuo Miura
In memoriam

CONTENTS

The technical community of offshore drilling has its own jargon and field units that may change depending on the region or operator. We will try to explain some of the jargon along with this chapter and will show physical units using both field units and SI units.

There are many types of wells in the oil and gas industry such as: wildcat well (a new well in an area or block) or exploratory well; stratigraphy well (a well for which the stratigraphy is confirmed and geological data is collected); appraisal well (a well for which the extension of discovery of oil or gas is confirmed); development well (a producer well to produce an oil and gas field and injector well to inject gas, water, CO_2, or vapor); a pilot well (a well, for example, where the top of the reservoir has been detected, then part of this well is plugged and abandoned to drill a horizontal well) or a relief well. The wildcat well is constructed to gather geological and geophysical data, identify the presence of hydrocarbon in a specific sedimentary rock formation to confirm a new discovery. If there is a discovery, then a production test, a.k.a. *DST* (Drill Stem Test), to collect dynamic data from the reservoir may be performed. The appraisal well verifies the extension of a recently discovered reservoir and/or performs also a DST.

After the discovery through a wildcat well and the *appraisal wells* (two to three wells in general) of a recently discovered field, an *operator* (petroleum company) may decide to develop such field if it is economically and technically feasible. It is when the development wells are constructed to produce petroleum and/or natural gas. Water, more specifically brine, may also be produced as a by-product. Besides the production well itself, injection wells may be also constructed. Such wells inject water and/or gas (CO_2 or natural gas) into the reservoir. There are several reasons to inject fluid into a reservoir, namely, to dispose of the produced water, to maintain the reservoir pore pressure keeping the production flow rate, and to increase the recovery factor. Eventually, an exploratory well can be converted into a development well. After the required data and samples are gathered, the exploratory well may be temporarily abandoned for future completion, becoming a development well.

The *relief wells* (a minimum of two are always constructed for higher success) are drilled to intercept the bottom of another well in a *blowout* (uncontrolled flow of fluid from the rock formation into the well and then to the environment) to carry out the *bottom kill*. In other words, heavy drilling fluid is injected from the relief well into the bottom of the blowing out well. When such a well is filled, the hydrostatic pressure of the drilling fluid is higher than the reservoir pore pressure, stopping the uncontrolled flow.

In this chapter, we will introduce the main challenges and current practices to construct a subsea well in ultra-deepwater. Our focus is on the drilling itself, discussing the challenges that the increasing water depth brings to the well design, drilling operation, and equipment. Then, we present the state-of-the-art regarding the drilling, completion, subsea equipment, floating platforms, etc. Furthermore, we present mud logging, which together with well logging and coring, gathers important data from subsurface formations (gas detection and drilling cutting analysis), helping to identify reservoirs and increase the safety of the operation with well monitoring and pore pressure assessment. Another important point is the well control, which are the activities to monitor the occurrence of a kick and its removal from the wellbore. Last, we present some new technologies that shall be used in the near future.

5.1 CHALLENGES OF THE DEEPWATER DRILLING

5.1.1 Advancement into Ultra-Deepwater and Immediate Implications

Besides the type of well according to its application, as we mentioned before, the offshore development wells may be constructed in two ways depending on the water depth. The dry completion wells are constructed with the wellhead on the platform, thus above the sea surface. Its drilling process is like the onshore drilling as explained in Chapter 4. The use of the dry completion well depends also on the region, as it is explained next comparing the Gulf of Mexico (GoM) and offshore Brazil. The second way to construct is as a subsea well, in which the wellhead is located on the seafloor. This chapter covers the challenges, required equipment, and drilling process of the subsea well.

5.1.1.1 The US Gulf of Mexico

Table 5.1 features the definition of deepwater and ultra-deepwater. The lower limit of deepwater, 610 m (2000 ft), is used to refer to the maximum water depth in which a fixed platform (complaint tower) can be deployed. The lower limit of the ultra-deepwater, 1830 m (6000 ft), is used as the maximum water depth where a spar platform[1] can be deployed.

Along with the Tension Leg Platform (TLP), the compliant tower and spar platform are widely used in the US Gulf of Mexico (GoM). In addition, these three types of production platform have in common a very low heave response (vertical motion) in waves. Despite being a fixed platform, the compliant tower may oscillate horizontally under action of waves and bend due to the sea current.

This characteristic of reduced vertical motion allows the use of dry completion wells for hydrocarbon production. Dry completion means that the wellhead is located above the sea surface on the platform. Usually, such wells are constructed using a modular drilling rig assembled on the platform in a very similar way of an onshore well. Once the drilling campaign is completed, the modular rig is disassembled and transported to other location using the platform supply vessels (PSV). For the platforms where electrical submersible pumps (ESP) are used, the drilling rig should be permanently assembled because the ESP must be replaced frequently (MTBF or mean time between failures of 2 to 3 years).

5.1.1.2 Offshore Brazil

Table 5.2 features the definition of water depth used in Brazil following the rules of the Brazilian National Agency of Petroleum, Natural Gas, and Biofuels (ANP).

Compared to the GoM, Brazil has a single TLP installed in 2014, which uses the ESP as the artificial lift method. Thus, such installation has a permanent drilling rig onboard for the workover (maintenance) of its dry completion well.

For deep- and ultra-deepwater, beside the TLP mentioned earlier, there are only two types of floating production units (FPU), namely, FPSO (Floating, Production, Storage, and Offloading) with 46 units in operation and 11 units under construction (Barton et al., 2018), followed by the semi-submersible with 14 units (Barton et al., 2018). In both types of FPUs, the subsea wells are connected to platforms using flexible risers in most of cases.

TABLE 5.1
Definition of Water Depth

Deepwater	610–1830 m (2000–6000 ft)
Ultra-Deepwater	>1830 m (>6000 ft)

Source: ISO (2005).

TABLE 5.2
Definition of Water Depth used in Brazil

Shallow Water	<400 m (1310 ft)
Deepwater	400–1500 m (1310–4920 ft)
Ultra-Deepwater	>1500 m (>4920 ft)

Source: ANP (2017).

TABLE 5.3
Maximum Water Depth of each MODU

Hull	Maximum Water Depth	Positioning
Jack-up	190 m (625 ft)	Fixed (legs)
Semi-submersible	3660 m (12,000 ft)	DPS/Moored
Drillship	3660 m (12,000 ft)	DPS/Moored*

The only dry completion wells in Brazil are those located in a conventional fixed platform located in shallow water and in a single *Papa-Terra* TLP.

5.1.2 System and Hardware Impacts for Drilling Installations

Because the subsea wells are in deep- and ultra-deepwaters, the first requirement is a floating platform with a drilling rig on board. Such platforms are classified as Mobile Offshore Drilling Units (MODU). Table 5.3 features the three different MODU available and the maximum water depth where each MODU may operate.

In offshore locations deeper than 190 m (625 ft), a floating MODU will be necessary. Drilling from floating MODUs required new technologies to be developed, such as subsea BOP composed by the BOP stack and Low Marine Riser Pack (LMRP), drilling riser, riser tensioning system, heave compensation system, etc.

Initially, in an anchored semi-submersible with a guideline drilling system, as the name says, four wires must connect the floating platform to the wellhead and remain connected during the whole operation. The wires guide all subsea equipment to the wellhead. Such a system has a limitation of 450 m of water depth.

For water depth deeper than 450 m (1500 ft), the guideline-less drilling system was developed. It is very similar to the previous system, except that the guidelines were removed. The guideline-less drilling system with the anchored platform is limited to a water depth of 800 m. Then, for deeper waters, the mooring system is replaced by a dynamic positioning system (DPS). However, it is possible to use polyester rope to moor a drilling platform in water depth down to 1800 m.

* Such a combination is limited only for drilling in shallow water in the Arctic. The majority of drillships operate with DPS.

5.1.2.1 Jack-up

Jack-up is a MODU that operates in shallow water. This platform is tugged to the location where its legs are lowered down to the seafloor (Figure 5.1a) and its hull is suspended above the sea surface.

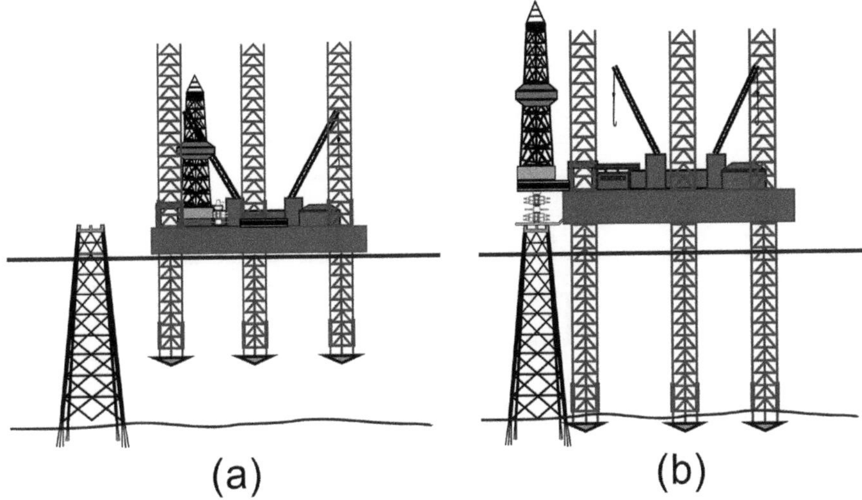

FIGURE 5.1 Jack-up is a MODU that operates in very shallow waters: (a) its legs are lowered down to suspend its hull above the sea surface, (b) then the well construction can be carried out over a fixed conventional platform (jacket).

Then the drilling rig is displaced over the well location, usually a conventional fixed production platform (Figure 5.1b). The well construction is like onshore construction; even the BOP is the same.

Despite jack-ups having floating capability, they are classified as fixed platforms. Its truss legs are transparent for the wave action and the hull is suspended high enough that the waves cannot reach its bottom. Thus, this MODU has no heave motion during its operation.

5.1.2.2 Guideline Drilling

Guideline drilling is used in a floating-type MODU in a maximum water depth of about 450 m (1500 ft). In such depth, all MODU shall be anchored platforms. There is an only economic reason to deploy an anchored platform rather than a MODU equipped with DPS. In addition, the radius of a watch circle of a DPS rig is about 1% of the water depth to keep station around the wellhead; therefore it will be hard to keep position in water depth shallower than 450 (radius of 4.5 m).

Before the spud-in, a temporary guide base (TGP) is lowered down and landed on the seafloor using drill pipes and the proper running tool. The TGP has 4 wires, the guidelines that are connected to the floating MODU.

After the 0.91 m (36 inches) bore is drilled, the conductor is run into the well together with the permanent guide base (PGB), which has the 4 guideposts. As a more time-efficient alternative, the conductor may also be jetted instead of drilled if the seafloor sediments are soft enough. Once the surface casing is run into the well with the high-pressure housing, the subsea BOP is lowered down hanged by the drilling riser (Figure 5.2). The guideposts help to center and align the BOP to seat over the high-pressure housing. During the BOP installation process, the choke and kill lines, which are high-pressure pipes/hoses connected to the BOP, are attached around the riser to be lowered down together.

5.1.2.3 Guideline-less Drilling

Guideline-less drilling is used in deep- and ultra-deepwaters and it requires a Remote Operated Vehicle (ROV). Because there are no guidelines, the conductor, the surface casing (Figure 5.3), and the subsea BOP (Figure 5.4) are lowered down above the wellhead. Then the ROV films the

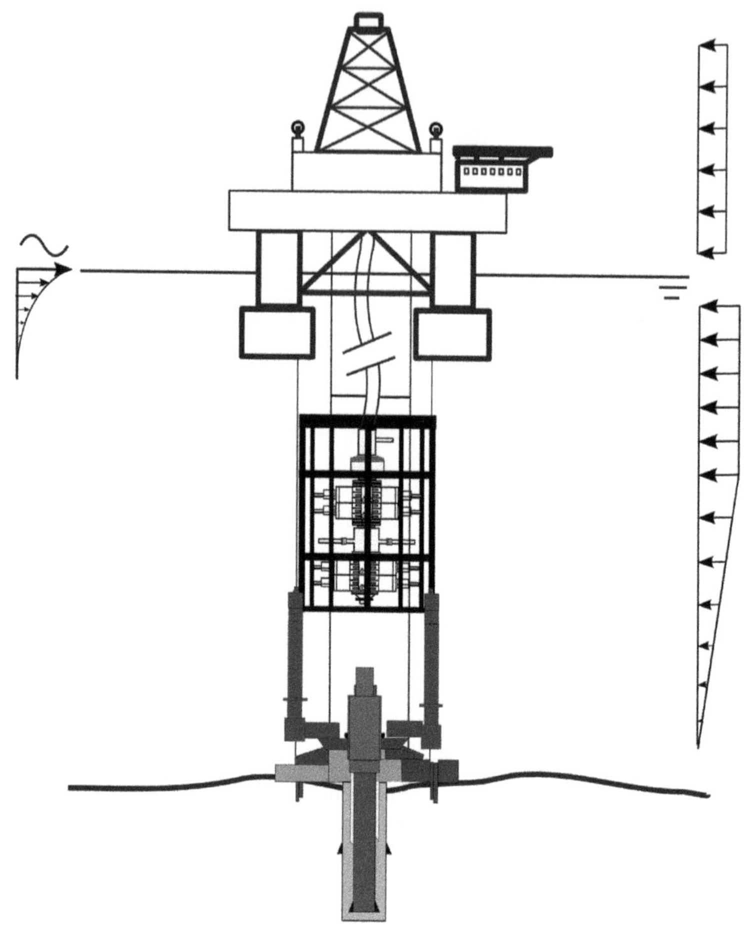

FIGURE 5.2 Guideline drilling system lowering down the subsea BOP.

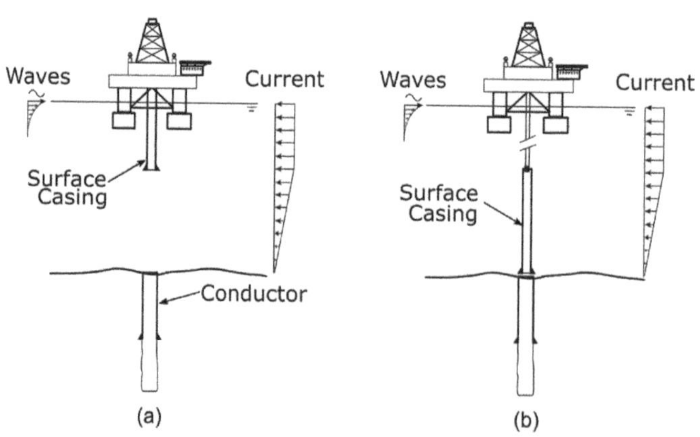

FIGURE 5.3 Running surface casing into the open-hole: (a) the surface casing is assembled joint by joint using threaded connectors; (b) it is run into the wellbore using a landing string.

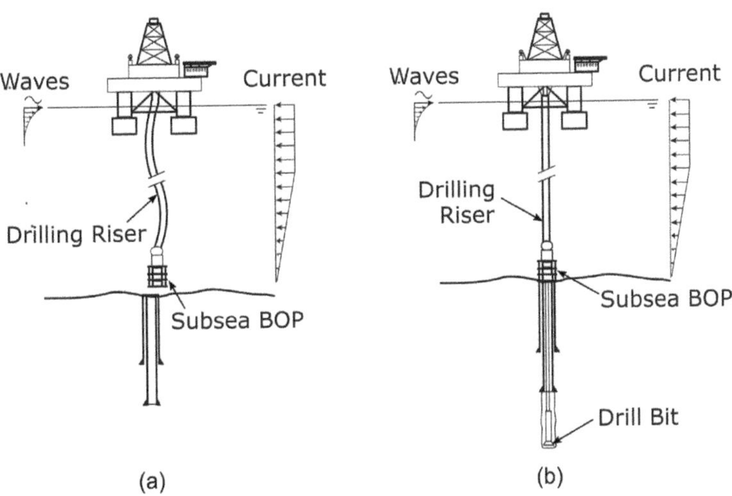

FIGURE 5.4 The subsea installation: (a) the subsea BOP is lowered down and hanged by the drilling riser; (b) the BOP is connected to the high-pressure housing and the drilling operation restarts.

distance between the equipment to the wellhead. If necessary, the whole floating platform is moved by changing the length of the mooring lines or the reference position in the DPS. It is a trial-and-error operation, but an experienced crew can carry out this operation smoothly.

This chapter will present more details about guideline-less drilling, explaining the challenge of a narrowing drilling operating window between pore pressure and fracture gradient, drilling processes (spud-in and drilling with fluid return), drilling equipment, mud logging, and well control.

5.1.3 Below the Mudline Conditions and Drilling Operating Window

5.1.3.1 The Industry's Practices

It is necessary to remark that the drilling engineering community prefers to use gradient pressure, instead of the pressure itself. The understanding of high pressure or low pressure is relative to the hydrostatic pressure of water, which depends on the depth. For example, if a closed vessel is pressurized at 10.7 MPa (about 1550 psi) on the Earth surface, then it is under high pressure. However, this same pressure inside a reservoir 3000 m below the surface is normal pressure. If the reservoir is deeper than 3000 m, the same pressure is assumed as low pressure.

Rather than using two separate data (pressure and depth), the petroleum community uses only the density in ppg (pounds per gallon). Thus, anything that is about 8.3 ppg[2] (1000 kg/m^3 or the water density) can be assumed normal pressure. Below this value is low pressure and above is high pressure. To make this pressure-depth vs. gradient conversion, it is necessary to use either Eq. (5.1) or Eq. (5.2).

$$P_{psi} = 0.052\,\rho\,D_{ft} \tag{5.1}$$

$$P_{psi} = 0.17\,\rho\,D_m \tag{5.2}$$

where P_{psi} is the pressure in pound per square inch (psi); ρ is the fluid density in ppg; D_{ft} is the well depth in feet; D_m is the depth in meters. These equations compute the hydrostatic pressure, where the coefficient at the right side includes the gravity acceleration and the proper unit conversion to result in psi.

Another important common practice in the drilling engineering community is to assume that the depth of any well offshore or onshore is measured from the rig floor (rotary table). Depending on the region or operator, the depth can be measured in feet (ft) or meters (m). Thus, when we refer to well depth in this chapter, we are always assuming the depth from the rig floor.

5.1.3.2 The Operating Drilling Window

The main challenge for operators, drilling contractors, and service companies in deep- and ultra-deepwaters is the narrowing of the "Drilling Operating Window" or "Mud Weight Operating Window," which means the difference between the pore pressure and the fracture pressure of a given formation. The main reason is due to the increase of the water depth, which reduces the overburden pressure and, consequently, reduces the fracture pressure.

Figure 5.5 shows an example of how a well is designed and its drilling fluid schedule is calculated based on the drilling operating window. The vertical axis is the well depth and the horizontal axis correspond to the pressure converted into density as defined before (Eqs. 1 or 2). Further, the origin of the total depth is located on the rig floor, which is located about 30 m (100 ft) above the sea surface (blue line in Figure 5.5). This "air gap" (the distance between the rig floor or the rotary table and sea surface) changes slightly depending on the drilling platform.

Drilling of a subsea well may be divided into two distinct stages: spud-in and drilling with the fluid return. Furthermore, each stage is divided into several phases. A phase is completed when a corresponding casing or liner is properly set in place. The spud-in is carried out on the sediments close to the seafloor, which behaves like soil. Thus, the geotechnical engineering practices shall be used for the design and planning of this phase. On the other hand, the drilling with fluid return occurs on the sedimentary rocks.

The use of the drilling operating window is required for the design of phases in the drilling with a fluid return below the surface casing (Figure 5.5) when the subsea BOP[3] and drilling riser are already installed (Figure 5.4b). Therefore, the well, subsea BOP, and the drilling riser are filled with the drilling fluid, which exerts a hydrostatic pressure on the wellbore.

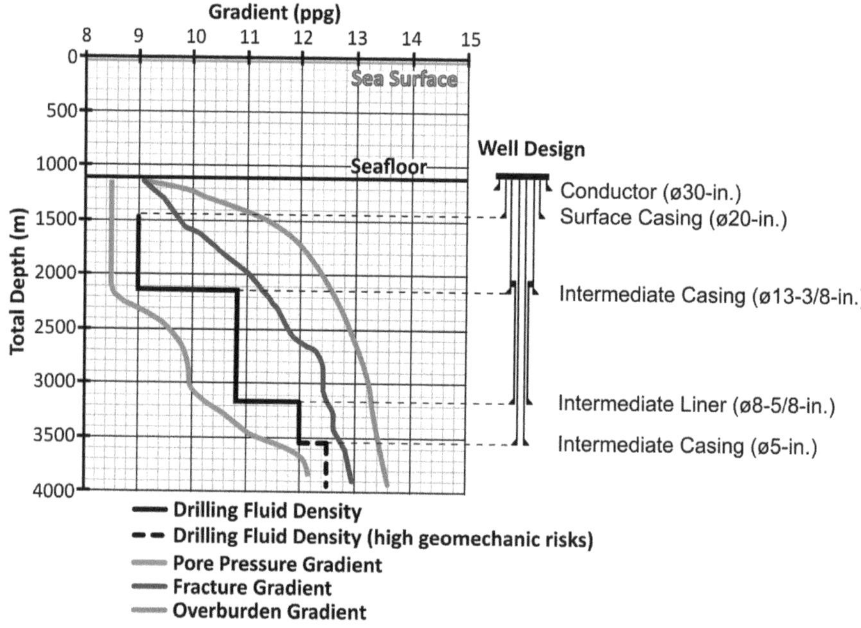

FIGURE 5.5 Example of a drilling operating window plotted together with drilling fluid schedule and the well design.

Along the uncased portion of the well, the hydrostatic pressure of drilling fluid must be kept within the pore pressure and fracture pressure interval. In addition, some formations have the wellbore collapse pressure lower than the fracture pressure. Thus, the interval is such cases is defined the difference between pore pressure and the collapse pressure.

If the hydrostatic pressure is less than the pore pressure, an undesired influx of formation fluid, a.k.a. kick occurs, which will result in non-productive time (NPT) and risks to the operation, personnel, environment, and equipment. When the kick occurs, the drilling operation is suspended, and this undesired influx must be removed from the wellbore using the proper well control operation. Failures to identify and to remove the kick properly may result in the uncontrollable flow of formation fluid into the environment, known as blowout.

On the other hand, if the drilling fluid hydrostatic pressure is higher than the fracture pressure, this overpressure will open fractures in the formation and drilling fluid will flow into such fractures. The loss of drilling fluid also will result in NPT and risks to the operation. The drilling operation is suspended to stop the circulation loss injecting the Loss Control Material (LCM) or in the worst case a cement plug on the fractured wellbore. Even a contingency casing may be set in place and properly cemented. In case of severe circulation loss, the level of drilling fluid drops so fast, reducing the hydrostatic pressure within the wellbore below the pore pressure, which may result in a kick. In the recent years, it is a common practice to use Managed Pressure Drilling (MPD) in a known loss of circulation zone, such as the Pre-Salt carbonate reservoir off Brazil (see Section 5.1.4). Even then, it is possible to drill using a Pressurized Mud Capping Drilling (PMCD) mode in a severe loss of circulation scenario where the mud is injected with no return to surface, and the drilled cuttings are injected in the loss of circulation zone.

It is a common practice to keep the gradient of the drilling fluid with a safety margin of 0.5 ppg from both the pore pressure gradient and fracture gradient. A good reason for this margin is because the drilling fluid gradient refers only to the hydrostatic pressure. However, during the circulation of the drilling fluid, the borehole faces a higher pressure, which is composed of the hydrostatic pressure and frictional pressure drop due to the circulation.

In the example in Figure 5.5, the drilling operating window is narrower due to the ultra-deepwater. The window is especially narrow in the depths below 3500 m where the window has less than 1 ppg. It means that the safety margin of 0.5 ppg cannot be respected, thus the drilling operation will face a high risk of a kick or fracture. The kick may occur when the mud pump is off during the connection of a new drill pipe stand to the drillstring. On the other hand, the fracture may occur when the mud pump is turned on and the pressure within the wellbore has a peak of pressure.

For better understanding, the drilling window shown in Figure 5.5 is converted to pressure units. Figure 5.6 shows, besides the window, the drilling fluid hydrostatic pressure within the wellbore. The drilling fluid hydrostatic pressure starts at zero on the rig floor and increases linearly according to its gradient.

To drill the last phase in the depth 3550–4000 m (Figure 5.6), the drilling window is very narrow. The pink lines represent the pressure range of the drilling fluid. As mentioned before, the window is so narrow that when the mud pump is turned on, the increased pressure due to frictional pressure drop may cause the formation to fracture. On the other hand, when the pump is off (e.g. to connect a new stand or drill pipe), the wellbore pressure drops and a kick may occur.

The petroleum industry has developed several technologies to address the narrowing window problem, especially the reduced fracture gradient. To reduce the risk of formation fracture due to the cement slurry high density, foamed cement has been developed (Davies and Hartog, 1981; Loeffler, 1984; Kutchko et al., 2016). During the drilling operation, several technologies have been tested including the foamed drilling fluid (Hall and Roberts, 1984; Liu et al., 2010), Dual-Gradient Drilling (Ziegler et al., 2013; Stave, 2014), and Managed Pressure Drilling (MPD). More details about the MPD technology are presented in the next subsection.

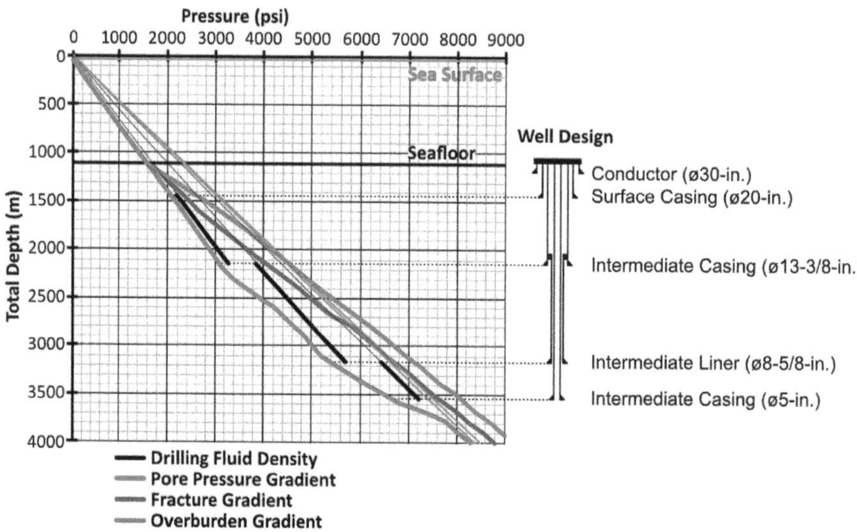

FIGURE 5.6 Example of a drilling operating window converted into pressure and plotted together with drilling fluid schedule and the well design.

5.1.4 The Fundamentals of Managed Pressure Drilling Methods

Managed pressure drilling (MPD) is a technology that has been developed to address several drilling issues, such as drilling through a highly permeable formation in which severe lost circulation is an issue, using the mud cap approach (Al-Awadhi et al., 2014; Bysveen et al., 2017). Furthermore, MPD systems have the capacity of controlling with accuracy the wellbore pressure, keeping it constant even when the mud pump is off. The main components of a MPD system will be presented next.

Figure 5.7 shows a schematic of a circulation system drilling a subsea well. The components highlighted with the red squares compose the MPD system that is assembled in parallel to the conventional circulation system of the drilling rig. Table 5.4 features a short explanation of each of such components.

In conventional drilling, the drilling riser is connected to the diverter that permits the drilling fluid and the rock cuttings to be conveyed to the shale shakers. The diverter can close the annulus similarly to an annular BOP in the case of a blowout, diverting the uncontrollable flow from the well to the overboard of the floating platform. During normal drilling operations, the diverter is fully open, which means that only atmospheric pressure is acting on the top of the drilling fluid column. Usually, the atmospheric pressure is neglected and, therefore, the pressure on the top of this column is considered zero.

In the MPD, the RCD seals the annulus between the drillstring and drilling riser, allowing the top of the drilling fluid column to be pressurized. There are two ways to generate pressure on the top of the drilling fluid column. The first one is shown in Figure 5.7 when the mud pump is circulating the fluid in the whole circuit during the drilling operation. In this case, the MPD Choke Manifold generates a back pressure due to the flow restriction. An automatic control system keeps the pressure in the wellbore, setting the choke opening (Yilmaz et al., 2013). This control receives the feedback signal of flow rate, measured by the Coriolis flowmeter, and pressures the well upstream and downstream.

The second way to pressurize the top of the drilling fluid column is used when the mud pump is turned off for the connection of a stand of drill pipes to the drillstring, for example. In this case,

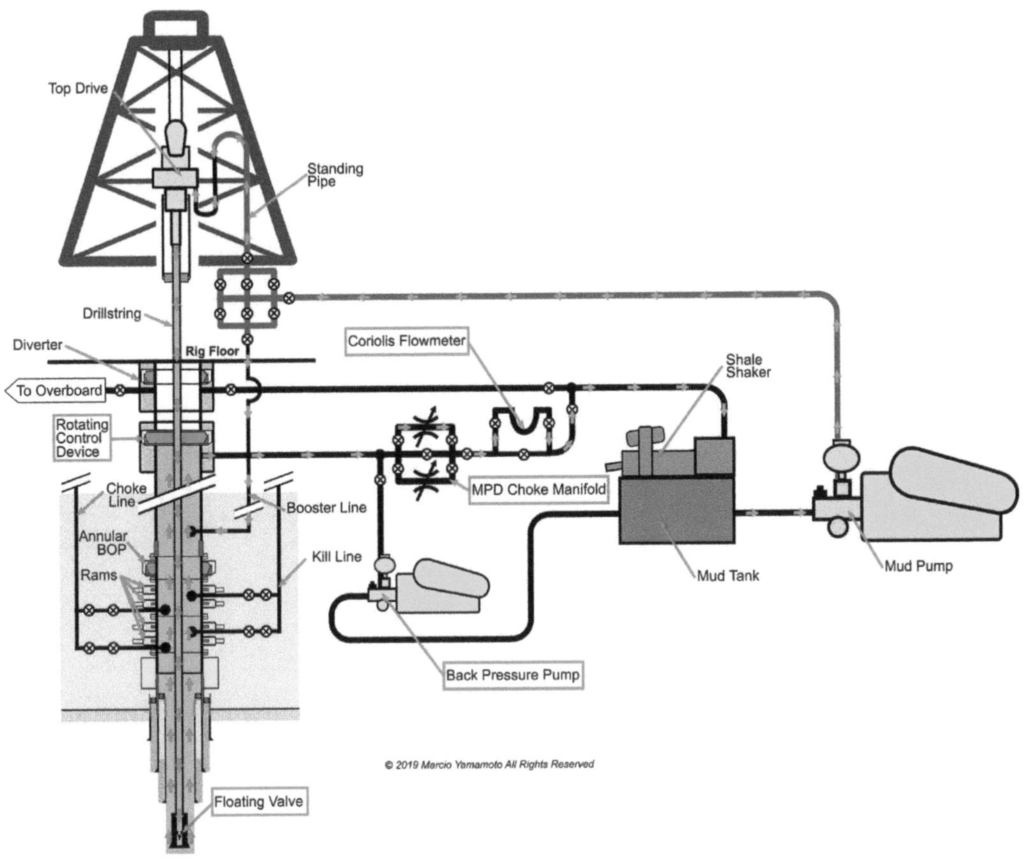

FIGURE 5.7 Schematic of the circulation system of a subsea well including the MPD components, which are highlighted with a red box.

TABLE 5.4
Main Components of an MPD

Component	Application
Rotating Control Device (RCD)	A piece of equipment installed below the diverter. It works as a shaft seal that allows pressurizing the well annulus during the drilling operation.
MPD Choke Manifold	It is not the choke manifold used for the well control operation. The MPD choke manifold has an automatic back-pressure control to keep the wellbore pressure.
Coriolis Flowmeter	It measures with accuracy the flow rate of the drilling fluid that returns from the well. The MPD choke manifold control receives the flow rate signal to actuate on the back pressure to compensate for pressure variation due to flow rate variation.
Back-Pressure Pump	When the mud pump is off, this pump is turned on to generate enough back pressure to keep the wellbore pressure.
Floating Valve	The floating valve has been a standard piece of equipment in the drillstring design. It avoids any fluid backflow into the drillstring. It is essential in the MPD system because, when the back-pressure pump is turned on, the floating valve shuts the drillstring allowing the connection of a stand of drill pipe.

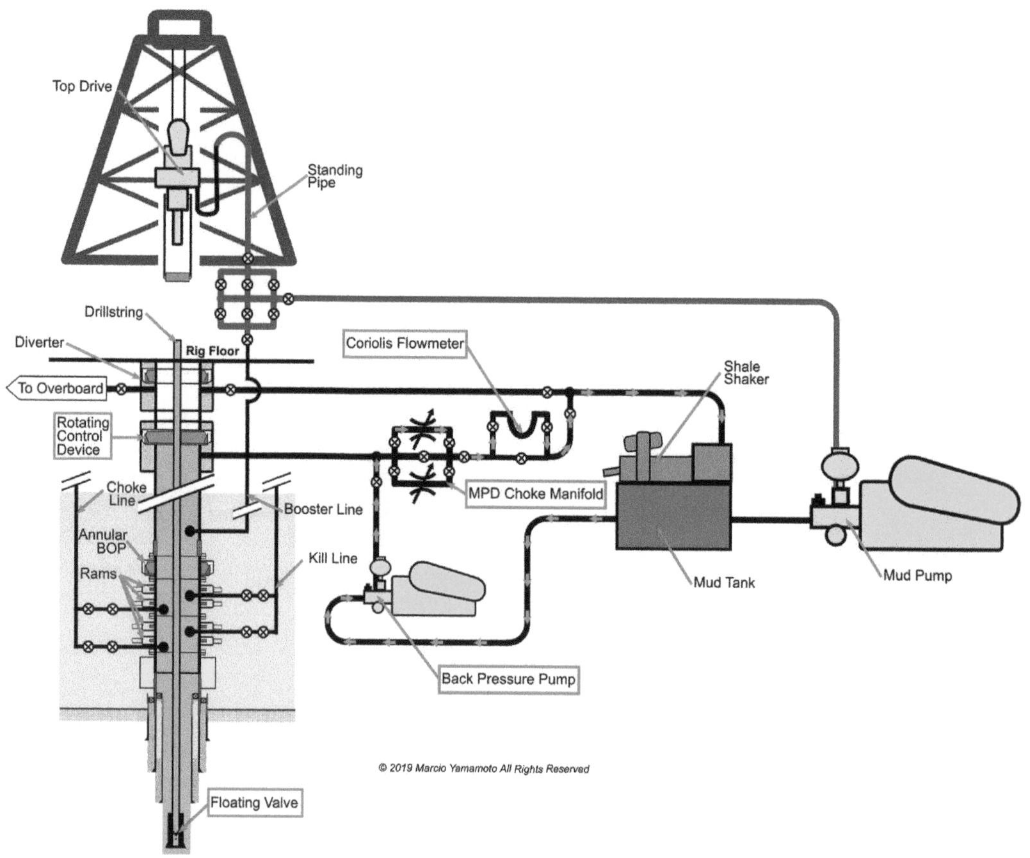

FIGURE 5.8 Schematic of the circulation system with MPD when the mud pump is off.

the backpressure pump is turned on, circulating the drilling fluid only through the MPD Choke Manifold and Coriolis flowmeter (Figure 5.8). Once again, the flow restriction will generate a backpressure on the top of the well. It can be noted here how important the floating valve is, installed within the drillstring, which will block the drilling fluid from return to the rig floor.

During the drilling operation using the MPD system, the drilling fluid gradient is slightly underbalanced, which means the hydrostatic pressure in the wellbore is lower than the pore pressure. The kick does not occur because the back pressure on the top of the fluid column drifts the hydrostatic pressure in the wellbore keeping the overbalance. Figure 5.9 shows how the wellbore pressure drifts when it is changed to the set point of the backpressure. The MPD's control keeps the wellbore pressure by adjusting the opening of the MPD choke manifold. When the mud pump is turned off, the backpressure pump is activated to generate the back pressure that keeps the pressure in the wellbore above the pore pressure.

5.2 EXISTING ARCHITECTURES AND METHODS FOR DEEPWATER DRILLING

The exploration and production (hereafter E&P) activities are carried out only in sedimentary basins as, for example, offshore Brazil, West Africa, and the Gulf of Mexico (GoM). In an offshore sedimentary basin, the seafloor is composed of completely loose sediments, especially clay. This condition of loose sediments continues several hundred meters below the seafloor.

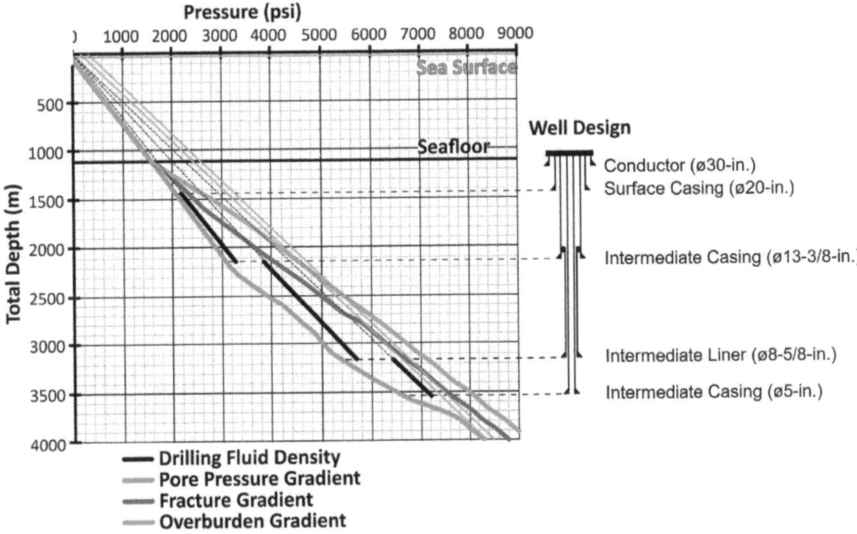

FIGURE 5.9 Example of a drilling operating window converted into pressure and plotted together with a drilling fluid schedule. The orange lines represent the range of pressure controlled by the MPD.

Increasing in depth, the sediments have already been exposed to several geological processes, such as lithification. Thus, the sediments start to become consolidated, transforming into sedimentary rock.

Drilling of a subsea well in deepwater may be divided into two distinctive stages: spud-in and drilling with the fluid return, with each stage divided into further phases.

The spud-in refers to drilling into loose sediments in very shallow depths. However, sometimes, instead of drilling, the spud-in can be carried out with jetting or hammering down a casing. On the other hand, the drilling with the fluid return is carried out into deeper sedimentary rocks, followed by the running and cementing a casing in place.

Because the geotechnical properties of the shallow sediments and the geomechanical properties of sedimentary rocks change with depth, it is necessary to drill in phases as mentioned before. Figure 5.10 shows the schematic of a typical subsea development well in the Santos Basin Pre-Salt Cluster off Brazil. The well has a telescopic structure, with the initial casings having the largest diameters. The following phase is always drilled within the previous casing using a bit with a smaller diameter.

In Brazil, there is a specific name for each spud-in casing. The first and second casings are called "conductor" and "surface casing." Other operators may call them "structural casing" and "conductor" for the same respective strings or call both surface casings.

In addition, the deepest phase of the well in Figure 5.10 has a special casing named "Liner," a casing string that does not extend back to the wellhead. The liner's top end is seated within the previous phase's casing. This shorter casing is used to replace long casings reducing operational time and cost. In addition, the number of casings is limited to the number of casing hangers that can be seated within the high-pressure housing. In some regions with very complex geology, the well design requires more casings than the high-pressure housing may afford. In such cases, the use of liners is indispensable. Furthermore, due to the weight of long casing string, it is possible to set a liner and then after, do a tie back with a casing up to the wellhead.

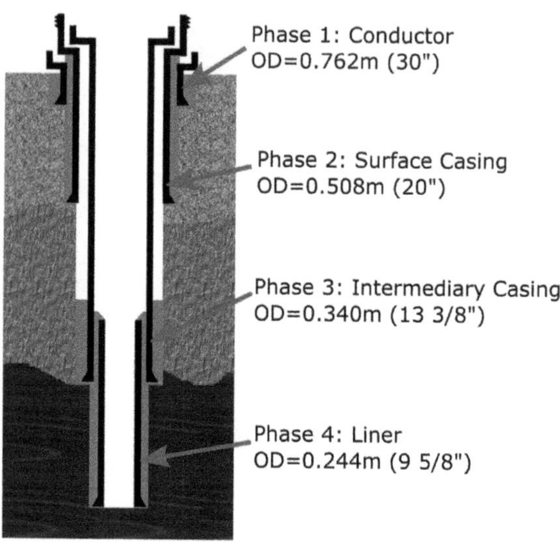

FIGURE 5.10 Schematic of the subsea well in the Santos Basin Pre-Salt Cluster (drawing is not to scale).

5.2.1 Spud-in

To complete the setting of the conductor in place, the well engineer has available a few methods, namely, (1) to jet the conductor and drill ahead to the second phase; (2) to drill, to run the conductor, and to cement in place; (3) free-fall installation using a torpedo-base conductor (Nogueira et al. 2005); or (4) to use a hydraulic hammer to drive down the conductor (Silvio et al., 2013) chiefly in a jack-up or a fixed platform for a safety reason (e.g., cratering near the well if it is drilled). The first method is the most widely used today because of the significant reduction in the operational time.

Figure 5.11 shows the steps of jetting the conductor, drilling ahead to the second phase, and running and cementing the surface casing. First, the conductor is assembled on the rig floor connecting the conductor joints using threaded connectors. Each joint has about 12.2 m (40 ft) length and external diameter of 0.762 m or 0.914 m (30 in or 36 in). About 3 to 10 joints can be used to assemble the conductor. But this number depends on the seafloor sediments characteristics and the combined submerged weights of the conductor itself and the surface casing.

The conductor's top end connects the "Low-Pressure Housing," which has a special internal geometry where the surface casing's "High-Pressure Housing" will be seated. Once the conductor is assembled, the drilling crew keeps it hanged by the low-pressure housing on the rig floor. Then a temporary or "false table" is placed over the housing in order that the Bottom Hole Assembly (BHA) can be assembled within the conductor.

For the jetting operation, a typical BHA is composed of a 0.660-m (26-in) drill bit, crossover subs, mud motor, drill collars, stabilizers, Logging While Drilling (LWD[4]), and Measurement While Drilling (MWD[5]) tools. It may also have a rotary steerable system for the directional drilling of the drill-ahead phase. After the BHA, the conductor's running tool is connected to the drillstring. A running tool is a generic term to refer to any specialized tool that makes the connection between the drill pipe string and any part or piece of equipment that will be installed inside or above the well. All casings, production tubing, Wet Christmas Tree (WCT) have their own respective running tool.

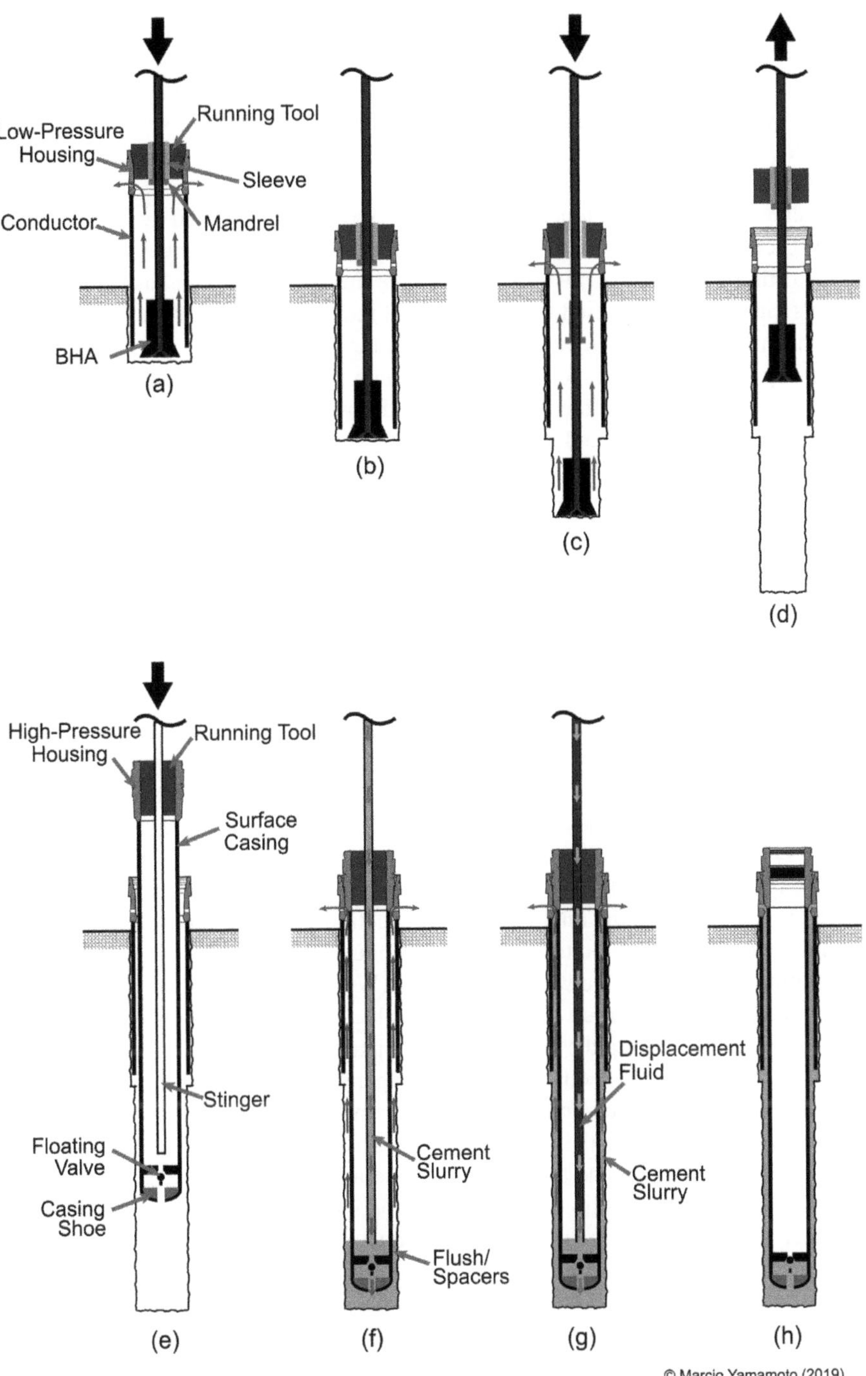

FIGURE 5.11 Schematic of the conductor's jetting operation, followed by the drill-ahead operation. The spud-in is completed after the surface casing is run and cemented into the well (drawing is not to scale).

Next, the running tool is connected to the low-pressure housing using a latch that can be released using the drillstring rotation. The whole conductor and BHA assembly are lowered down to the seafloor (Figure 5.11a). The conductor must be spud within the square- or triangle-formed target buoys on the seafloor, which were installed previously by a geodesic-specialized vessel. A ROV must dive from the floating drilling platform to make sure the conductor is set in the correct position and to inspect the whole jetting operation.

When the conductor assembly touches the seafloor, it penetrates the soil by its own weight (Beck et al., 1991). There is no rotation of the whole drillstring. During the jetting operation, the seawater is pumped as drilling fluid. The drillstring conveys the fluid downward and at the BHA, part of the drilling fluid's kinetic energy is used to drive the mud motor, generating the rotation of the drill bit only. Then, the drilling fluid accelerates when it flows throughout the nozzles of the bit.

These drilling fluid jets are responsible to dissociate the sediments located below the bit. The mixture of drilling fluid and sediments are conveyed upward within the conductor. At the top of the conductor, the drilling fluid and sediments are discharged through drain holes located on the low-pressure housing (Figure 5.11a) and vents located at the running tool. High viscous sweeps must be pumped periodically to clean the sediments inside the annulus between the conductor and the BHA.

The accumulation of sediments within the conductor may form a packoff (the annulus is clogged). The packoff forces the drilling fluid to flow over the conductor's outside surface resulting in a broaching, which means an annulus is created between the conductor outside surface and sediments that avoids the build-up of skin friction, which holds the conductor in place. In case of a severe broaching, the well must be abandoned.

According to Akers (2008), the drilling fluid[6] is pumped at a low flow rate to avoid the plugging of the drill bit's nozzles when the conductor touches the seafloor. When the weight on the bit (WOB) overcomes 22.24~44.48 kN (5~10 kips), the flow rate is increased with the depth. The maximum flow rate is typically achieved at the depth of 45.7 m (150 ft) below mud line (BML). The maximum flow rate may change according to the penetration rate and the conductor's diameter, but a typical maximum rate is $63 \times 10^{-3} \sim 94 \times 10^{-3}$ m^3/s (1,000–1,500 gal/min). Then, the flow rate is reduced about 3–5 m (10~15 ft) above the final depth of this section to reduce the turbulence and avoid the wash-out of the sediments at the bottom end of the conductor.

Another jetting parameter is the weight on the bit (WOB) that is also increased with the depth. In case of the actual WOB increases, faster than the values programmed, it is necessary to carry out the reciprocation, i.e., pulling up the conductor while jetting to break the building up skin friction. The length of each stroke depends on the personnel experience in a specific offshore area. But some operators prefer, instead to pull up, to lock the heave compensation system and let the floating platform heave motion to reciprocate the whole conductor string.

Once the conductor achieves the final depth, all operations are suspended, and the conductor is held in the position for the build-up of the skin friction with time (Figure 5.11b). Jeanjean (2002) presented a methodology to assess the increased conductor's bearing capacity with time.

Before continuing, some operators may pull the string up to a precalculated tension to test the skin friction on the conductor. If the conductor holds itself in place, the drill-ahead operation shall start. First, it is necessary to release the drillstring from the running tool. The conductor's running tool has a sleeve where a mandrel is seated inside (Figure 5.11). The mandrel is a piece of pipe with the same pin-box threaded connector of a drill pipe. A pin latches the mandrel to the sleeve during the jetting. For the drill-ahead, such pin may be removed using an ROV or it may also be sheared using the self-weight of the drillstring.

The second phase of spud-in is then drilled using the 0.660-m (26-in) bit. Because the mandrel is released of the running tool, it is possible to drill rotating the whole drillstring (Figure 5.11c). The combination of rotations of mud motor and the whole drillstring may result in a higher rate of penetration (ROP). The seawater is pumped as drilling fluid carrying the sediments upward to be discharged through the drain holes and vents located at the low-pressure housing and running tool. When necessary a "pump and dump" method (pump: an environment-friendly drilling fluid instead of seawater) can be used, such as drilling through an instable formation or gaining angle in a directional well.

When the drilling of the second phase is completed, the whole drillstring must be tripped out from the well and the conductor running tool must be recovered. The mandrel fits again inside the running tool's sleeve. Another J-slot and pin coupling allows transferring the drillstring rotation to the sleeve. The clockwise rotation of the drills string releases the running tool allowing its recovery (Figure 5.11d).

Next, the surface casing is assembled on the rig floor. Its bottom end is called a casing shoe where the casing end pipe is reinforced with cement and has a round shape to avoid any dent due to accidental impacts during the casing running. At the casing shoe, there is also a floating valve, which does not allow any fluid to flow back into the inside of casing. On the top end of the surface casing, there is the high-pressure housing, which must be seated within the low-pressure housing. All subsea pieces of subsea equipment, namely, subsea BOP and WCT must be installed on the high-pressure housing. In addition, the high-pressure housing has an internal geometry where the subsequent casing hangers should be seated.

The surface casing has its own running tool that is latched to the high-pressure housing. Below the running tool, there is a string of drill pipes named "stinger." The surface casing is lowered down by drill pipes (Figure 5.11e).

Some operators prefer to use a shorter stinger with the traditional two cementing plugs attached to it. The plugs separate the bottom and top of cement from the drilling fluid when it is pumped downward throughout the casing, avoiding the mixture of the cement slurry with the fluid contained inside the casing. A contaminated cement slurry may result in "bubbles" in the annulus. Such bubbles may affect the well integrity because they may become the source of leakage of fluid from the formation into the well. A bad cement job requires a secondary remedial cement operation.

However, an argument against the use of the two plugs is whenever it is necessary to circulate drilling fluid and reciprocate the casing string, it may cause the accidental release of the plugs from the stinger. When it happens, it is necessary to trip out the casing again to reset the plugs. On the other hand, the long stinger conveys the cement all the way down through the casing, just above the floating valve. After the cement slurry flows out the stinger, it cannot flow upward within the casing because the running tool is sealing the top end. Furthermore, the fluid trapped inside the casing is always lighter than the cement slurry. This difference of density keeps this fluid separated at the top avoiding the mixture. Thus, the only way for the cement slurry flow is through the floating valve and casing shoe.

Using the long stinger, the cement slurry may get mixed, at the very beginning, with the drilling fluid located inside the casing shoe. But once the cement slurry flow becomes steady, there is a low risk of contaminating the slurry with the fluid located at the top part of the casing.

After the surface casing is seated inside the well, the cement job starts pumping flush and spacer fluids (Figure 5.11f). These fluids must displace the drilling fluid that remained in the annulus between the well and the casing. They also clean the surface of the casing and sediments for better adherence of the cement slurry. For the surface casing, the cement must be pumped to reach the low-pressure housing (Figure 5.11g). The ROV inspects the cement operation to make sure that cement slurry reaches the drain holes. A common practice is to use a black die in the spacer so the spacer coming out of the well, at the sea floor, can be seen through the ROV.

When the annulus is filled with the cement slurry, the hydrostatic pressure at the bottom of the annulus is higher than the hydrostatic pressure inside the casing since the cement density is heavier than the displacement fluid. Thus, the cement tends to flow back into inside the casing. This backflow does not happen because the floating valve blocks the backflow keeping cement in the annulus until the cement is cured.

The last operation of the spud-in is the recovery of the drill pipe string used for the cementing job. The running tool is released using the clockwise rotation of the whole drill pipe string. Then the spud-in stage is completed (Figure 5.11h). Some cured cement and floating valves remain at the casing shoe. However, all these materials, such as rubber and aluminum of the floating valves and cement are easily drilled by the drill bit of the next phase.

5.2.2 Drilling with Fluid Return

After the spud-in is completed as shown in Figure 5.11h, it is necessary to re-entry the well with the installation of the subsea BOP and drilling riser. In Brazil, the operators lower down the whole subsea BOP hanged by the drilling riser (Figure 5.4). In the GoM, the operators prefer to install the BOP stack first using the proper running tool hanged to a land string. Then, the LMRP is lowered down and hanged by the drilling riser and connected into the top of the BOP stack.

A VX ring provides the sealing between the high-pressure housing and BOP (Figure 5.12a). Once installed, the BOP system must be pressure tested to ensure that all the components function properly. The BOP stack must be tested periodically (each two or three weeks) during a drilling-with-fluid-return operation.

Before the drilling operation starts, it is necessary to install a wear bushing within the high-pressure housing to protect its inner surface. Any dent or wearing caused by the drillstring on the inner surface of the high-pressure housing may cause problems when the casing hanger is seated inside in the next phase. The wear bushing is run and connected to the proper running tool hanged by a string of drill pipes (Figure 5.12a).

Next, it is necessary to carry out the Leak-off Test (LOT) to verify the strength of the rock layer located just below the casing shoe depth. The drillstring to be used to drill this phase is run into the well. It drills the floating valve, casing shoe, and about 10 m (32.8 ft) of sedimentary rock. Then the annular BOP is closed around the drillstring, sealing the annulus. Drilling fluid is then injected using the cement pump rather than the mud pump. The pressure on the rock increases while drilling fluid is pumped (Figure 5.12b). During the test, the surface pressure vs. volume injected is plotted. This graph follows a straight-line behavior up to the point where this line starts to bend. This indicates a fracture initiation at the rock and the test is stopped. In addition, the Formation Integrity Test (FIT) may be performed, instead of LOT, to a certain maximum pressure or equivalent mud density. Usually in an area where the pore pressure regime is known, thus the maximum necessary mud density to drill the section or phase is known, then the FIT may be performed lowering the risk to break down or to fracture the formation at the casing shoe depth. In addition, the LOT or FIT serve as an integrity test of the casing and cement bond because if it leaks during the test the cement job is not good.

When the LOT is completed, the drilling operation starts (Figure 5.12c). The drilling fluid is injected from the top of the drillstring. The fluid is conveyed downward to the bit and returns to the surface throughout the annulus carrying the drilling cuttings.

During the drilling operation, there are three main drilling parameters (Table 5.5).

When the bit reaches the designed final depth, the well is circulated until the shale shaker is clean from the cuttings (usually one to three "bottom-up"[7]). After this, the drillstring is tripped out of the

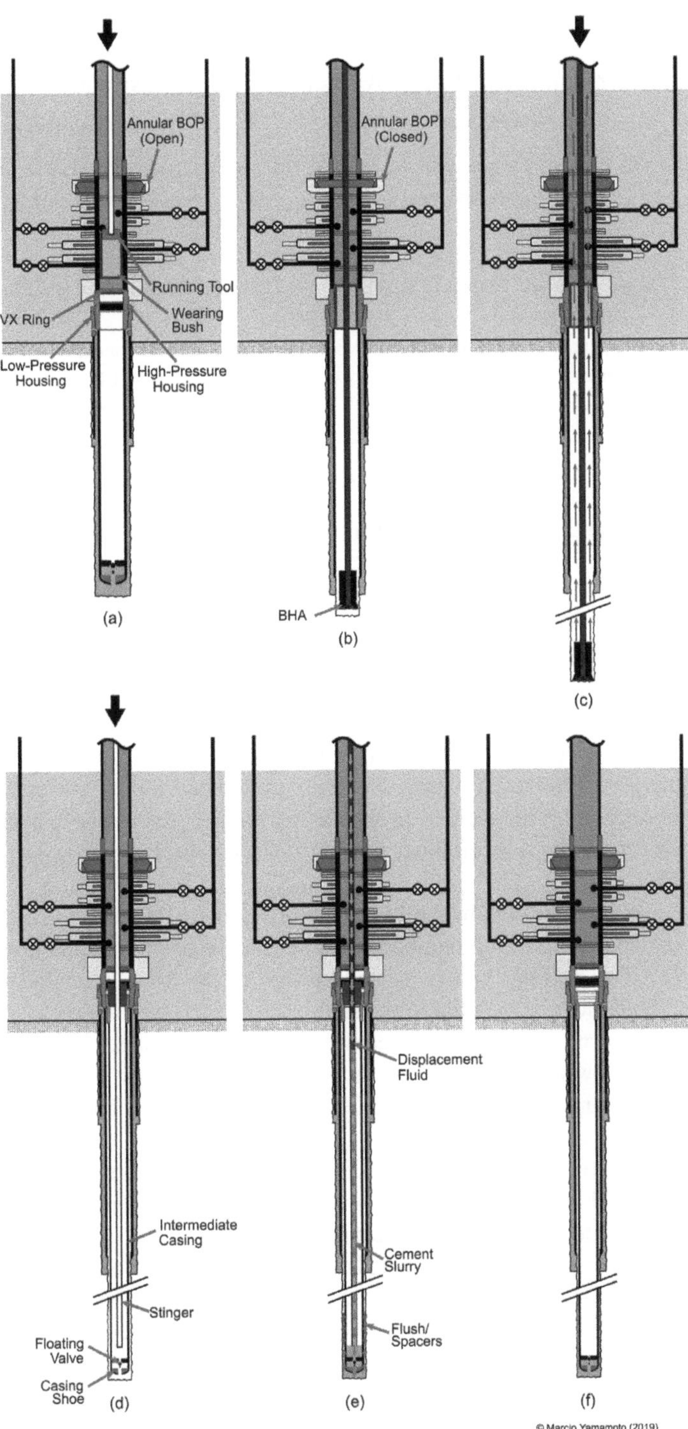

FIGURE 5.12 Schematic of the drilling with the fluid return. Once the drilling is completed, the casing is run and cemented in place (drawing is not to scale).

TABLE 5.5
The Drilling Parameters and their Respective Actuator and Measurement Method

Drilling Parameter	Actuator	Measurement
Weight on Bit (WOB)	The brakes of the draw-works in synergy with the hoisting system keep the WOB	The weight difference between the drillstring suspended off the bottom and drillstring seated on the bottom. The weight is assessed measuring the tension of the dead line
Rotation	Top drive drives the rotation on the drillstring	Rotation sensors embedded in the top drive
Pump rate	Mud pump's reciprocating pistons drives the flow rate	Flow rate is assessed multiplying the pump speed by the "pump factor" (the volume displaced in a pump cycle) and its efficiency

hole for the casing running operation. However, it is necessary to remove the wear bushing first using the proper running tool attached to the drill pipe string.

Then, if a dual activity drilling rig (single derrick with dual hoisting system and dual rotating system) is available, the intermediate casing joints can be assembled in stands of two or three joints offline and hanged on the rig floor to save connection time. The intermediate casing's bottom end has the casing shoe with the floating valve, and the top end has the casing hanger. Two or three joints above the casing shoe, a collar with a float valve is installed so it can hold the cement plugs (top and bottom). Like the surface casing, a temporary or "false" table is placed over the casing hanger to make the string assembly composed by the stinger and the running tool. After the running tool is connected to the casing hanger, the temporary table is removed, and the casing is lowered down throughout the drilling riser, BOP, and previous casing into the well (Figure 5.12d).

The casing hanger is seated within the high-pressure housing. Then, the cementing operation starts. The flush and spacer fluids are pumped first to clean the open-hole and displace the drilling fluid (Figure 5.12e). The drilling fluid is pushed upward to the drain holes located at the casing hanger. Such drain holes convey the fluid to the annulus above the running tool.

Compared to the surface casing, the intermediate casing is much longer. Its casing shoe is located thousands of meters below the seafloor. Thus, the cement pump does not have enough power to pump the cement slurry up to the wellhead. The top of the cement shall stay in a predetermined depth.

After the cement is cured, the running tool is once again unlatched from the casing hanger, and the whole string is tripped out (Figure 5.12f). A well logging tool may be run into the well to verify the top of cement depth and the integrity of the cement barrier.

When the phase is completed (Figure 5.12f), the same drilling operation cycle for the next phase starts: wear bushing is set in position, LOT or FIT, drilling, run and cement casing in place (Figure 5.12a–f). The number of phases depends on the sedimentary basin. Brazilian East Margin, where the most prolific basins are located, is a passive margin with relatively simple geology. As shown in Figure 5.10, a well is usually drilled comprising 4 phases. By comparison, the subsea wells in the GoM, which presents a very complex geology, are drilled with more than 8 phases.

Differently from the spud-in stage, when seawater was used as drilling fluid, the drilling-with-fluid-return stage uses more complex and expensive drilling fluid, usually called only as mud, which shall have higher density and viscosity than the seawater. The drilling fluid has several applications as shown in Table 5.6.

TABLE 5.6
Main Applications of the Drilling Fluid

• Cool down the bit and lubricate the drillstring	• Keep the open-hole stability
• Carry the drilling cuttings to the surface	• Overcome the pore pressure

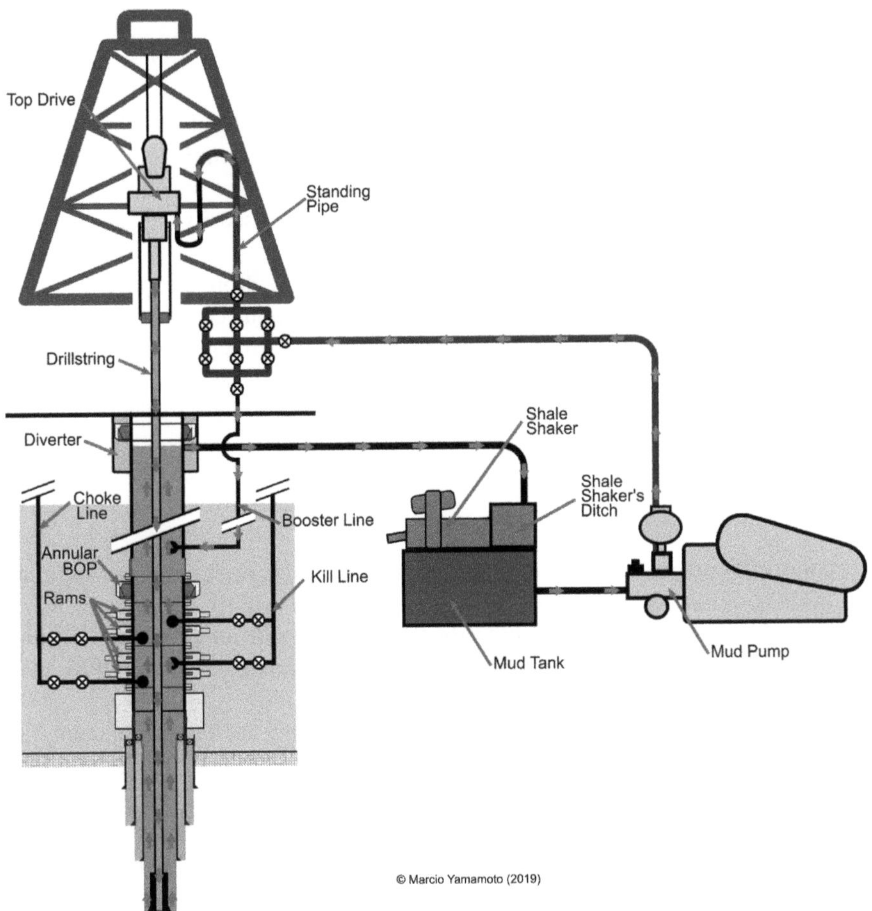

FIGURE 5.13 Schematic of the circulation system during the drilling with fluid return (drawing is not to scale).

The drilling fluid circulation system is a closed loop circuit as shown in Figure 5.13. Following the mud pump, the fluid is conveyed through a manifold to the standing pipe. A hose connects the standing pipe to the top drive where the drillstring top end is connected. The fluid is conveyed to the bottom of the well, flowing throughout the internal part of the drillstring (drill pipes plus the BHA) and throughout the bit nozzles and returning to the surface through the annulus up to the diverter. At the diverter, the drilling fluid is at atmospheric pressure. Then, it is conveyed down to the shale shaker through the flowline by gravity.

The drilling fluid and cuttings must be separated over the vibratory screens of the shale shakers. The mixture fluid and cuttings arrive in the ditch and then flow over the screen. The screen vibration displaces the mixture forward over the screen and, at the same time, helps to separate the solid and liquid phases. Drilling fluid flows down to other components of the solid control equipment, such as desander, desilter, and mud cleaner and then into the tanks where it may receive further treatment, with addition of chemicals, depending on its desired properties, and then is reinjected into the well.

The drilling cuttings discharged from the shale shakers are soaked with drilling fluid. They must pass then through a centrifugal cuttings dryer, where some drilling fluid is removed and pumped back into the tanks to be reinjected. The dry drilling cuttings are stored to be properly discarded onshore. If a water-based mud is used, then the drilling cuttings might go overboard into the sea depending on the local legislation.

The drilling cuttings are carried out to the surface by the drilling fluid. The cuttings upward velocity is the difference of the drilling fluid annular velocity and the cuttings slip velocity. The slip velocity is a function of the drilling fluid viscosity and the difference of specific weight between the cutting and drilling fluid (Bourgoyne et al., 1991). Drilling fluid viscosity and specific weight may be increased adding bentonite and barite, respectively.

Regarding the drilling fluid annular velocity, a large velocity drop is expected to occur when the fluid reaches the drilling riser because of the increased annulus cross-area. The drilling riser always has an inner diameter larger than the open-hole and last casing installed. To avoid any problem of drilling cutting accumulation within the riser, additional flow of drilling fluid is pumped through an auxiliary pipe named Booster Line that is connected just above the bottom part of the riser (Figure 5.13).

5.2.3 Drilling Installations for Deepwater

This subsection will compare the drilling installation used in deepwater and that used in onshore rigs. Such installations are usually called drilling rigs and also drilling packages, which are composed of all systems and equipment required for the well construction. Thus, the floating mobile drilling unit (MODU) itself, its living quarters, Dynamic Positioning System (DPS), etc., are not included in the drilling rig.

A drilling rig is composed by the derrick, its rig floor including ancillary equipment, such as the hoisting system (draw-works, crown block, traveling block, and drilling cable), top drive, drilling fluid circulation system, well control system, drillstring, etc. (Bourgoyne et al., 1991).

The floating drilling unit is a no-inertial frame because it has motion in all degrees of freedom due to the waves, current, and wind that generate acceleration on the drilling unit. Further, the platform's heave movement literally stretches and shortens the well depth. Even though the drilling unit motion can be measured, such data is not used to adjust the measurement sensors. Instead, the fluctuation on the sensors reading is assumed as a noise, which reduces the accuracy of all measurements.

A very important issue in well control is to identify a kick as soon as possible. One way to identify the kick is by comparing the drilling fluid pump-in and pump-out flow rates. In addition to the platform motion, especially heave, the drilling riser, which conveys the returning drilling fluid to the surface, is exposed to the vortex-induced vibrations (VIV) generating more fluctuation to the returning flow rate. Thus, because the reduced accuracy of the return flow rate, prompt identification of a kick may not be conclusive or require a flow check, which increases non-productive time (NPT) and, consequently, the costs of the well construction.

Next, another difference is that the most of offshore rig has a top drive, instead of a Kelly and rotary table which are still widely used in onshore rigs. The top drive can drill, adding a stand of three drill pipe joints into the drillstring. This capability reduces the connection time to connect a drill pipe by three drill pipes with a single drill pipe that can be added to the drillstring.

Further, the offshore rig has a heave compensation system to reduce the heave motion on the top of the drillstring. The heave can generate compression on the drillstring, which may result in the buckling of the drill pipes, and/or cause the impact of the drill bit on the well bottom resulting in bit failure. In either case, the final results are increased NPT and costs.

In the last decades, the heave compensator technology has evolved. Initially, the Passive Heave Compensator (PHC) was designed as an automobile suspension with a spring-damper system. Then the Active Heave Compensator (AHC) was developed together with active-passive hybrid compensators (Woodacre et al. 2015). The AHC is a feedback control system where a controller receives a signal of the platform heave and drives an actuator.

The PHC was used in the Drillstring Compensator (DSC), where the hydraulic pistons are installed between the top drive or traveling block and the top of drillstring (Butler and Larralde, 1971; Woodall-Mason and Tilbe, 1976). The top drive moves together with the floating unit while the drillstring is kept suspended by the pistons. In this case, compressed air works as a spring. The stiffness can be changed by adding or removing compressed air accumulators to the system. The restriction valves located at the hydraulic cylinders can also adjust the damping.

Later on, the PHC started to be used in the Crown Mounted Compensator (CMC), where the hydraulic pistons keep the crown block, traveling block, top drive, and drillstring suspended, while the rest of the platform is heaving. Further development includes AHC (Haaø et al., 2012) and hybrid compensators (Cuellar Sanchez et al., 2017) used in the CMC.

The AHC is used in the Active Heave Drilling Draw-works (AHD), in which the actuator is the draw-works itself (Offshore Magazine, 1999). There are no hydraulic piston or compressed air accumulators. A controller drives the draw-works based on the heave motion. The draw-works keeps the traveling block, together with top drive and drillstring, stationary in the Earth reference frame.

The last difference between the onshore and offshore rigs is the derrick. Because of the floating platform motions, the offshore rig's derrick is more robust to withstand the dynamic loads. Furthermore, the last generation of offshore rigs is the dual-activity rig, which means that in the same derrick, there are two hoisting systems and, sometimes, two top drives. It does not mean that the dual-activity rig can drill two wells simultaneously. Instead, it is possible to carry out some parallel activities improving the performance. For example, assuming a spud-in scenario, the surface casing can be assembled and run to mudline using a hoisting system, while the other hosting system is jetting the conductor and drilling ahead (Keener et al., 2003).

5.2.4 Drilling Platform

Semi-submersible rigs and drillships are the floating-type MODU that can operate in deep- and ultra-deepwaters. The semi-submersible rig has two pontoons, which are connected to the deck using columns (Figure 5.14). The pontoons together with the columns provide the buoyancy. Furthermore,

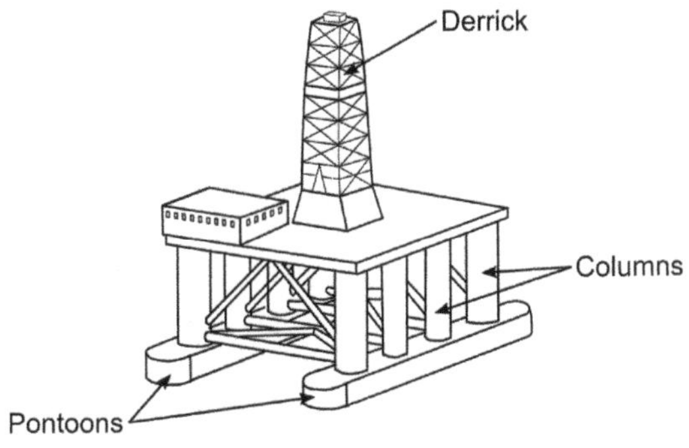

FIGURE 5.14 Schematic of a semi-submersible drilling platform.

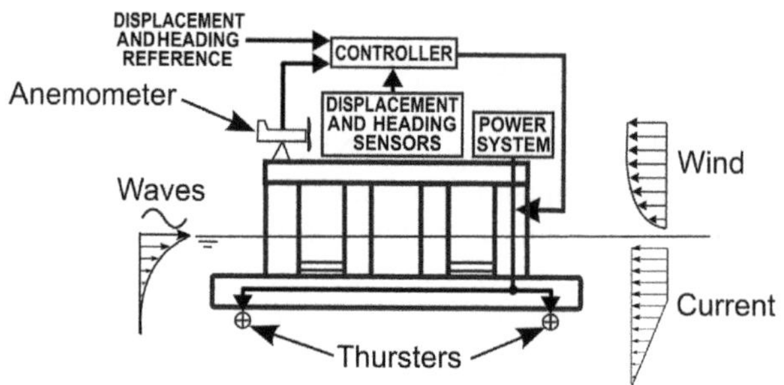

FIGURE 5.15 Schematic of a DPS. *Source:* Courtesy of Yamamoto and Morooka (2005).

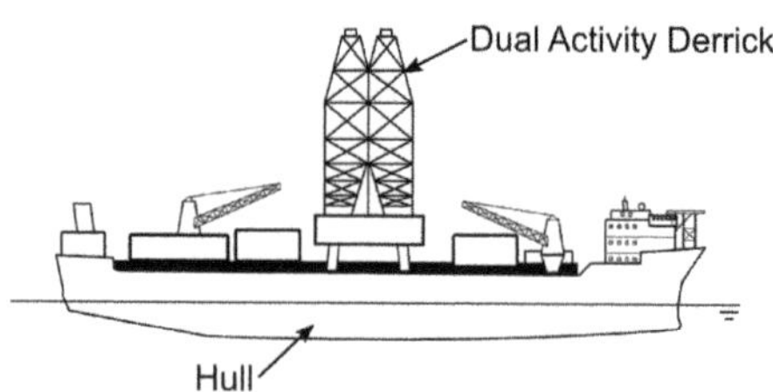

FIGURE 5.16 Schematic of a drillship.

the pontoons are located more than 20 m below the sea surface where the wave effects are reduced. According to the Airy wave theory, the velocity and acceleration of water particles due to the waves reduce with increasing water depth (Newman, 1977).

Besides the reduced wave effects, because the pontoons are horizontal buoyant members, they increase the added mass and viscous drag in the vertical direction. Thus, the heave motion is reduced compared to a drillship allowing the semi-submersible to operate in harsher sea conditions. In high-heave amplitude motion, the well construction operation must be suspended.

As to positioning, the semi-submersible rig can have a mooring system to operate in shallow and deepwaters. For ultra-deepwater, the DPS is used for positioning. The DPS is a feedback control system in which a controller receives the data of platform position and heading. If the platform drifts from a reference position, the controller drives the thrusters installed on the bottom of the pontoons to restore position (Figure 5.15). There is no technical restriction for a semi-submersible equipped with DPS to operate in shallow water precluded that a watch circle of 1% of water depth is obeyed due to a bending moment caused by the riser column in the wellhead. This means that a minimum of around 500 m of water depth is required to operate a semi-submersible or drillship using a DPS. However, because the daily rate of DPS semi-submersible rigs is higher than moored one, the latter option is preferred wherever is possible. Although, the anchor handling tug supply (AHTS) vessel's daily rate can be also very expensive. In fact, it is possible to upgrade an anchored semi-submersible to operate in deeper locations. In this case, the weight of the longer mooring lines and drilling riser are compensated attaching more buoyant devices on the pontoons and columns (Guo et al., 2006; Tsukada et al., 2007).

The second possible MODU to operate in deepwater is the drillship (Figure 5.16). It is necessary to point out that a drillship is designed and built to be a drillship. Some people confuse the drillship with an FPSO (floating, production, storage, and offloading); the latter is a production-floating platform with a ship-shaped hull. Most of the FPSO is converted from an old large crude carrier ship.

Most of the drillship has a DPS to operate in ultra-deepwater. A few drillships, which operate in the shallow waters, have a mooring system instead.

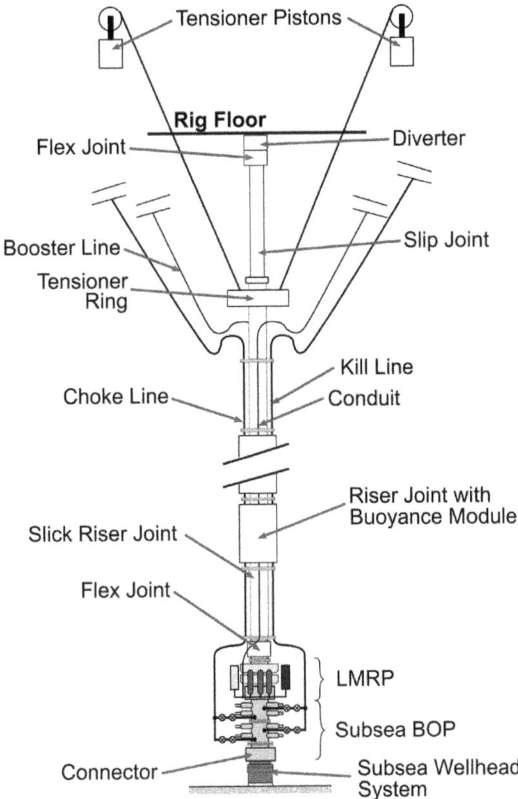

FIGURE 5.17 Schematic of a subsea system composed by the BOP stack (LMRP and subsea BOP) and drilling riser (the graph is not to scale).

5.2.5 Drilling Riser, Subsea BOP, and LMRP

The drilling riser, BOP stack, and LMRP are the subsea equipment required to drill a subsea well (Figure 5.17). These subsea pieces of equipment are installed after the spud-in when the surface casing is run and cemented.

The subsea BOP is composed by the BOP stack and the LMRP (Low-Marine Riser Package). The subsea BOP is safety equipment with a primary use during the well control operation. More details about the well control, namely, the occurrence of a kick and its removal from the wellbore are presented in Section 5.4. If the kick develops towards a blowout (the uncontrollable flow of the formation fluid to the surface), the drilling personnel can shut the well using the available rams (Figure 5.17).

There are two sequences to shut the well, depending on whether there is a drillstring inside the well or not. When there is no drillstring, the well is shut by closing the blind ram. If the drillstring is in the well, in an emergency disconnection procedure before the disconnection at the LMRP, the drilling personnel shall first close the pipe ram to hold the drillstring in a proper position. Then, one shall close the shear ram to cut the drill pipe. It is important to cut the drill pipe in its body and not in the tool joint that has a large diameter and wall thickness. The upper part of the drillstring is recovered, while the pipe ram holds the lower part of the drillstring. With the bore of the blind ram cleared, one can close the blind ram, shutting the well.

It is important to highlight that most of the shear rams can only cut the body part of the drill pipe and, depending on the ram specification, some casing strings (some up to 9-5/8" casing). Shear rams usually cannot cut the drill pipe tool joint, heavy weight drill pipe (HWDP), drill collars, and other thick pieces of tubular structures. It is also important that the shear ram has the capability to cut the wireline of well logging tools, coiled tubing, etc. Thus, the BOP can be used during the drilling-with-fluid return stage, completion stage, and heavy workover.

The BOP has two redundant control pods, the yellow one and the blue one, that have solenoid hydraulic valves to drive the annular BOPs and rams. The BOP's main control panel is located in the "dog house" (the driller's control cabin on the rig floor) and a secondary BOP control panel may be located in the bridge. At the control panel, one can select one control pod and then actuate any annular BOP, ram, and/or valve. The control pods on the BOP are continuously confirming the connection to the control panel on the surface. If the connection is lost, the control pod safety protocol automatically shuts the well.

All the valves, rams, and annular BOPs are hydraulically driven. The hydraulic fluid, which is pumped from the floating unit, is adjusted using the proper additives such as antifreeze, etc. The hydraulic fluid is pumped from the floating platform and conveyed downward to the BOP throughout an ancillary pipe named "conduit." It is one of the four ancillary pipes attached around the drilling riser. Further, the BOP stack has accumulators where hydraulic potential energy is stored compressing an inert gas located inside each accumulator's high-pressure vessel.

The accumulators enhance the BOP availability; all hydraulic components may be driven even if no hydraulic fluid is pumped from the floating platform. In addition, the accumulators increase the speed to drive hydraulic components because the gas expansion is faster than the fluid pumped through the conduit.

The secondary use of the BOP stack is to allow emergency disconnection of the floating platform from the wellhead. The disconnection is required when the floating platform is drifting off the position above the wellhead. In Brazil, about 30% of the drift-off that resulted in emergency disconnection is due to environmental conditions and natural events (e.g. ionospheric scintillation that prevented the GPS receiver from locking on the satellite signals). "Environmental conditions" refers when the DPS could not withstand harsh sea conditions and drifted off position. A "natural event" refers to losing a GPS satellite signal due to scintillation or when the ionosphere suffers an ionization, thus blocking the GPS signals from passing through. The other 70% of emergency disconnection refers to all sort of technical failures including human error (Paula Jr. and Fonseca, 2013).

In the emergency disconnection situation, the drillstring has to be hung first on the pipe ram of the BOP and it has to be cut using the shear ram, then the well is secured or shut in with the blind ram, and the hydraulic connector that fixes the LMRP to the BOP is released. This operation is automatic (cut the pipe and release at LMRP) and it takes less than a minute after the drillstring is hanged, just pushing one button. In addition, the BOP stack configuration changes widely according to the manufacturer and drilling contractor specifications. In some BOP stack configurations, a single ram has the blind and shear ram capabilities together.

The last device of subsea equipment is the drilling riser, which is top tensioned vertically with 4 ancillary pipes. Table 5.7 features the main application of each one of these ancillary pipes. The riser guides the drillstring and other equipment to the wellhead and conveys the drilling fluid to the surface.

The whole drilling riser is assembled using, for example, 22.86-m (75 ft-) long joints. The joints may have a buoyancy module to reduce their submerged weight. There are also slick joints (without buoyancy modules), which are used at the riser's bottom or close to the sea surface to reduce the effects of the waves. In addition, it has flex joints and/or ball joints installed on both ends to reduce the bending moment concentration.

On its top end, there is a slip joint that avoids that the platform's heave motion to be transferred to the drilling riser. The whole riser is kept tensioned by the tensioner ring connected to a wire-pistons system (Figure 5.17). The riser tensioning system also is a passive-type heave compensator.

TABLE 5.7
The Main Applications of the Drilling Riser's Ancillary Lines

Ancillary Line	Application
Choke Line	It is used during the well control operation to control the circulation of the drilling fluid with a kick to the choke manifold (see details in Section 5.4). The kill line can substitute it.
Kill Line	It is used to monitor the pressure at BOP during the circulation of a kick or to inject heavier drilling fluid into the wellhead during well control operations.
Buster Line	It conveys drilling fluid to the bottom end of the drilling riser to increase the flow velocity within the drilling riser, avoiding the clogs from the drilling cuttings.
Conduit	It conveys hydraulic fluid to keep the BOP accumulators pressurized.

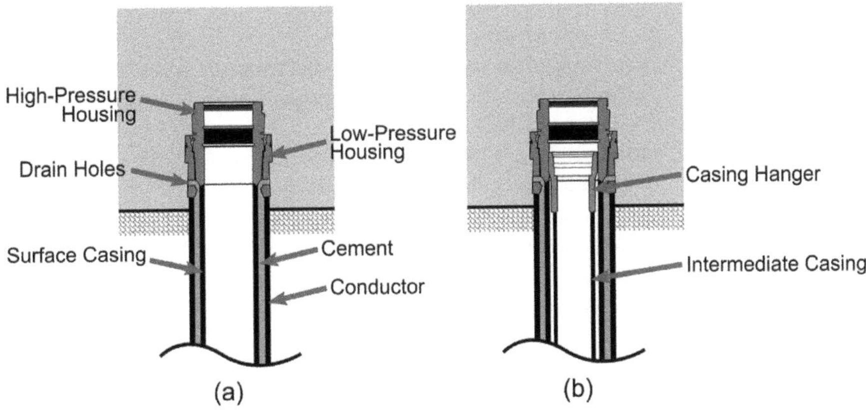

FIGURE 5.18 Schematic of a subsea wellhead system: (a) the high-pressure housing is seated within the low-pressure housing and cemented in place; (b) a casing hanger is seated within the high-pressure housing holding the intermediate casing.

5.2.6 Subsea Wellhead System

The subsea wellhead system is composed of a low-pressure housing (LPH) and a high-pressure housing (HPH) as shown in Figure 5.18. The bottom of the LPH is welded to a piece of conductor joint and it is assembled at the top end of the conductor string. In the same way, HPH is welded to a piece of a surface casing joint and assembled on the top of the respective string (Kaculi and Witwer, 2014; Williams, 2015).

As explained in Section 5.2.1, the conductor is installed first. Then the second phase is drilled to receive the surface casing. The surface casing is run into the wellbore and the HPH seats within the LPH (Figure 5.18a). The following casings must be run into the wellbore with a casing hanger on their top end, and all casing hangers are seated within HPH (Figure 5.18b). However, the number of casing hangers seated inside the HPH is limited. For a complex well, which requires several phases of drilling with the fluid return, liners are used to replace the casing because of such limitation.

In some offshore regions, the tubing hanger is also installed within the HPH above the casing hanger. For the major Brazilian operator, the tubing hanger is installed within subsea equipment called production adapter base (PAB), which is installed between the subsea wellhead system and the Wet Christmas Tree (WCT).

TABLE 5.8
Main Components of a BHA

Equipment	Application
Drill Bit	It has several types (fixed cutters, cone type, and hybrid) and it is selected according to the formation to be drilled.
Drill Collar	Tubular component with a very thick wall making it very heavy. Its weight is responsible to put weight on the bit (WOB is a drilling parameter) and to keep the drill pipes above tensioned.
Stabilizer	A component with the same outer diameter of the bit and with a spiral surface, which is contact on the wall of the open well. It keeps the BHA centralized into the wellbore, reduces the drillstring vibration, and avoids the well deviation.
Cross-over subs (XO's)	Any small piece of tubular to connect two BHA components with different threaded-connector types.
Rotary Steerable	Equipment that changes the direction of the tool face for the directional drilling with the rotation the whole drillstring.
Measurement While Drilling	Equipment to measure the inclination from vertical, magnetic direction from the North, gyroscope for direction, azimuth, gamma ray, pressure inside and outside the drillstring (PWD).
Logging While Drilling	Basic well logging tool (gamma ray, resistivity, magnetic resonance (NMR), etc.) to measure the wellbore during the drilling.
Non-Magnetic Drill Collar	A drill collar manufactured with a non-magnetic material to not affect the LWD measurement (magnetic direction).
Drilling Jar	Equipment that accumulates energy and releases an impact load to release the stuck drillstring.

5.2.7 Drillstrings

Drillstring is a tubular structure, which has the drill bit at its lower end. It can be separated into two parts: Bottom Hole Assembly (BHA) and drill pipes. Each part is composed of several components, which are connected to each other by threaded connectors. The male connector is called a pin, while the female connector is called a box.

Table 5.8 features common pieces of equipment that compose the BHA. Then the rest of the drillstring is composed of the drill pipes, which are tubular with very narrow walls compared to the drill collars (Bourgoyne et al., 1991). Thus, the drill pipes must always be tensioned in the axial direction to avoid buckling.

Due to the large difference of stiffness between BHA and drill pipes, a stress concentration is expected in the connection above the BHA. In order to have a smooth transition between BHA and drill pipes, the American Petroleum Institute (API) recommends the use of nine to ten joints of Heavy Weight Drill Pipe (HWDP) to connect the BHA to the drill pipe string to reduce the risk of failure of the drill pipe due to the stress concentration just above the BHA (API, 1998).

5.2.8 Completion

In this section, some details are presented about the completion, that is the operation and equipment installed within the wellbore and on the wellhead to allow the production or injection of the development wells.

At the end of the drilling operation, the well is cased and filled with drilling fluid. Then the well conditioning operation starts to prepare the well for the completion. A drillstring equipped with special scrapers are run into the wellbore. Brine is injected throughout the drillstring to displace the drilling fluid while any dust adhered to the casing inner surface is scrapped out. The completion cannot be carried out with drilling fluid because it has solids in suspension, which may cause the failure or leakage in the completion equipment. The brine used during the completion operation has the density adjusted to keep the well overbalanced to avoid a kick.

The completion equipment and installation may vary depending on the reservoir characteristic, well design, operator, etc. However, the installation is usually carried out in two phases. The lower completion phase refers to the operations and equipment installed below the packer including the packer itself. A packer is a piece of equipment anchored on the casing that seals the annulus and holds the weight of other pieces of equipment.

Furthermore, there is a variety of reservoir-well interfaces available that must be selected mainly according to the reservoir characteristics. For consolidated reservoirs, the well is cased and perforations are deployed whereas unconsolidated reservoirs may require an open-hole gravel pack or a slotted liner solution (Bellarby, 2009).

Figure 5.19 shows the schematic of a cased completed well with perforations. To make such perforations, a wireline tool named a perforating gun, which is equipped with several explosive charges, is lowered down into the well bottom. This tool shoots a high-pressure high-temperature jet against the well wall that penetrates the casing, cement, and several meters into the formation creating a channel where the hydrocarbon will be produced.

Once the low completion is set in place, the production tubing is run into the wellbore. The tubing lower end is connected to the packer, while its top end is connected to the tubing hanger. Along the production tubing string, beside the pipe joints, several ancillary parts are attached such as the surface controlled subsurface safety valve (SCSSV) and the gas lift valve

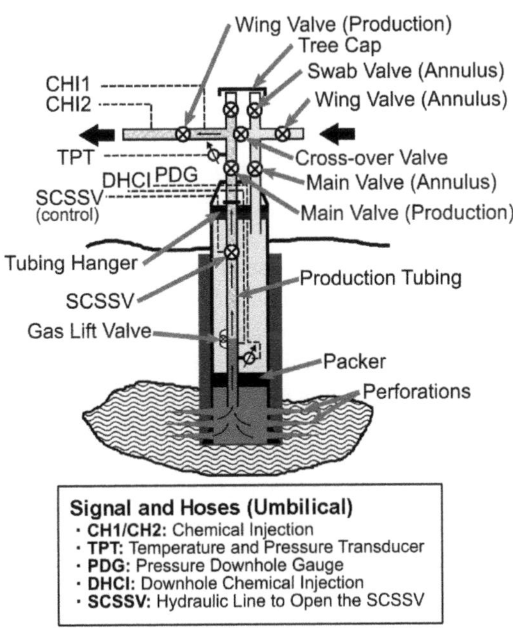

FIGURE 5.19 Schematic of a completed well for hydrocarbon production. Gas is injected in the annulus for the gas lift (artificial lift method). The dashed lines represent the control and sensors lines (description is in the box).

(Figure 5.19). In addition, it may be equipped with an electrical submerged pump (ESP) to artificially lift the oil.

In Brazil, the major operator seats the tubing hanger within the PAB. In other offshore regions, the tubing hanger may be seated within the HPH. The tubing hanger seals the annulus while holding the weight of the production tubing. Further, the tubing hanger is the boundary/interface between subsurface and subsea. An oil field contractor may be responsible for the subsurface completion equipment as shown in Figure 5.19 and a subsea contractor is responsible for the subsea equipment above the tubing hanger, namely, WCT together with flowlines, risers, control umbilical, etc.

The set of valves above the tubing hanger refers to the WCT functionalities (Agostini et al., 2017). A WCT has three subsea lines attached, namely, the production line, annulus line, and the umbilical for remote operation from the FPU. The production line conveys the produced fluid to the FPU. The annulus line conveys gas that is injected in the well annulus for the gas lift. When the natural gas is not yet available, for example, during the production startup, the nitrogen generated on board the FPU can be injected in the gas lift. When the production is in a steady flow, the produced natural gas is injected for the gas lift. The air never can be used in the gas lift because the oxygen in the air with the hydrocarbons within the wellbore in high pressure and high temperature shall result in an explosion and/or fire inside the well.

When the tree cap is installed, in addition to covering the swab valves it activates the remote operation of the WCT functionalities. When the tree cap is removed, the WCT functionalities must be driven locally using an ROV.

5.3 MUD LOGGING

During the well construction, it is necessary to collect data of subsurface formations to assess the reservoir rock (depth, porosity, permeability, presence of shale, etc.), to identify the presence of hydrocarbons and their saturation within the reservoir and among other things. Then, the geologist, geophysics and reservoir engineers may compound the well data with their seismic model to assess the reserves, to define a development plan, etc.

Chapter 4 introduced the well logging tools which are run into the well to measure geophysical properties of the recently drilled formations. For example, the resistivity log induces an electrical current through the reservoir to assess the formation resistivity. A reservoir rock with low resistivity means it may be filled with water (brine), while a high resistivity means the reservoir may contain hydrocarbon. Further, the most common well logging tool can be embedded in the BHA in what is named LWD.

Before the Schlumberger brothers had introduced well logging, mud logging had been used to identify the presence of natural gas in the drilling fluid that returned to surface. The presence of gas has been used as a correlation to identify economical reservoirs.

Mud logging contractors have developed this service over the decades. Nowadays, mud logging has three main services: gas detection, drilling cutting analysis, well monitoring and drilling parameters recording (Whittaker, 1987). In addition, they can assess the pore pressure of the drilled formation and to monitor the mud tank volume that are crucial data for the kick identification. All services are gathered in the mud logging cabin that is installed on board the drilling platform (Figure 5.20).

Another characteristic of mud logging is that the results are reported at the well depth: gas detection, rock samples characterization, drilling parameters.

5.3.1 GAS DETECTION

The gas detection is carried out in the drilling fluid that returns from the well. A piece of equipment named a gas trap (Figure 5.21) is installed in the drilling fluid return. The gas trap has an agitator

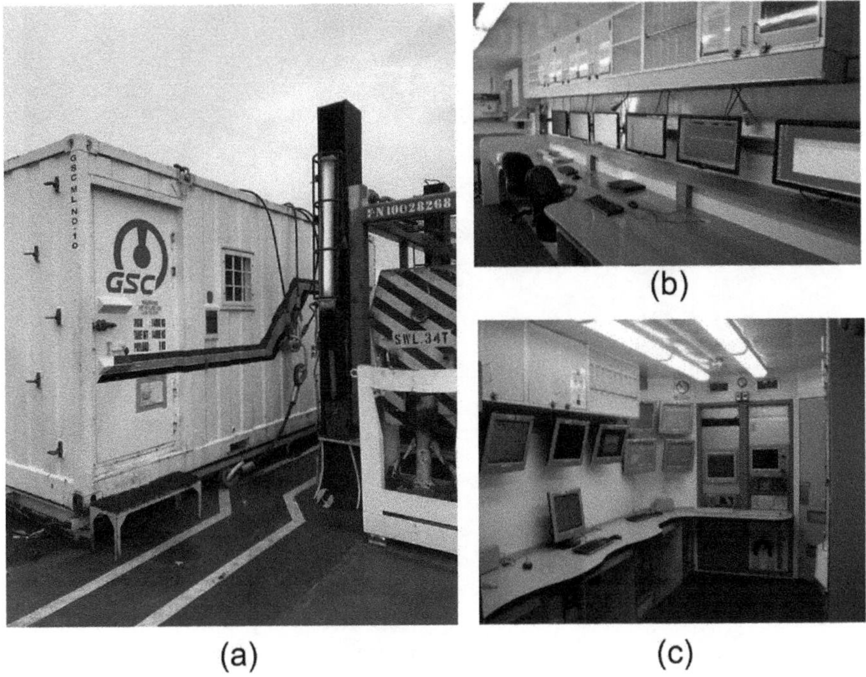

FIGURE 5.20 Several views of a mud logging cabin installed on board a drilling platform, (a) external view, (b–c) internal views. *Source:* Courtesy Geophysical Surveying Co., Ltd.

to separate the gas from the drilling fluid, and a vacuum line to convey the gas to the mud logging cabin (Whittaker, 1987).

After the proper conditioning, the gas sample is analyzed. The traditional method is the Catalytic Combustion Detector, a.k.a. "hot wire" (Pixler, 1961). In this methodology, the gas is continuously analyzed, the sample is injected in the detector chamber where a detector wire filament is located. The filament is connected to a Wheatstone bridge. The filament is heated due to electrical current from the bridge, the heat causes the combustion of the gas on the filament surface. The heat from the combustion alters the filament resistance, which could be measured by the Wheatstone bridge's voltmeter (Whittaker, 1987).

Despite the gas injected in the detector chamber is a mixture of hydrocarbons, the hot wire cannot assess the concentration of each hydrocarbon concentration. Thus, the instrument is calibrated assuming only methane and its results are showed as "equivalent methane in air" (EMA).

To overcome the inability to measure the concentration of different hydrocarbon present in the gas sample, the industry has been using gas chromatography. This methodology can measure the sample composition (usually C1~C5) and concentration. The limitation of this methodology is that it cannot make continuous measurements because the sample analyses are carried out in batches (Whittaker, 1987).

The modern mud logging service includes the interpretation of the gas chromatography measurement. This interpretation may indicate if the drilled formation is a non-productive zone, a gas-bearing reservoir, or an oil-bearing reservoir (Loermans et al., 2005).

Beside hydrocarbons, the formation gas may contain CO2 and traces of H2S. Thus, the mud logging gas detection device must identify and measure the concentration of both gases. The CO2 may be accumulated in a large range of concentration. Usually, higher concentration of CO2 is

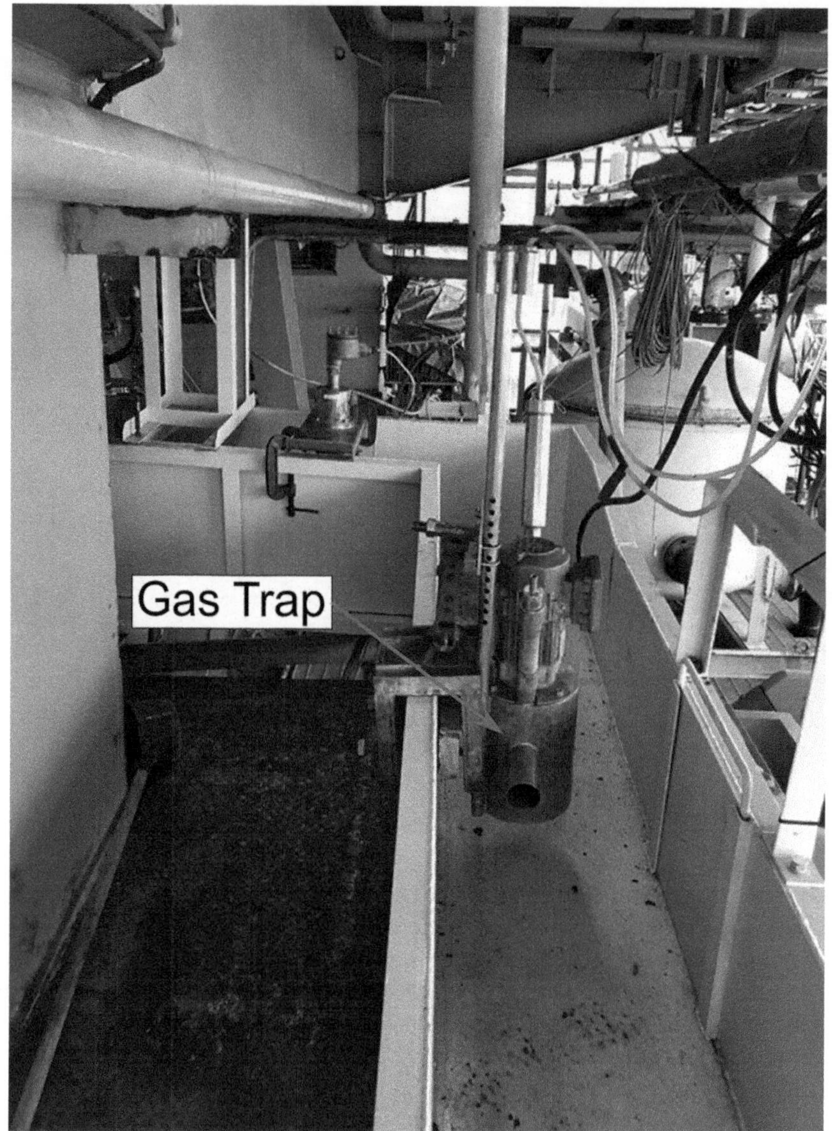

FIGURE 5.21 Gas trap located in the return line for the drilling fluid. *Source:* Courtesy Geophysical Surveying Co., Ltd.

present in carbonatic reservoirs, such as the Pre-Salt reservoirs off Brazil. The H2S is a poison gas even in at low concentrations, which may require special safety measures for the personnel on board.

5.3.2 Drilling Cutting Characterization

5.3.2.1 Collection of the Drilling Cuttings

The drilling fluid is pumped and injected on the top of the drillstring. The fluid is conveyed to the bottom hole, and then it flows through the bit's nozzles, cleaning the drilling cuttings in front of the bit. The drilling cuttings mixed with the drilling fluid are conveyed first through the annulus between the drillstring and casing or well, and later, through the annulus between the drillstring and riser. At the surface, the drilling fluid and cuttings must be separated at the shale shakers (Figure 5.13).

The sample of drilling cutting shall be collected at the shale shaker, before the cuttings dryers. Especial care for the collection is required depending on the type of the shale shaker. Samples from the desander and desilter should also be collected. The samples are collected from different depths following the contract between the operator and mud logging contractor. Usually, the samples are collected every 3 m (3x3 sampling rate) at the reservoir depth and every 15 m (15x15 sampling rate) at other formation.

According to Whittaker (1987), the samples must be collected in time intervals no longer than 15 minutes. For example, during one hour of drilling operation in a 3x3 sampling rate depth, only 3 m had been drilled, so the mud logger should collect 4 samples to be representative of this depth. In addition, the extra sample must be collected whenever the background gas measured by gas chromatography changes or after a drilling break[8] occurs.

The drilling cuttings generated at the well bottom will take several minutes until arrival at the shale shakers. This period is called "Sample Lag Time" calculated using Eq. (5.3),

$$t_{lag} = \frac{D}{v_t} \tag{5.3}$$

where t_{lag} is the sample lag time, D is the well depth, and v_t is the cuttings velocity at the annulus defined by Eq. (5.4).

$$v_t = v_{an} - v_{sl} \tag{5.4}$$

In Equation 5.4, v_{an} is the fluid velocity at the annulus and v_{sl} is the cuttings' particle slip velocity. The fluid velocity is calculated dividing this flow rate by the annulus area.

This computation must be considered with care since the area calculation does not take into account variation of the well diameters, which always occurs. In addition, the drilling cuttings may be contaminated with debris from shallower formations.

Thus, compared to well logging tools, the mud logging's drilling cutting characterization does not have a resolution in terms of depth. A sample assumed depth may be a few meters drifted from its actual depth, while the well logging tools have a precision of centimeters. Keeping the limitations of both tools in mind, the geologists and geophysics shall integrate both results.

Another recent issue is that diamond bits (impregnated bits) are used at very high rotation to increase the ROP. The problem is that this way of drilling pulverizes the drilling cuttings and/or the high temperature due to the high rotation causes metamorphism of the drilling cuttings, which may cause physical and/or chemical changes of the samples.

5.3.2.2 Characterization of the Samples of Drilling Cuttings

Once the drilling cuttings are collected, they are cleaned and classified in terms of the type of sedimentary rock, qualitative analysis of porosity and permeability, and possible presence of hydrocarbons, especially oil, in the porous media.

Inside the mud logging cabin (Figure 5.20) the samples are washed and sieved through a 5-mm mesh. Usually, the samples that remain on the sieve are discarded; only samples smaller than 5 mm are characterized (Whittaker, 1987).

Despite the oil-based drilling fluid being in disuse nowadays due to its environmental impact, in case the samples are covered with oil-based fluid, an industrial detergent or nonfluorescent solvent

TABLE 5.9
List of Geological Characteristics with which Each Constituent of the Samples Collected from Drilling Cuttings Should be Described

1.	Rock type
2.	Color
3.	Hardness (induration)
4.	Grain size
5.	Grain shape
6.	Sorting
7.	Luster
8.	Cementation or matrix
9.	Structure
10.	Porosity (qualitative)
11.	Inclusions

may be used to wash the samples (Whittaker, 1987). Then part of the samples are put in an oven to dry.

Inside the mud logging cabin, a trained geologist must characterize the samples using a stereoscope. The sample's constituents must be classified in terms of percentage, and then each constituent must be described as shown in Table 5.9. In the case of samples composed of carbonates, in addition to the visual description, a physical-chemical analysis may be required to identify the type of carbonate.

The next step is to identify the presence of hydrocarbon inside the porous space of the samples. Because the samples can be contaminated with grease, lubricants, or even the drilling fluid, it is necessary to make a crosscheck to confirm the presence of hydrocarbon (Whittaker, 1987).

First, the geologist may identify the presence of hydrocarbon inside the sample's pores during the visual characterization using the stereoscope. Second, a significative amount of unwashed sample is placed with water inside a hermetically closed blender. Then the sample is pulverized releasing the porous fluid in the water. After waiting for decantation of hydrocarbons, first, the gas inside the blender is conveyed to gas chromatography to identify and calculate the amount of each gas component. Then the water is inspected for the presence of oil. The third and last crosscheck is the ultraviolet light inspection box where the geologist looks for fluorescent oil droplets on the samples of drilling cuttings and diluted drilling fluid samples (Whittaker, 1987).

Recent researches are trying to make the characterization of the samples more quantitative using new technologies. Loermans et al. (2005, 2012) introduced the Advanced Mud Logging (AML) concept, which is a very comprehensive project, aiming to provide a complete, preliminary formation evaluation based only on mud logging data. The stereoscope shall be replaced by a microscope with high-resolution cameras to allow the use of image processing software and to share the images in real time with the onshore operator's office (Salim & Lagraba, 2018). In addition, several laboratory measurement techniques should adapt for the use in a mud logging cabin. For example, the measurement of the grain density with a gas displacement pycnometer or the use of X-ray Computed Tomography and Nuclear Magnetic Resonance to measure the porosity and permeability of samples. Even the mineralogy of samples can be measured using X-ray Diffraction and X-ray Fluorescence (Loermans et al., 2012; Salim & Lagraba, 2018).

TABLE 5.10
Sensor's Location and The Drilling Parameters Measured Directly

Location of the Sensor	Measured Drilling Parameters
Mud pump (Figure 5.22)	Quantity and rate of the strokes
Hook (Figure 5.23)	Position
Dead line (Figure 5.24)	Drilling cable tension
Top drive	Rotation and torque
Stand pipe manifold	Pressure
Choke line manifold (Figure 5.25)	Pressure
Cement line	Pressure
Drilling fluid return line connecting the diverter to the shale shakers (Figure 5.26)	Drilling fluid return flow rate
Drilling fluid tanks (Figure 5.27)	Drilling fluid density, temperature, resistivity, level position
Heave compensation system	Platform's vertical displacement due to waves

TABLE 5.11
Calculated Drilling Parameters and Required Data

Calculated Drilling Parameters	Required Data
Bit and well depth	Length of the drillstring and position of the hook
Weight on the bit (WOB)	Hoisting system configuration and tension on the drilling cable
Rate of penetration (ROP)	Hook position
Drilling fluid injection flow rate	Mud pump's configuration and stroke rate
Volume of drilling fluid in the tank	Tank configuration and level position
Lag time	Well and drillstring configurations, and the drilling fluid flow rate

5.3.3 Well Monitoring and Drilling Parameters Recording

5.3.3.1 Measurement of Drilling Parameters

The drilling rig has a set of sensors that measures the drilling parameters. The driller receives all these measured data in the driller's cabin (dog house) where the whole operation is controlled, and the well parameters are monitored to identify a kick or any other problem.

There are some parameters that are directly measured by the sensors (Table 5.10), while other parameters must be calculated using the system configuration and measured parameters (Table 5.11).

The drilling fluid measurements are carried out in both tanks; the one which receives the returning fluid, and the one that stores the fluid before it is pumped.

5.3.3.2 Well Monitoring

The mud logging cabin receives the same sensors signals and the drilling parameters available for the driller. Initially, the drilling parameters are plotted in the mud log report together with drilling cutting

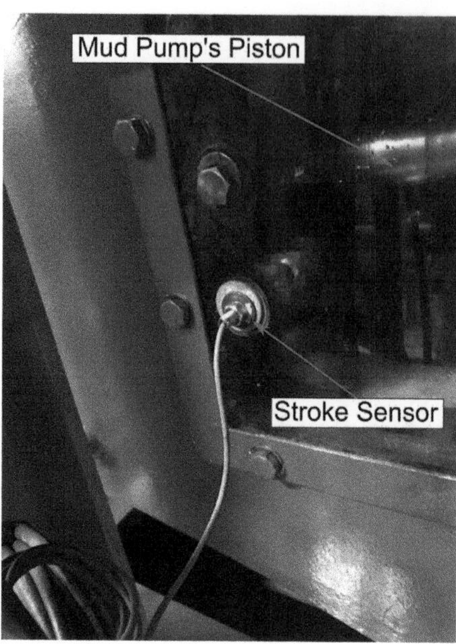

FIGURE 5.22 Sensor to measure the stroke of a single piston of mud pump. The sensor is integrated to the mud pump. *Source:* Courtesy Geophysical Surveying Co., Ltd.

FIGURE 5.23 The geolograph is an encoder-type sensor that measures the hook position measuring the length of a narrow wire which is connected to the top of the derrick. *Source:* Courtesy Geophysical Surveying Co., Ltd.

samples description and gas detection. However, the mud logging contractors have also started to offer other services, such as the well control monitoring and the assessment of pore pressure.

Well control is always the main issue during well construction. A kick, the undesired influx of formation fluid into the wellbore, must be detected and circulated out of the wellbore quickly and properly. Failure to detect and remove the kick may develop into a blowout which is the uncontrollable leakage of fluid formation into the environment. The blowout is the scariest and most severe top event of the industry.

In the past, the driller was the only one who monitored the drilling parameters to detect the occurrence of a kick. To reduce the risk of failure to detect the kick, a second human operator, the mud logger, monitors the same drilling parameters (Whittaker, 1987).

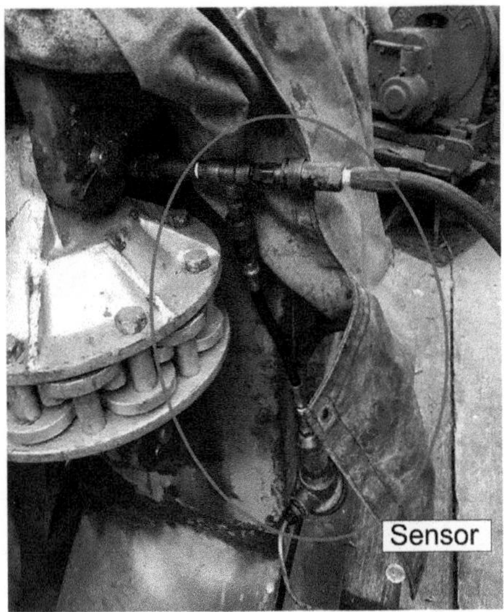

FIGURE 5.24 A load cell attached to the anchor of the deadline (anchored point of a drilling cable) measures the drilling cable tension. *Source:* Courtesy Geophysical Surveying Co., Ltd.

FIGURE 5.25 Pressure gauges are largely used in a drilling rig. The picture shows the pressure gauge installed in the choke manifold. *Source:* Courtesy Geophysical Surveying Co., Ltd.

FIGURE 5.26 Flowmeter to measure the returning drilling fluid flow rate. *Source:* Courtesy Geophysical Surveying Co., Ltd.

Another service of the mud logging is the assessment of pore pressure, which can be evaluated using the d-exponent (Jorden, Shirley, 1966), as shown in Eq. (5.4).

$$\mathrm{d} = \frac{\log\left(\dfrac{R}{60\,N}\right)}{\log\left(\dfrac{12\,W}{10^{6}\,D}\right)} \tag{5.4}$$

FIGURE 5.27 Sensors assembled on the drilling fluid tank to measure density, temperature, resistivity, and level of the drilling fluid. *Source:* Courtesy Geophysical Surveying Co., Ltd.

In this equation, d is the non-dimension drilling exponent, R is the penetration rate in ft/h, N is the rotation per minute (RPM), W is the weight on the bit in pounds, D is the bit diameter in inches.

The d-exponent is a normalized penetration rate, which was developed for the drilling performance analysis. It was realized that the d-exponent log plot (d-exponent vs. total depth) has a similar shape of the resistivity log. Thus, this log could be used to assess the pore pressure gradient using Eaton's method (Eaton, 1975).

The assessment of pore pressure in real time is especially important during the drilling of exploratory wells. For these "wildcats" wells, the geopressures (pore pressure, fracture pressure, and overburden stress) are assessed before the drilling with limited data. Thus, it is a good practice to have an expert team onshore who receives the pore pressure, LOT, etc. data in real time to update the geopressure model. If necessary, this team can change drilling parameters, drilling fluid parameters, or even the well design to reduce the risk of geomechanical problems, kick occurrence, etc.

5.4 WELL CONTROL

All personnel engaged directly in the well construction operation or supervision personnel, such as the drilling supervisor, must have a WellSharp certification granted by the International Association of Drilling Contractors (IADC) or an International Well Control Forum (IWCF) equivalent certification. The certification is issued upon completing successfully a training regarding the well control operation including simulator practice.

Well control refers to identify the occurrence of a kick promptly and carry out the operations required to remove the kick from the well and take measures to avoid a new kick. This section will cover the difference between the kick and blowout, how to identify a kick and how to remove it from the well by the Driller's Method and the Wait and Weight Method. However, it does not cover underbalanced drilling, which has a different paradigm, where the well is drilled with a continuous kick along the permeable formations.

5.4.1 Kick and Blowout Definition

One of the functions of the drilling fluid is to exert a hydrostatic pressure on the formation wall in an open-hole section of the well. This hydrostatic pressure has a geomechanical application avoiding

TABLE 5.12
Classification of Blowouts

Classification	Description
Underground Blowout	The formation fluid flows uncontrollably from a reservoir into the well. Because the BOP is shut in at the wellhead, the fluid flows into a shallower permeable and/or fractured formation. Sometimes, because the invaded formation does not have a competent cap rock or if there is any fault, the invader fluid can move upward generating a seepage on the surface.
Subsea Blowout	The formation fluid flows uncontrollably from the reservoir into the well, then the well conveys the fluid until the seafloor. In this case, there is no drilling riser and the BOP was not able to shut the well or the blowout can flow behind the casing in case of a poor cement job.
Surface Blowout	The formation fluid flows uncontrollably from the reservoir into the well. The influx is conveyed to the surface through the drilling riser. In this case, the BOP was not able to shut the well. The drilling personnel can shut the diverter annulus diverting the blowout to the platform overboard. Such action avoids the influx of natural gas and oil into the rig floor, reducing the risk of explosion and fire.

Source: Santos (2013).

the collapse of the well. In addition, the most important is that this hydrostatic pressure must overcome the pore pressure of the formation (overbalanced drilling) to avoid the influx of formation fluid into the wellbore.

When the wellbore pressure is lower than pore pressure and the formation has enough porosity and permeability, the formation fluid will flow into the well. The formation fluid can be natural gas, oil, brine, or any mixture of such fluids. If this influx is undesired, it is called a kick. It is important to highlight that the kick is undesired in most of well construction phases. However, during a Drill Stem Test (DST), the wellbore pressure is reduced to allow the influx. Furthermore, the data flow rate vs. pressure is measured during this test. In the case of DST, the influx is desired and the fluid from the formation is brought up to the surface in a controllable way.

As explained before, the drilling crew is trained to identify the occurrence of an undesired influx of formation fluid, or kick. Once the kick is identified, the well construction operation is suspended, and the well control operation starts. Thus, the occurrence of a kick implies in NPT increasing the costs of the well construction. If the crew fails in the kick identification, the kick flows upward, usually through the annulus between the drillstring and the casing or well wall. In the case of natural gas kick, the gas accelerates when flowing upward due to the gas expansion.

If the kick reaches the environment uncontrollably, it is called blowout. Because the environment has a broad meaning, we can classify three different blowouts as featured in Table 5.12.

5.4.2 Kick Identification

It is important to emphasize that the drilling crew must identify the kick promptly. The delay in its identification results in a larger volume of kick that means the well and well control equipment will face higher pressures.

Table 5.13 features the symptoms that indicate the occurrence of a kick. If any of these symptoms are confirmed, the crew must suspend the normal operation and start the well control operation. The problem is that all symptoms require the measurement of volume or flow rate of drilling fluid, and this measurement is carried out on board of a floating unit, which is a non-inertial reference frame

TABLE 5.13
Primary Symptoms of a Kick

Symptoms	Description
Increased volume of drilling fluid in the tanks	Before the drilling operation starts, the total volume of drilling fluid in the tanks is measured. It is normally expected that the volume drops when the well is drilled. An increasing volume of drilling fluid in the tanks means that more fluid is flowing into the circulation system, which means a kick is occurring. A special attention is required because the mud transfer between active tank and reserve pit are very common so this transfer has to be accounted.
Increased flow rate of the drilling fluid returning from the well	During the drilling operation, it is normally expected that the drilling fluid flow rate injected on the top of the drillstring added to the flow rate injected on the top of the booster line should be close enough to the flow rate of the drilling fluid returning from the well. A return flow rate higher more than 10% of the injection flow rate is interpreted as a kick occurrence. However, the measurement of the return flow rate, using a simple pad, is not accurate unless a radioactive sensor is installed.
Returning drilling fluid with pump off	When the pump is off (e.g., making a connection of a stand of drill pipe to the top end of the drillstring), if there is returning drilling fluid, usually seen in the trip tank, it means another fluid in the subsurface is pushing up the drilling fluid.
During the tripping, mismatch of the volume of drill pipe removed from drillstring and the volume of drilling fluid used to refill the well	During the tripping (disassembling the drillstring), the mud pump is not connected to the drillstring. The pipes are pulled out using the elevator. Thus, when the drilling crew removes a stand of drill pipe from the drillstring, it is necessary to refill the well with the same volume of this stand to keep the hydrostatic pressure within the wellbore. Figure 5.28 shows a schematic of the trip tank system, which refills the well during the tripping. If the volume drop of the trip tank is smaller than the stand volume, it means an influx is filling the well.

Source: Santos (2013).

because of its oscillatory motion in all degrees of freedom mainly caused by the ocean waves, current, and wind.

For example, the volume of drilling fluid in the tanks is continuously measured using an ultrasound sensor that measures the position of the fluid surface. Due to a poor performance of such ultrasound sensor in a scenario of fluid emulsion, another type of sensor is used in parallel to measure the drilling fluid height. However, the tank has a wide area, which means a gain of volume may result in a very small variation of the surface position. Further, the whole platform is oscillating due to the ocean waves, which also generates waves on the surface of the drilling fluid in the tanks. Both conditions make it harder to identify the kick occurrence when its volume is still low.

The monitoring of the drilling parameters related to the kick symptoms (injection flow rate, returning flow rate, volume in tanks, etc.) is so important that there are always two human operators onboard watching such parameters, namely, the driller and the mud logger (see Section 5.3). Sometimes, a third human operator is monitoring the same parameters from a base onshore.

When the kick diagnosis is inconclusive due to a small variation in any of such parameters, a flow check is required. In the flow check, the normal operation is suspended, the drillstring is stopped, the mud pumps are off, the returning drilling fluid is diverted into the trip tank, and the volume in the trip tank is observed for, at least, 15 minutes. It is similar to the schematic in

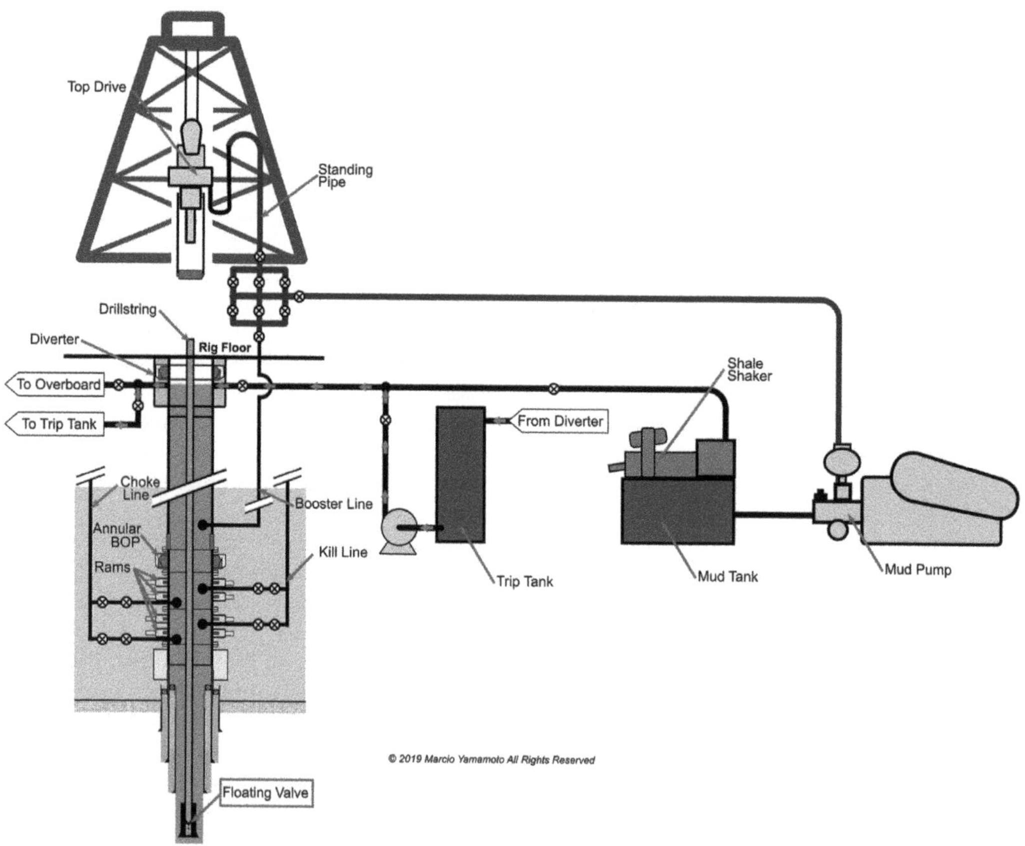

FIGURE 5.28 The schematic showing the trip tank keeping the well filled with drilling fluid during the tripping operation. The same trip tank is used for the flow check to identify the occurrence of a kick.

Figure 5.28. Because the trip tank has a much smaller surface area, a small gain of volume can be identified easier.

5.4.3 Well Control Operations

Once the kick is detected, the BOP must be shut down, pressures be read (SICP—shut in casing pressure and SIDPP—shut in drill pipe pressure), the gained volume of tank determined, and the kick must be circulated out of the wellbore. However, there are several methodologies of well control. For example, the well can be shut "hard" (with the choke valve closed) or "soft" (choke valve open) (Jardine et al., 1993). Further, there are several methodologies to circulate out the kick, but the Driller's Method and the Wait and Weight Method are widely used (Grace, 2003). The operator must choose which methodologies shall be used during the well control operations, which usually are included in the clauses of a contract with the drilling contractor. Further, all personnel involved in the drilling operation must be trained regarding the actions to take for the chosen methodologies. Following, the two main methodologies, the Driller's Method and the Wait and Weight Method, will be presented.

5.4.3.1 Driller's Method

In this explanation, let assume that the kick was confirmed after a flow check and the choke valve is already closed (hard shut-in) and the mud pump is off.

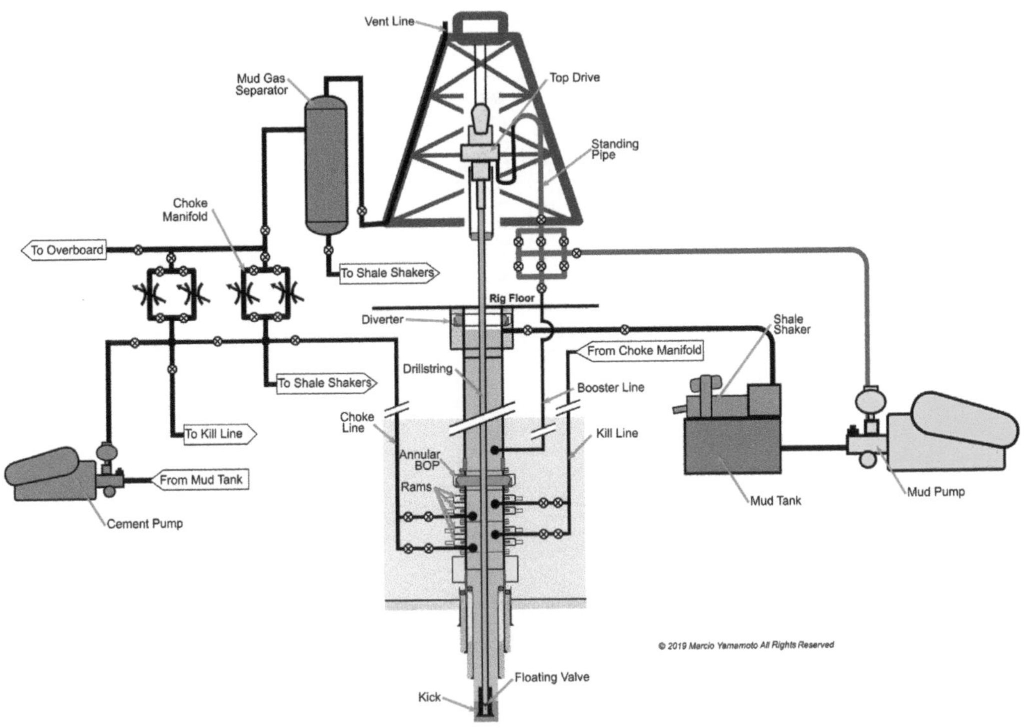

FIGURE 5.29 The schematic showing the moment when the annular BOP shuts the well.

First, the annular BOP must be closed and valves to access the choke line must be opened. The kick will keep flowing into the well pressurizing the wellbore (Figure 5.29). When the bottom wellbore pressure matches the pore pressure, the inflow stops and it necessary to record the standing pipe pressure (SIDPP), choke manifold pressure (SICP), and the gained volume of the trip tank. All these data are used to assess the actual pore pressure and the volume of the kick (Grace, 2003; Santos, 2013). The actual pore pressure is important information which is used to calculate the density of heavier drilling fluid to overbalance the pore pressure. The whole well, including the drilling riser, must be filled with this new drilling fluid to restart the drilling operation avoiding a new kick.

The next step is to circulate out the kick as shown in Figure 5.30. The choke valve is opened and the flow from the choke manifold is diverted to the mud gas separator. This separator has two outputs, the gas output is connected to the vent line that discharges the gas on the top of the drilling derrick, and the drilling fluid output is connected to the shale shakers. Then, the mud pump is turned on, circulating at a reduced flow rate, which will be constant during the whole well control operation.

During the circulation, the bottom wellbore pressure must be kept above the actual pore pressure to avoid another kick, controlling the opening of the choke valve located in the choke manifold (Figure 5.30). The bottom pressure is assessed using the measured pressure at the standing pipe because the fluid within the drillstring is single-phase with a known density. Since the flow rate is kept constant, the bottom pressure is controlled actuating on the choke valve.

From the BOP, the kick is conveyed through the choke line until the choke manifold located on the platform. In Figure 5.30, we can observe that the kill line is parallel to the choke line and has its own choke manifold interconnected to the mud gas separator. The kill line is used to inject a heavier drilling fluid into a well when there is no drillstring. However, we can assume the kill line as a redundancy to the choke line and it serves as pressure control line during the well control operation.

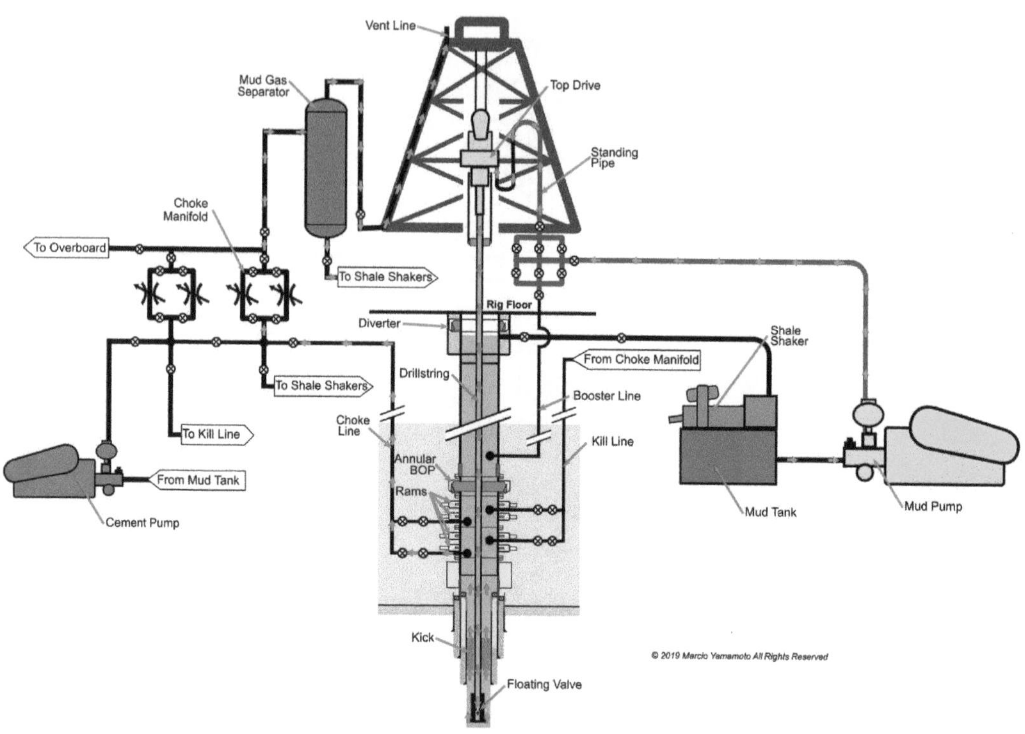

FIGURE 5.30 The schematic showing the kick being circulated out of the well using the Driller's Method.

Once the kick is conveyed into the mud gas separator, the gas is released into the atmosphere through the vent line and the fluid returns to the shale shakers. The last part of the Driller's Method is to circulate and refill the whole well, including the drilling riser, with heavier drilling fluid (Figure 5.31). The density of this heavier fluid is calculated to overbalance the measured pore pressure.

5.4.3.2 Wait and Weight Method

This method is also called Engineer's Method. Similar to the previous, let us assume that the kick was confirmed after a flow check and the choke valve is already closed (hard shut-in) and the mud pump is off.

When the kick is identified, the proper valves are closed as explained in Section 5.4.3.1. The kick inflow stops when the bottom wellbore pressure matches the pore pressure, then the standing pipe pressure (SIDPP), choke manifold pressure (SICP), and the gained volume of the trip tank are recorded. The pore pressure is assessed using the recorded data and the density of the new heavier drilling fluid is calculated.

Next, the new drilling fluid is prepared while the kick waits in the wellbore (Figure 5.32). Usually the kick has a lower density than the drilling fluid. Therefore, the kick starts to flow upward in a relative slow velocity due to the drilling fluid viscosity.

After the heavier drilling fluid is prepared, it is pumped into the well bottom throughout the drillstring (Figure 5.33). The kick is conveyed up to subsea BOP using the reduced flow rate. During the circulation, the well bottom pressure is monitored using the pressures measured at the standing pipe and choke manifold. Whenever is necessary, the choke valve must be adjusted to control the well bottom pressure.

From the subsea BOP, the kick is conveyed by the choke line up to the choke manifold located on the drilling platform. Then, the kick is conveyed to the mud gas separator. The gas is discharged through the vent line and drilling fluid is conveyed to the shale shakers.

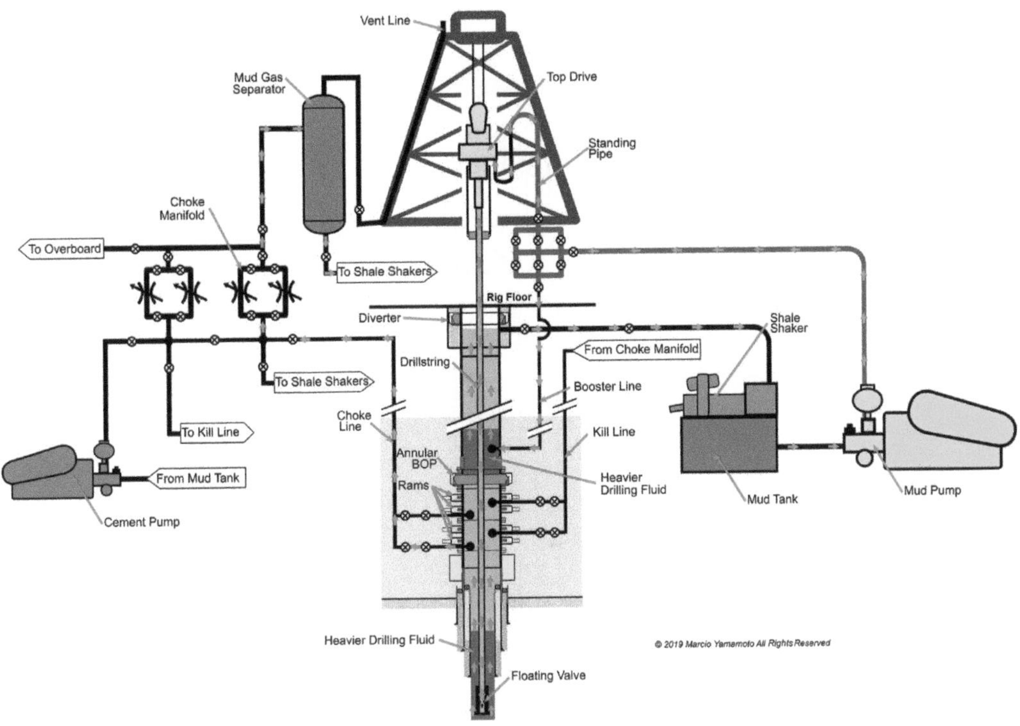

FIGURE 5.31 Filling the well and drilling riser with a heavier drilling fluid to avoid another kick.

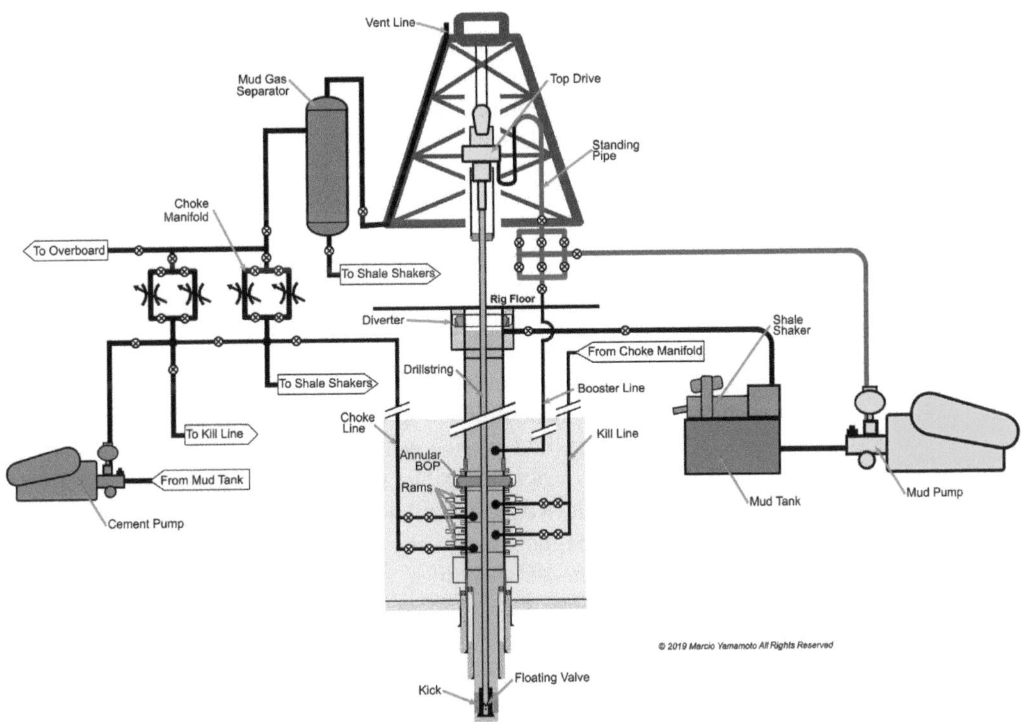

FIGURE 5.32 The kick waits in the wellbore while the heavier drilling fluid is prepared to circulate the kick out of the well.

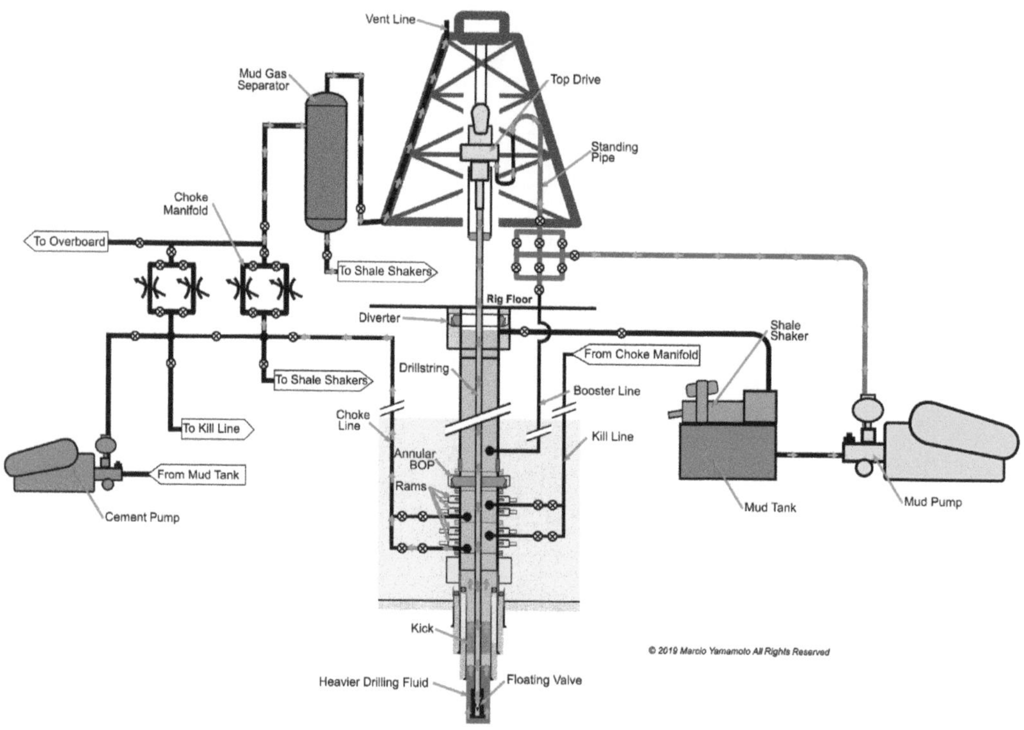

FIGURE 5.33 The schematic showing the kick being circulated out the well using the Wait and Weight Method.

When the kick reaches the mud gas separator, the whole well, subsea BOP, and choke line are already filled with the heavier drilling fluid. Thus, the last step of the Wait and Weight Method is to fill the drilling riser with the heavier fluid. Figure 5.34 shows the drilling fluid is pumped at the bottom of the drilling riser though the Booster Line.

5.5 FUTURE TECHNOLOGIES

5.5.1 Riserless Drilling Using Reverse Circulation Technology

Riserless drilling technologies have a limited capacity due to the subsea pump that needs to pump drilling fluid with cuttings (solids), and it is not a proven a technology yet. However, using current available proven technologies riserless drilling can become possible. One possibility is using the reverse circulation drilling. This way there is no problem to bring cuttings to the surface (drilling rig) by using a flexible line to pump drilling fluid into the well annulus and retrieve it with cuttings from the internal part of the drillstring. To make it feasible, two technologies shall be developed. One is a subsea version of rotating control device introduced with MPD to seal the well annulus. The other is the rotating fluid circulating device (gooseneck; kelly hose; and stand pipe manifold, or top drive) on the surface (rig) that supports both solids (cuttings) and well control with flow back fluids.

5.5.2 Hydraulic Jet Drill Bit

Current bit technology relies only on mechanical cutters and fluid hydraulics to clean up the bit cutters. This current drill technology has problems when drilling very abrasive formations (like those that contain silex in their composition) due to very fast wear of bit cutters. In this environment, hydraulic jet cutting technology available in other industries can be transferred to drill an oil and gas well.

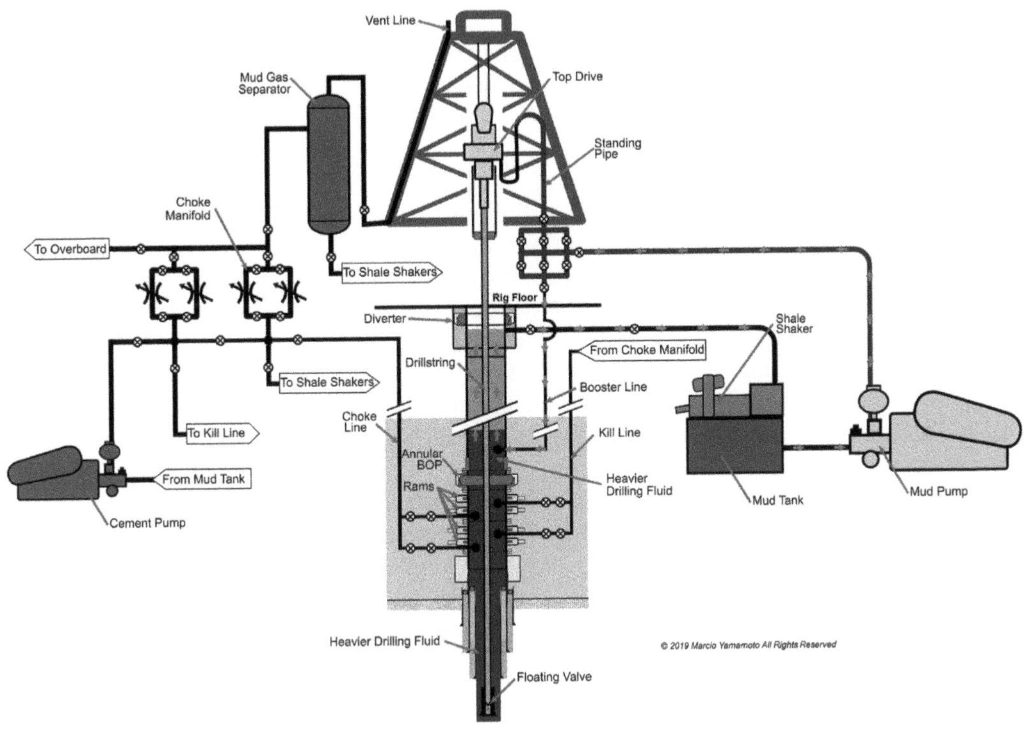

FIGURE 5.34 Filling the drilling riser with a heavier drilling fluid.

5.5.3 Subsea Well Design for Well Intervention Robot

Currently "intelligent" or "smart" completion technology is the focused of the oil and gas industry but looking it from a field maintainability point of view, it is not the solution for the majority of the production loss problems. It solves only two problems (water cut and gas cut problem) of 21 mapped production loss problems. Also, as it relies on mechanical equipment, it needs constant verification test and maintenance that, in most cases, shall lead to costly heavy workover.

The well intervention robot technology introduced in the North Sea relies on riserless subsea well interventions that are considered the cheapest way to maintain well productions. But for this technology to dominate as an approach of well maintenance, the well itself must be designed to enable intervention by these robots.

5.6 SUMMARY/CONCLUSIONS

The construction of a subsea well in deepwater is a complex endeavor that requires interdisciplinary knowledge including geology, geophysics, drilling engineering, offshore engineering, subsea engineering, etc. Further, several important topics were not included in the text, such as drilling fluid, dynamic behavior of rotating drillstrings, etc.

The major challenge in ultra-deep drilling is the narrow operating drilling window or the difference between fracture and pore pressure gradients. The industry has invested a lot of money to develop new technologies to address this challenge. We presented one of these technologies, the MPD that allows more accurate control of well bottom pressure, including compensating the pressure when the mud pump is off.

Furthermore, we presented the current state-of-the-art of the subsea well construction giving details about the spud-in and drilling with the return of fluid, equipment, mud logging (a very important topic that is never included in textbooks), and well control.

ACKNOWLEDGMENTS

The authors would like to thank Heitor Lima, Texas A&M University, College Station, TX; João Carlos Ribeiro Plácido, Mechanical Engineering Department, DEM PUC-Rio, Brazil; and Shiniti Ohara, Vice President Operations, Barra Energia, Rio de Janeiro, RJ, Brazil, for reviewing this chapter and providing valuable technical comments and suggestions.

The Acknowledgments extends to Tsuyoki Fujii, Geophysical Surveying, Co., Ltd., who shared the pictures of mud logging equipment, and Mr. Breno Carbognin, a former student who prepared the figures regarding jack-up MODU and guideline drilling. We also acknowledge Dr. Katsuya Maeda (National Maritime Research Institute, Japan) and Mr. Norihito Inada (Japan Oil, Gas and Metals National Corporation) for their support.

The first author also acknowledges the experienced Brazilian well engineers who have shared their knowledge with the academia, helping the academic training of a new generation of petroleum engineers.

IN MEMORIAM

Dr. Kazuo Miura passed away in April 2020, after fighting several years against cancer. Dr. Miura's career started in Petrobras at the beginning of the 1980s where he served as Completion Engineer and Researcher. For more than 20 years, Dr. Miura worked in the most challenging completion projects in the Campos Basin. He also trained several generations of Petrobras young petroleum engineers. Late in his career in Petrobras, he was a researcher in the Petrobras Research Center where he advocated for BIS (Safety Barrier Integrated Set), a new philosophy for well safety barriers. The framework of the BIS was established in his PhD studies. His activism on BIS reverberated outside Petrobras and, in 2017, the Brazilian National Agency of Petroleum, Natural Gas, and Biofuel (ANP) demanded the use of BIS for the whole service life of all wells in Brazil. After his retirement from Petrobras, he held academic positions at the University of Tokyo (Japan) and University of Campinas (Brazil). His long career and service to Petroleum Engineering, together with his devotion to the training of the younger generation of professionals in Brazil and Japan, deserve our recognition. He is survived by his wife Teresa, daughters Olivia and Mariana, and son Bernardo.

NOTES

1. In 2010, the Perdido platform was installed in 2383 m (7817 ft) of water depth, which remains the world's deepest spar (Barton et al., 2018).
2. The 8.3 ppg is a reference value of fresh water density. In practice, each sedimentary basin may have a different density as normal pressure.
3. In this chapter, we refer to subsea BOP as the set of the BOP stack and the LMRP.
4. LWD is a set of geophysical sensors that measure the formation properties during the drilling operation.
5. MWD is a set of sensors that measure the wellbore path during the drilling. It is used to measure the well deviation from the vertical and the wellbore geometry in the directional and horizontal drilling.
6. During the spud-in, seawater is used as drilling fluid.
7. Bottom-up: the operation that pumps a drilling fluid volume, which corresponds to the annulus volume, in the way that drilling fluid at the well bottom flows back to the platform on the top.
8. Drilling break: when the rate of penetration (ROP) has a rapid increase after the bit penetrates a different formation. The drilling break usually occurs at the boundary between a cap rock formation and a reservoir. The cap rock, which is located above the reservoir, is a compacted formation, meaning it is harder to drill (low ROP). On the other hand, the reservoir, especially sandstone, is less compacted and easier to drill (high ROP).

REFERENCES

Agostini, C., Yamamoto, M., Miura, K. (2017), "The application and functionalities of a wet Christmas tree applied in Santos Basin Pre-Salt Cluster", *Proceedings of the JASNAOE Annual Autumn Meeting*, Hiroshima, November 27–28, 2017.

Akers, T. J. (2008), "Jetting of structural casing in deepwater environments: Job Design and Operational Practices", *SPE Drilling & Completion, SPE-102378-PA*, March 2008, pp. 29–40.

Al-Awadhi, F. K., Al Ameri, F., Kikuchi, S., Afifi, H. A. H. (2014), "Making un-drillable HPHT well drillable using mud cap drilling", *Proceedings of the Abu Dhabi International Petroleum Exhibition and Conference, Paper SPE-171865-MS*, Abu Dhabi, November 10–13, 2014.

ANP (2017), "ANP Resolution No. 670/2017".

API (1998), "API-RP-7G Recommended Practice for Drill Stem Design and Operating Limits", American Petroleum Institute, Washington, DC.

Barton, C., Hambling, H., Albaugh, E. K., Mahlstedt, B., Davis, D. (2018), "Deepwater Solutions & Records for concept selection", *Offshore Magazine*, Houston, TX, May 2018.

Beck, R. D., Jackson C. W., Hamilton, T. K. (1991), "Reliable Deepwater Structural Casing Installation using controlled jetting", *Proceedings of the 66th Annual Technical Conference and Exhibition, Paper SPE-22542-MS*, Dallas, TX, October 6–9, 1991.

Bellarby, J. (2009), "*Well Completion Design*", Elsevier, Amsterdam.

Bourgoyne, A. T., Jr., Millheim, K., Chenevert M. E., Young, F. S., Jr. (1991), "Applied Drilling Engineering", *Society of Petroleum Engineers*, Richardson, TX, ISBN 1-55563-001-4, 2nd print, 502 pages.

Butler, B., Larralde, E. (1971), "Motion compensation on drilling vessels", *Proceedings of the Offshore Technology Conference, Paper OTC-1335*, Houston, TX, April 19–21, 1971.

Bysveen, J., Fossli, B., Stenshorne, P. C., Skärgård, G., Hollman, L. (2017), "Planning of an MPD and Controlled Mud Cap Drilling CMCD Operation in the Barents Sea using the CML technology", *Proceedings of the IADC/SPE Managed Pressure Drilling & Underbalanced Operations Conference & Exhibition, Paper SPE/IADC-185286-MS*, Rio de Janeiro, March 28–29, 2017.

Davies, D. R., Hartog, J. J. (1981), "Foamed cement—a cement with many applications", *Proceedings of the Middle East Oil Technical Conference, Paper SPE 9598*, Manama, March 9–12, 1981.

Eaton, B.A. (1975), "The equation for geopressure prediction from well logs", *Proceedings of the Fall Meeting of the Society of Petroleum Engineers of AIME, SPE-5544-MS*, Dallas, TX, September 28–October 1, 1975.

Grace, R. D. (2003), "*Blowout and Well Control Handbook*", Gulf Professional Publishing, Burlington, ISBN 978-0128126745, 472 pages.

Guo, Y., Ji, S., Ye, A. (2006), "New device for drilling in deepwater under semi-submersible: ABS unit", *Proceedings of the IADC/SPE Asia Pacific Drilling Technology Conference and Exhibition, Paper SPE-101349-MS*, Bangkok, November 13–15, 2006.

Haaø, J., Vangen, S., Tyapin, I., Choux, M., Hovland, G., Hansen, M. R. (2012), "The effect of friction in passive and active heave compensation of crown block mounted compensators", *Proceedings of the 2012 IFAC Workshop on Automatic Control in Offshore Oil and Gas Production*, Trondheim, May 31–June 1, 2012.

Hall, D. L., Roberts, R. D. (1984), "Offshore drilling with preformed stable foam", *Proceedings of the SPE California Regional Meeting, Paper SPE 12794*, Long Beach, CA, April 11–13, 1984.

ISO (2005), "Petroleum and natural gas industries—Design and operation of subsea production systems—Part 1: general requirements and recommendations", ISO 13627-1.

Jardine, S. I., Johnson, A. B., White, D. B., Stibbs, W. (1993), "Hard or soft shut-in: which is the best approach?", *Proceedings of the SPE/IADC Drilling Conference, Paper SPE-25712-MS*, Amsterdam, February 22–25, 1993.

Jeanjean, P. (2002), "Innovative design method for deepwater surface casing", *Proceedings of the SPE Annual Technical Conference and Exhibition, Paper SPE-77357*, San Antonio, TX, September 29–October 2, 2002.

Jorden, J. R., Shirley, O. J. (November 1966), "Application of drilling performance data to overpressure detection", *Journal of Petroleum Technology*, 18(11), 1387–1394.

Kaculi, J. T., Witwer, B. J. (2014), "Subsea wellhead system verification analysis and validation testing", *Proceedings of the Offshore Technology Conference, Paper OTC-25163-MS*, Houston, TX, May 5–8, 2014.

Keener, C., Keji-Ajayi, I., Allan, R. (2003), "Performance gains with 5th generation rigs", *Proceedings of the SPE/IADC Drilling Conference*, Paper SPE-79833-MS, Amsterdam, February 19–21, 2003.

Kutchko, B., Crandall, D., Spaulding, R., Gill, M., Moore, J., Gieger, C., Haljasmaa, I., Benge, G., Maxson, J. (2016), "A look at processes impacting foamed cements", *Proceedings of the SPE Deepwater Drilling & Completions Conference, Paper SPE-180337-MS*, Galveston, TX, September 14–15, 2016.

Liu, C., Zhang, N., Guo, B., Ghalambor, A. (2010), "An investigation of heavy-foam properties for offshore drilling", *Proceedings of the SPE Annual Technical Conference and Exhibition, Paper SPE-132464-MS*, Florence, September 19–22, 2010.

Loeffler, N. R. (1984), "Foamed cement: a second generation", *Proceedings of the 1984 Permian Basin Oil & Gas Recovery Conference, Paper 12592*, Midland, TX, March 8–9, 1984.

Loermans, T., Kanj, M., Bradford, C. (2005), "Advanced mud logging: from ARCHIE'S DREAM to reality", *Proceedings of the SPE Technical Symposium of Saudi Arabia Section, Paper SPE-106324*, Dhahran, Saudi Arabia, May 14–16, 2005.

Loermans, T., Kimour, F., Bradford, C., Bondabou, K., Marsala, A. (2012), "Successful pilot testing of integrated advanced mud logging unit", *Proceedings of the SPWLA 53rd Annual Logging Symposium, SPWLA-2012-184*, Cartagena, Colombia, June 16–20, 2012.

Nogueira, E. F., Borges, A. T., Junior, C. J. M., Machado, R. D. (2005), "Torpedo Base—a new conductor installation process", *Proceedings of the Offshore Technology Conference, Paper OTC-17197-MS*, Houston, TX, May 2–5, 2005

Offshore Magazine, (1999), "Active heave drilling drawworks system goes to work", https://www.offshore-mag.com/drilling-completion/article/16757687/active-heave-drilling-drawworks-system-goes-to-work.

Paula, R. R., Fonseca, D. R. (2013), "Emergency Disconnection Guidelines", *Proceedings of the Offshore Technology Conference Brazil, Paper OTC-24418-MS*, Rio de Janeiro, October 29–31, 2013.

Pixler, B. O. (April 1961), "Mud Analysis Logging", *Journal of Petroleum Technology*, Vol. 13, No. 4, 323–326.

Salim, A., Lagraba P. J. O. (2018), "Utilizing drill cuttings to enhance characterization and description of tight carbonate reservoirs". *Proceedings of the SPE Annual Technical Conference and Exhibition, SPE-191662-MS*, Dallas, TX, September 24–26, 2018.

Santos, O. L. A. (2013), "*Well Safety in the Drilling*", Sao Paulo: Blucher and Petrobras (*in Portuguese*).

Silvio, A., Ruiz J. C., Pereira M. F. (2013), "Deepwater conductor pre-installation for first TLWP in Brazil", *Proceedings of the Offshore Technology Conference Brazil, Paper OTC-24291-MS*, Rio de Janeiro, RJ, October 29–31, 2013

Stave, R. (2014), "Implementation of dual gradient drilling", *Proceedings of the Offshore Technology Conference, Paper OTC-25222-MS*, Houston, TX, May 5–8, 2014.

Tsukada, R. I., Morooka, C. K., Yamamoto, M. (2007), "A comparative study between surface and subsea BOP systems in offshore drilling operations", *Brazilian Journal of Petroleum and Gas*, Vol. 1, No. 2, pp. 88–94.

Whittaker, A. H. (1987), "Mud Logging", in *Petroleum Engineering Handbook*, Chapter 52, SPE, Richardson, TX.

Williams, G. D. (2015), "Saving time and reducing risk with subsea wellhead system running and test tools", *Proceedings of the Offshore Technology Conference, Paper OTC-25906-MS*, Houston, TX, May 4–7, 2015.

Woodacre, J. K., Bauer, R. J., Irani, R. A. , "A review of vertical motion heave compensation systems"(2015), *Ocean Engineering*, Vol. 104, pp. 140–154.

Woodall-Mason, N., Tilbe, J. R. (1976), "Value of heave compensators to floating drilling", *Journal of Petroleum Technology*, Vol. 28, No. 8, pp. 938–946.

Yamamoto, M., Morooka, C. K. (2005), "Dynamic positioning system of semi-submersible platform using fuzzy control", *Journal of the Brazilian Society of Mechanical Sciences and Engineering*, Vol. 27, No. 4, pp. 449–455.

Yilmaz, M., Mujeeb, S., Dhansri, N. R. (2013), "A H-infinity control approach for oil drilling processes", *Procedia Computer Science*, Vol. 20, pp. 134–139.

Ziegler, R., Ashley, P., Malt, R. F., Stave, R., Toftevag, K. R. (2013), "Successful application of deepwater dual gradient drilling", *Proceedings of the IADC/SPE Managed Pressure Drilling and Underbalanced Operations Conference and Exhibition, Paper IADC/SPE 164561*, San Antonio, TX, April 17–18, 2013.

6 Recent Innovations in Drilling in Ice

Mary R. Albert,
Dartmouth College, Hanover, NH
Kristina R. Slawny, Grant V. Boeckmann, Chris J. Gibson, Jay A. Johnson,
University of Wisconsin, Madison, WI
Keith Makinson, and Julius Rix
British Antarctic Survey, Cambridge, UK

CONTENTS

6.1 INTRODUCTION

The need for scientific ice drilling in glaciers and ice sheets has been driven by many fields of science, including drilling ice cores for evidence of past environment and paleoclimate information, and drilling access holes through the ice to gather data relevant to glacial dynamics, history of glacier extent, sediment sampling, and discovery of ecosystems within and beneath the ice. Many nations have contributed to drilling technologies relevant to each of these fields, and developments in any one nation often build on prior designs from other nations. A description of the very early polar ice coring endeavors in Greenland and Antarctica is provided in Langway (2008). Ice drilling and coring technologies that were developed before 2008 are well described in Bentley et al. (2006), including a wide array of ice coring drills, drills designed to create holes in ice only, and autonomous instruments that melt their way through ice. The text by Talalay (2016) provides a review of mechanical ice drilling technology that includes design, parameters, and performance of an assortment of tools and drills for making holes in snow, firn, and ice. Described in detail are direct-push drilling, hand- and power-driven portable drills, percussion drills, conventional machine-driven rotary drill rigs, flexible drill-stem drill rigs, cable-suspended electromechanical auger drills, cable-suspended electromechanical drills with bottom-hole circulation, and drilling challenges and perspective for future development.

In this chapter our goal is to describe new ice drilling and coring technologies that have been designed, built, and used in the field in the most recent decade. Some of these technologies are improvements on prior drills, while other technologies such as a replicate ice coring drill, geologic drilling underneath many meters of glacial ice, and the rapid access isotope drill are the first of their kind. There are many additional ice drilling and sampling designs currently in the design or development stage that are not included in this chapter; rather our goal in this chapter is to describe proven ice drilling technologies that have been developed since 2009.

6.2 ICE CORING

The need to retrieve cores from ice is driven primarily by the natural archive of past climate and environment that is embedded in the physical, electrical, and chemical properties of the ice. In very cold polar and high-altitude regions where snow rarely melts, glaciers and ice sheets are the result of snowfall and ice crystals metamorphism over thousands of years. The snow deposition events result in layering that remains evident on seasonal time scales in high-accumulation areas to decadal time scales in low-accumulation areas. As snow continues to accumulate on the surface, underlying snow that is more than one year old, termed "firn," continually compacts, sinters, and undergoes metamorphosis until finally at depths of pore close-off, which vary from approximately 60–120 m, the interstitial pores have become closed to the atmosphere and the medium becomes solid ice with air trapped in bubbles within the ice. Deep in the ice the trapped air is forced into clathrates. The layering and the structural differences between firn, bubbly ice, and clathrate ice impacts the drilling and coring of ice. For retrieving ice cores for evidence of past environment and climate, it is almost always important that the drilling process retrieves full-diameter, intact cores.

Early and historical ice coring efforts have been documented previously, e.g. Bentley et al. (2006), Langway (2008), Talalay (2016), and improved techniques continue to be developed. This chapter describes recent innovations in ice coring for drills that have demonstrated success in the field and thus have field performance data. One advance within this decade has been the development of a system for doing replicate ice coring at multiple targeted depths while still allowing the main borehole to be amenable to borehole logging systems.

6.2.1 Replicate Coring

The driving scientific interest in retrieving replicate ice cores is to obtain additional ice from depths where the ice contains evidence from significant environmental or climatic events. Often the scientific demand for such ice is higher than for the ice from the overall core; replicate coring can satisfy the demand without the need to drill an entire duplicate ice core. Results from Russian drilling at Vostok in the 2008–2009 field season showed that by repeatedly running the drill up and down around a depth in a slanted (6°) borehole, a shelf-like indentation could be made on the downhill side of the borehole that could serve as the start of a replicate core. It is often the goal in ice coring to have the core drilled as close as possible to vertical, so there are instances where the inclination of the borehole alone may not be sufficient to initiate a replicate core at a specified depth. In rock coring sometimes whipstocks are used to initiate deviatory cores, but the presence of the whipstock may impede or prevent the travel of borehole logging sensors that are used in boreholes after the core has been drilled. In addition, the geologic and petroleum industry has shown the utility of "steerable" geological drilling. To address the need to drill a replicate core on the high side of the borehole to facilitate subsequent borehole logging, and also to avoid the expense of developing a fully steerable drilling system for ice coring, the use of actuators to force the sonde to the high side of the borehole was successfully pursued by the US Ice Drilling Program between 2010–2013. The drill to which replicate coring capability was developed was the existing Deep Ice Sheet Coring (DISC) Drill, an electromechanical system designed to take 122-mm-diameter ice cores from the main borehole and 108-mm-diameter replicate ice cores to depths of 4000 m. The DISC Drill performed very well in drilling the main ice core to a depth of 3405 m in five field seasons at WAIS Divide, Antarctica, setting a new US deep ice drilling record (see Bentley et al., 2006, p. 241–244 for description of the DISC Drill). Replicate ice coring capability using a controlled sonde angle for the DISC Drill was designed, constructed, and deployed for the US Ice Drilling Program at the University of Wisconsin-Madison. The Replicate Coring System was developed and built in 2010–2011, tested in Antarctica during the 2011–2012 WAIS Divide field season and tested further in Madison, WI, during summer–fall 2012 (Johnson et al., 2014a). During the 2012–2013 Antarctic field season, the system produced five azimuth and depth-controlled deviations at four target depth levels. A total of 285 m of replicate ice core was recovered in the first coring of its kind. The entire main/replicate ice core, including ductile, brittle, and warm ice, had excellent quality and satisfied the needs of the ice-science community (Shturmakov et al., 2014).

6.2.1.1 Science Requirements of Replicate Coring for the DISC Drill

The following science requirements were formulated by scientists associated with the WAIS Divide ice core project and were used by the engineers as the target of the design and construction of the Replicate Coring System for the DISC Drill.

Core Characteristics.

- A minimum core length of 1 m is required, though 2 m is desirable.
- A target core diameter of 100 mm. It is desirable that the diameter does not vary by more than 3 mm, i.e. 97–103 mm core diameter is acceptable.
- Total replicate core collection up to 400 m.
- Ice pieces to fit together snugly without any gaps.
- Ability to determine the in situ orientation (azimuth) of core segments to within ±10°.
- It is desirable that the replicate core is within 0 m to 20 m of the parent borehole; however, somewhat larger deviations than 20 m are tolerable though not desired.

- Core to be collected with an angle of deviation of less than 20° from vertical, 10° or less is desirable.
- An amount of rubble during the initial deviation from the parent borehole is permissible. Note: Deviation from the parent borehole will require some amount of drilling or reaming that does not produce science-grade core. This is necessary in order to start the deviating borehole.

Hole Characteristics

- The parent borehole must remain open and usable after replicate coring has occurred. No reduction in diameter of the parent borehole is allowed (e.g. no permanent rings or whipstock may be used).
- Reaming of the parent borehole is permissible but should be kept to a minimum. Reaming should be less than 20 m in length, and result in a borehole diameter increase less than 100 mm. This is allowed for each deviation from the parent borehole.
- Replicate coring should be possible at any location of the borehole, starting 100 m below casing end.
- Damage to the parent borehole wall should be kept to a minimum, outside the reaming interval.
- It is required that deviation is performed on the "uphill" side of the borehole. This is to ease logging of the borehole.
- The borehole inclination must be measured with an accuracy of ±5°.
- The Replicate Coring System must be operational at a lowest temperature of −55°C.
- If any permanent equipment is to remain in the parent borehole it must have an ID equal to, or larger than, that of the parent borehole and not obstruct borehole continuity.

Drilling Fluid

- The drilling fluid must be the same, or compatible with, that used in the parent borehole.

6.2.1.2 Mechanical Design of the Replicate Coring System for the DISC Drill

The Replicate Coring System added several new mechanical subsystems to the DISC Drill. These included two electromechanical actuators capable of pushing the sonde to any targeted azimuth, core and screen barrels that are reduced in diameter from the main DISC Drill, and new cutter heads optimized for the multiple stages of the replicate coring procedure. These mechanical subsystems employed as part of the complete Replicate Coring System were used to create five deviations from the 3405-m-deep parent hole collecting a total of 285 m of replicate core from the most interesting time periods in the WAIS Divide climate record, with depth locations identified in Figure 6.1.

A description of the major sections of the downhole portion of the drill, the sonde (as illustrated in Figure 6.2), is essential to understand the basic operation of the system. The cable interface section is the uppermost portion of the sonde and includes a freely rotating connection to the drill cable. Below this are the two actuator modules on either end of the instrument section. The actuator modules each include three levers to tip the drill in the borehole. Onboard electronics are contained within the instrument section, providing power and controls to the actuator modules as well as the pump and cutter motors just below the lower actuator. Chip barrels collect ice chips created from drilling. A core barrel is added as needed to collect and support the ice core. Multiple styles of cutting heads are employed for the various requirements of the replicate coring operations.

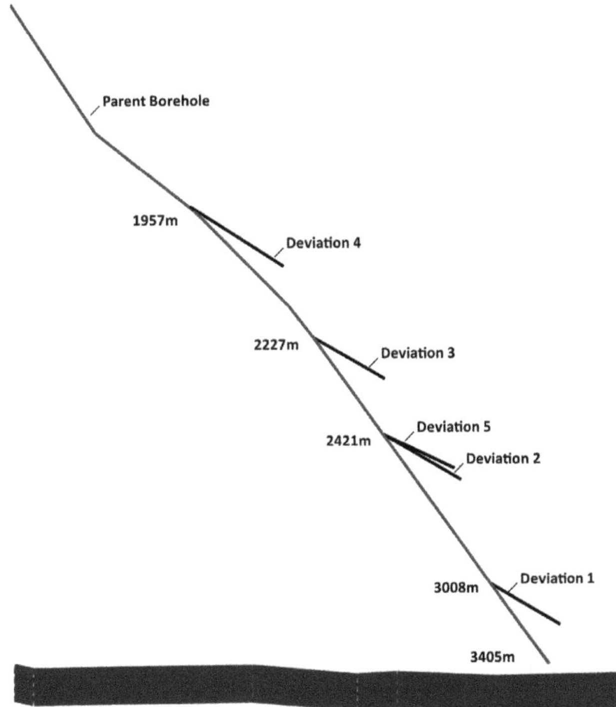

FIGURE 6.1 Replicate coring WAIS divide. Replicate core was collected from five intentional deviations from the parent borehole at four target depths. *Source*: The inclination of the parent borehole varied from vertical up to 5°. Deviations were created on the high side of the borehole at an angle of about 1° from the parent borehole. Scaling of the figure exaggerates the apparent angles. Courtesy of Gibson et al. (2014).

The architecture of the Replicate Coring System maximizes the amount of flexibility in operation while adjustable design parameters focus and simplify operation. In production operation, the operator inputs azimuthal direction and the amount of force to be exerted on the bore wall. Sensors onboard provide real-time feedback of rotation, torque, weight on bit, and inclination to the operator. Onboard controls make near-instantaneous adjustments to maintain the commanded force and direction.

Through lab and field testing and a ten-week production season, a detailed procedure was developed to efficiently create a deviation in three basic steps as shown in Figure 6.3 (Johnson et al., 2014a; Slawny et al., 2014).

First, a deviation of about 15 m in length is created by cutting in an upward stroke using a broaching cutting-head, as seen in Figure 6.3A and Figure 6.3B. The cut is expanded in repeated passes until it is deep enough to allow the sonde to move 75 to 100 mm in a radial direction out of the parent borehole. An inclination sensor with <0.1 degree accuracy is used to make this measurement. Although using a milling cutter to create the deviation may be possible, using an axial cutter

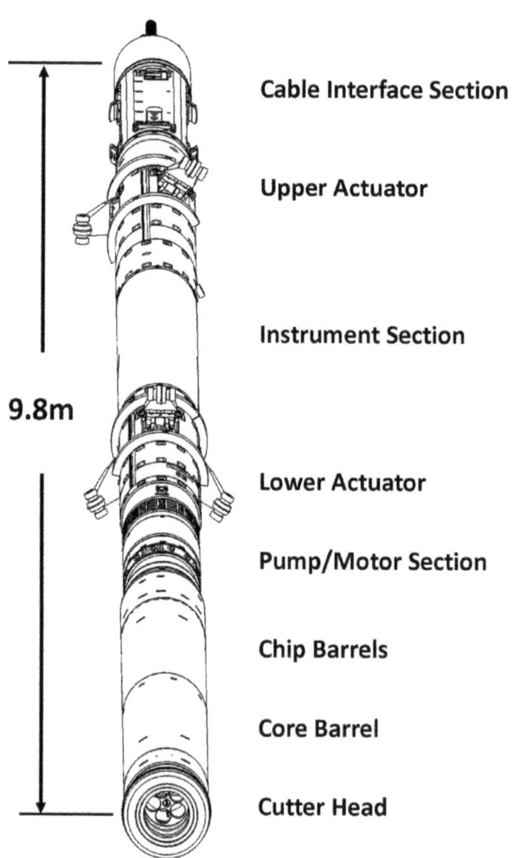

FIGURE 6.2 Replicate Coring System sonde. *Source:* Courtesy of Gibson et al. (2014).

like a broach efficiently removes the material of the deviation with an inherently straight cut. This approach, however, leaves a gradual slope at the lower end of the deviation not suitable for landing the coring head.

The second step in cutting the deviation is to create a landing for the coring head, and this is accomplished by rotary cutting with a milling head over an additional 1 m of depth, as seen in Figure 6.3C and Figure 6.3D. Finally, the coring operation can commence. The sonde is tilted into the deviation to an angle of 1 degree relative to the parent borehole. An inclination sensor with <0.1 degree accuracy is used to make this measurement and assures the cutter lands at the full radial depth of the deviation, 75 to 100 mm. The first core is 1 m in length and the cross section of the bottom of the core is nearly a full replicate core diameter, 108 mm; this assures that the core dogs can engage to break the core, as depicted in Figure 6.3E and Figure 6.3F.

Subsequent cores are recovered by reentering the deviation. Accurate inclination and depth readings are essential to assure smooth reentry. Radial force provided by the lower actuator is not necessary once engaged in the deviation to the depth of the base of the first core. At this point, the lower levers are retracted and the upper levers remain extended to provide anti-torque. Upper levers transition from the parent to the replicate borehole by first coring to the point of transition with a 2-m barrel. On the following run, the 1-m barrel is used allowing the upper levers to descend fully into the replicate borehole before coring is resumed.

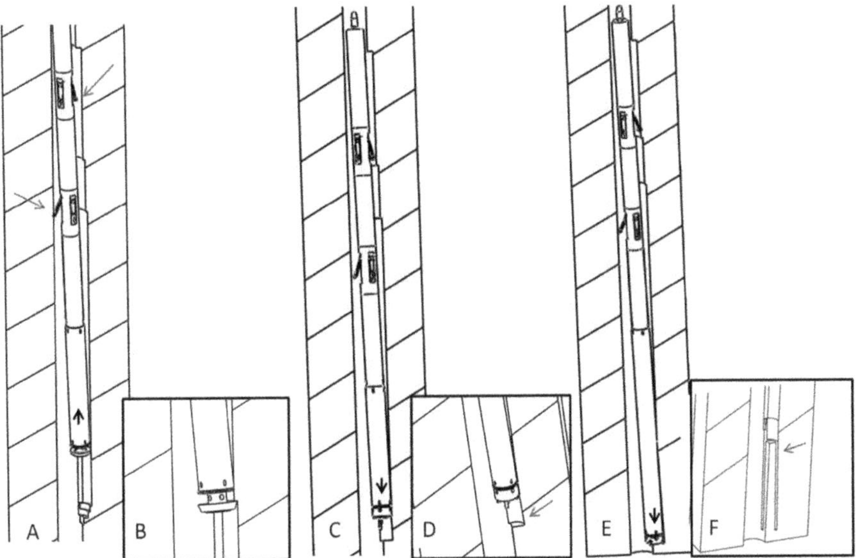

FIGURE 6.3 Replicate coring deviation procedure can be presented as three basic steps. First, the deviation is begun using the broaching head (A). Upper and lower actuators extend on opposing sides (blue arrows) to tip the sonde and engage the cutter. The broaching cutter engages the ice wall, removing material in repeated passes in an upward stroke (B). In a second step, the milling cutter is installed and a flat landing surface is created (C and D). In the third step, a coring cutter head and core barrel are added. The sonde is again tipped and coring starts on the flat surface provided by the milling cutter (E). The first partial core has a tapered geometry (F). This core is broken by core dogs and removed as the sonde begins to ascend. *Source:* Courtesy of Gibson et al. (2014).

The Replicate Coring System is a combination of actuators, motors, sensors, and a computerized control system that is critical to the system function. The control system requires information about the actual conditions of the system with means to correct the guidance system as drilling commences, and the actuator system requires capability of positioning and steering the system in the direction identified through the control system. Details of the system design are described in Mortensen et al. (2014), and subsequent updates identified through system testing are described in Johnson et al. (2014a).

The unique Replicate Coring System supports ice coring science by allowing rapid collection of large volumes of ice core from depths of interest without risking access to future borehole logging in existing boreholes. The system is a key advance because it allows scientists to take core samples at targeted depths while leaving the parent borehole open for future logging of information. Deployment of this technology to WAIS Divide for a ten-week season allowed for the retrieval of critical core samples from five deviations at four target depths. This would have otherwise required four or more seasons to recover. The expectation is that the technology could be applied with equal success on future DISC Drill projects and/or adapted for use with other drill systems in the future.

6.2.1.3 Electrical, Electronic, and Software Design for the Replicate Coring System for the DISC Drill

Science requirements developed for the DISC Drill Replicate Coring System dictated that the deviations must be completed on the uphill side of the borehole, as shown in Figure 6.4. This ensures

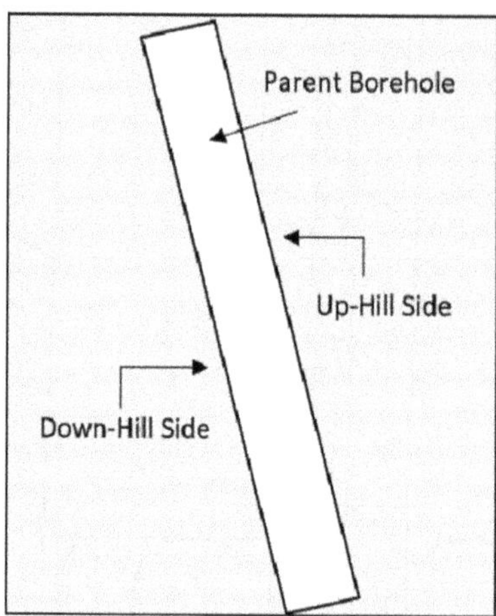

FIGURE 6.4 Depiction of uphill side of borehole. *Source:* After Mortensen et al. (2014).

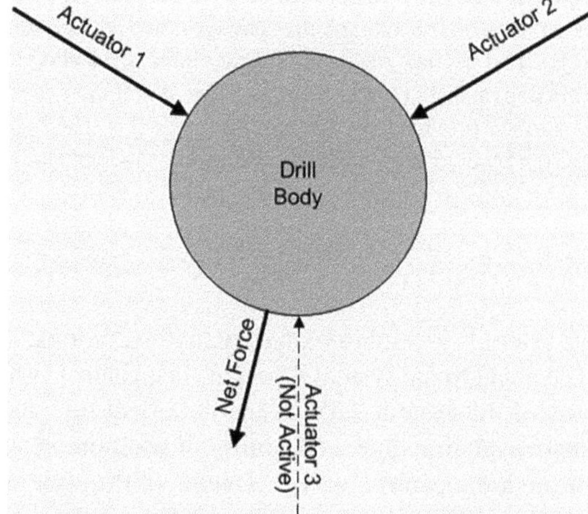

FIGURE 6.5 Two actuators work to produce one net force in a desired direction. *Source:* Courtesy of Mortensen et al. (2014).

passive logging tools will remain able to traverse the entire length of the parent borehole following replicate coring. Such a requirement necessitates complex electronics to enable the drill to know where it is in the borehole, where it should be and an ability to steer itself to the correct location. Repeatable accuracy of locating a certain position in the borehole is paramount to successfully creating and reentering a replicate borehole.

As noted in Section 6.2.1.2, two actuator sections are added to the DISC Drill sonde. Each section contains three actuator arms as illustrated in Figure 6.5, each of which is controlled by a brushed DC motor. Real-time feedback is essential for control and positioning of the drill.

A complex control module is necessary to control navigation and pushing forces. The control system block diagram is shown in Figure 6.6. Six linear variable differential transformers (LVDT)

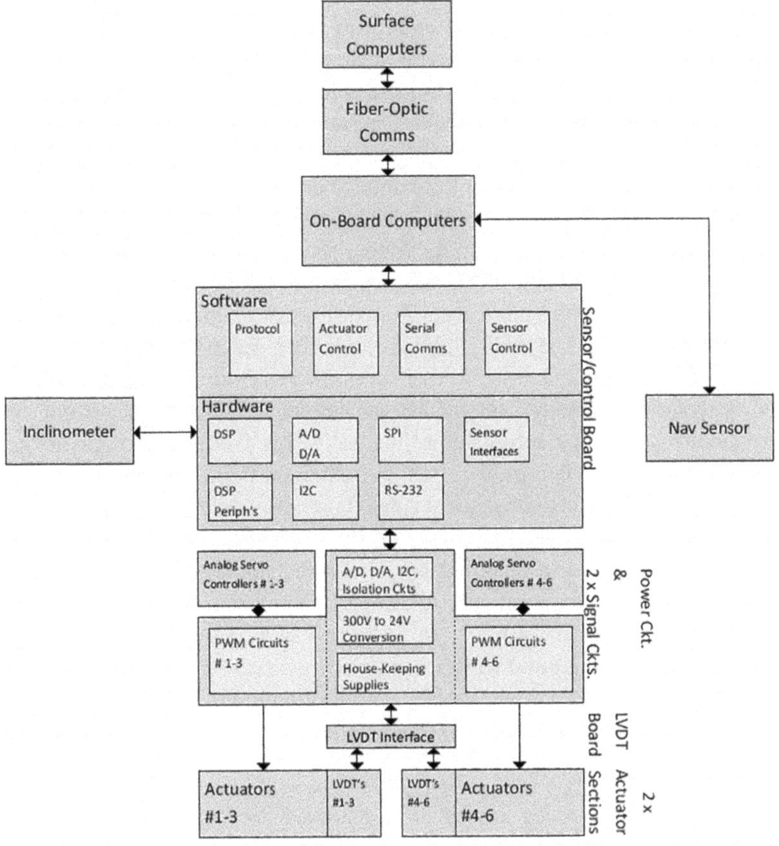

FIGURE 6.6 Replicate Coring System block diagram. *Source:* Courtesy of Mortensen et al. (2014).

continuously measure the location of each actuator arm. During descent into the borehole, all arms remain retracted.

Separate power supplies are implemented for the power board, LVDT board, and the sensor/motor control board to incorporate multiple grounds and control noise in the system. Commercially available Vicor power supplies were used. The Replicate Coring Module (RCM) consists of six circuit boards: one power board, two signal boards, one LVDT board, one sensor/control board, and one inclinometer board. A commercially available navigation module from 3DM is used to continuously read and report the drill's azimuth, that is, the drill's position relative to geographic north. Operators are able to enter a heading between 0° and 360°.

The RCM uses both integrated downhole software and modifications to the DISC Drill PC-based software.

The RCM software was designed as a "round robin," meaning the software executes the same function cyclically at a fixed rate of 20 cycles s^{-1}. Each cycle has the following top-level functions:

- Read all sensors
- Parse commands
- Process actuators
- Wait until time to start over.

Advanced error-checking software is implemented using checksum and other redundant information. Operators are able to set drill heading and effort levels of the actuator arms or a "SetEffort" value. To ensure commands sent downhole are accurately received and processed, operators can query the system for a "GetEffort" value. Any discrepancy between the two values signals that the previous command was not properly received. Operators control all RCM functions via a laptop computer on the surface that runs National Instruments LabVIEW as the interface. A second laptop also runs LabVIEW software and controls the winch operation.

Even the most carefully drilled boreholes are not perfectly vertically oriented. An important part of RCM system design is its ability to move to a pre-determined depth and at that point begin drilling a replicate core from the uphill side of the parent borehole. To initiate a new deviation, the drill is lowered to the target depth in the hole. Actuator arms are then activated to force the drill to one side of the borehole wall. The drill's inclination is then recorded. The actuator arms are then retracted and the drill is commanded to another set point 90° from the original point. The actuator arms are again activated and another inclination measurement is taken. Repetition of this process allows the operators to locate the uphill side of the hole. This process need only be completed one time when initiating a new deviation. Location and other set point information is then stored and automatically incorporated during subsequent runs. In creating a deviation, many runs are required to remove chips from the hole and to switch out drill heads depending on the technique being used. A broaching head is used to begin a new deviation. The winch is used to raise the drill to allow for upward cutting motion for approximately 15–20 m, known as stroke length. At the end of each stroke length, the drill is lowered again to the start point of the stroke, and the borehole is repeatedly widened, allowing the drill to increase its inclination and deviation out of the parent borehole. Following broaching, a downward cutting milling cutter is used to create a flat landing for the coring head. The design of the RCM, though complex, is in concept adaptable to other deep ice coring drills.

6.2.2 Intermediate Depth Ice Coring

Both climate history and glacial dynamics have driven the need for retrieving core from depths between 400 m and 1800 m. Existing quantitative reconstructions of the past two millennia are lacking in annual data prior to 1600 AD in many areas, and many cores are needed in order to understand the highly regional nature of many climate processes. In addition, the need to better understand ice dynamics processes for improved predictions of current ice sheet behavior in response to warming conditions requires a drill that can penetrate similar depths. Climate and environmental science issues relating to the past 500–1000 years require cores from depths of 400 m and deeper. The need to access areas with limited logistics requires that the drill be portable and versatile for use under a variety of conditions. Thus, a number of intermediate depth ice coring drills have been developed to be sufficiently portable that they can be used for coring at a wide variety of sites with production drilling in two field seasons or less, and they are able to retrieve core from depths of interest for a variety of science goals.

An overview of a number of intermediate depth drills in use by 2006 has been published (Bentley et al., 2006). The British Antarctic Survey electromechanical drill was used to retrieve a 948-m ice core and some sediment from beneath the ice (Mulvaney et al., 2007). Several drills developed at the Byrd Polar Research Center were successfully used in electromechanical dry drilling to 310 m depth in Tibet in 2015 (Zagorodnov, V., pers. comm. 2019), and deeper using an ethanol thermoelectric drill to 586 m depth in Svalbard in 1987 (Zagorodnov, 1988; Zagorodnov et al., 2005a). Early electrothermal intermediate depth drills described in Bentley et al. (2006) include the CRREL thermal drill, the Russian TELGA-14 drill, the CNRS drill, the Australian drill, and the Japanese JARE electrothermal drill. Motivated by the need to drill in temperate and polythermal glaciers in high altitude locations with minimal logistical support,

TABLE 6.1
Comparison of IDP Foro Drill Specifications

	Foro 400	Foro 1650 (IDD)	Foro 3000 Drill
Max. Depth (m)	450	1650	3000
Core Length (m)	1	2	3
Trench (L × W × D, m)	None	15 × 4.6 × 1.5	19.3 × 4.6 × 1.4
Slot (L × W × D, m)	None	3 × 0.9 × 3.5	5.5 × 0.9 × 5.3
Pilot Hole	Main borehole diameter	2-stage reamer	2-stage reamer
Casing	None	Polyethylene sewer liner	Polyethylene sewer liner
Tower	Fixed	Tilting	Tilting
Fluid Recovery from Chips	None	Centrifuge	Chip melter
Ventilation	None	Forced air/ERV	Forced air/ERV
Tent (lbs.)	755	4000	4000
Electrical Generator (kW)	5	30	35
System Weight (lbs.)	3500	30,000	35,700–52,700 (estimate; depends on shop option)

(Zagorodnov and Thompson 2014) identified design improvements for thermoelectric intermediate depth ice coring. These include injection of pure ethanol in the hole above the drill and circulation of an ethanol-water system through the kerf to produce good core quality with a high production drilling rate. With power requirements of only 1.5 kW, the system could be powered by solar panels.

In cold glaciers and ice sheets, essential concepts involved in ice coring apply to many electromechanical ice coring drills, from those designed to drill to tens of meters to drills designed to retrieve cores from several kilometers in depth. The characteristics of the firn and ice change with depth in an ice sheet or glacier, and also the ability of the drill to operate without a drilling fluid decreases dramatically with depth, so that any single drill is not optimal for use in reaching all depths. However, keeping as many characteristics of the collective system of drills as uniform as possible facilitates ease of maintenance and interoperability. Toward that goal, the US Ice Drilling Program is developing a series of drills in the "Foro" family of drills that have a common sonde diameter and similar design. The Foro 1650 is an existing, proven intermediate depth drill; originally targeted for drilling to 1500 m, it was later updated to 1750 m and successfully used to drill the "SPICE" ice core at the South Pole. The Foro 400 is a shallow drill targeted at reaching 400 m, and the Foro 3000 Drill is designed for use in deep drilling to 3000 m depth. A comparison of the system specifications is found in Table 6.1. These drills are described in more detail in the following sections.

6.2.2.1 Foro 400 Ice Coring Drill

Targeted to retrieving ice cores up to 400 m depth, the Foro 400 Drill is designed to have the same sonde diameter and similar design to other drills in the Foro line. As the next generation of the PICO 4-Inch Drill system, the Foro 400 Drill has a winch capacity of 450 m of 5.7-mm-diameter cable and recovers 98 mm diameter by 1-m-long ice cores. The Foro 400 Drill retains the modular fiberglass

FIGURE 6.7 Foro 400 winch sled and tower base.

FIGURE 6.8 Foro 400 Drill sonde.

tubular tower and sled-type base of the 4-Inch Drill, which have been desirable features of this system, but does so with a more refined design, as shown in Figure 6.7.

The anti-torque and motor section design is common with the larger Foro Drill systems, but does not contain an instrument section electronics package. However, the pressure tight motor section allows the drill to be used in wet conditions and with drilling fluids. The Foro 400 Drill sonde is shown in Figure 6.8.

The speed and direction of the 500 W, 5000 rpm, brushed DC motor in the motor section is controlled with a pulse width modulated (PWM) motor controller housed in the control box at the surface. The winch is driven by a 3600 W brushless servo motor, which provides fully controllable line speeds from 1 mm/s to 0.8 m/s and a maximum pulling force of 8 kN. The crown sheave is instrumented with a load pin and ring encoder to monitor line speed and payout. A spring-loaded cable guide on the downhole side of the sheave prevents the cable from being able to jump off the sheave (see Figure 6.9) and will stop the winch if the drill is raised too high on the tower.

Cable load, payout, and speed are displayed on an LCI-90i line control display manufactured by Measurement Technology Northwest, which is integrated into the control box, as shown in Figure 6.10.

The drill system packs into custom-made shipping cases that have been designed with specific compartments for each piece to maximize protection while in shipment and to minimize overall size and weight. Overall weight has been reduced more than 40% compared to the 4-Inch Drill system it will replace.

6.2.2.2 Foro 1650 Ice Coring Drill

The US Ice Drilling Program (IDP) generated the following science requirements for an intermediate depth drill through a series of iterative discussions between the scientific community and

FIGURE 6.9 Foro 400 crown sheave with spring-loaded cable guide.

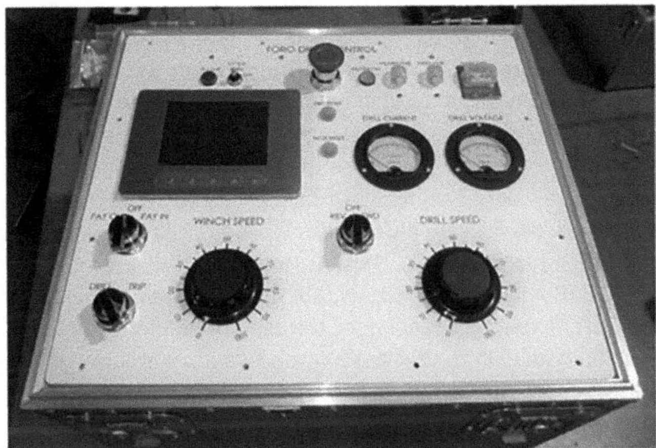

FIGURE 6.10 Foro 400 control box.

engineering staff of the program. The following are the original science requirements for the Foro 1650 Drill, also known in the US community as the Intermediate Depth Drill (IDD):

- Requirements:
  - Target depths: from the surface down to 1500 m.
  - Ice core diameter: 98 +/−3 mm.
  - Core length: 2 m.
  - Minimum 10-m temperature at the site: −55°C.
  - Air transport type: Bell 212 or similar helicopter and/or Twin Otter aircraft.
  - Replicate coring capability: no.
  - Drilling fluid: drill should be compatible with existing fluids, e.g. Isopar-K or butyl acetate.
  - Maximum field project duration: two field seasons.
  - Core quality requirements:

a. Complete core recovery over entire borehole, as close as possible, including brittle ice
b. Ice pieces to fit together snugly without any gaps
c. In non-brittle ice, the packed core should have no more than 12 pieces of ice per 10-m section of core
d. In brittle ice there may be a lot of pieces in a single ~1-m core segment, but the pieces must fit together retaining stratigraphic order; more than 80% of the ice volume must be in pieces that each have a volume >2 liters

- Absolute borehole depth measurement accuracy: 0.2% of depth.
- Sonde inclination will not exceed 5°.
- Field setup time: the minimum that is realistically possible with a three-person effort at a small remote camp.
- System complete with receiving area for core from core barrel and ability to cut into 1m sections.

Note that although the original science requirements were targeted at 1500 m, the system is currently capable of reaching 1650 m. The Foro 1650 is sufficiently portable for coring at a wide variety of sites, fits into a Twin Otter aircraft or Bell 212 helicopter, though numerous flights are required, and is capable of retrieving 98-mm-diameter core from the surface down to 1650 m depth in two field seasons by a ten-person field team for 24-hr per day drilling and core handling operations.

The proven design of the Danish Hans Tausen drill (Johnsen et al., 2007) was chosen as the basis for the Foro 1650 design, with a number of modifications made for the final Foro 1650 (Johnson et al., 2014b). The Foro 1650 has been designed to recover 2-m long cores to balance the trade-offs between the length of the drill, which directly affects the size of all equipment on the surface and the length of the ice cores drilled per run. This keeps the size of all equipment manageable and within the design goals. Table 6.2 presents a summary of the drill dimensions.

The major components of the drill are the anti-torque, motor section, chips chamber with internal drive shaft, the core barrel, and cutter head. The anti-torque features a three spring steel skate design surrounding an EVERGRIP winch cable termination inside a bearing housing and a multi-channel slip ring to allow the drill to spin independent of the cable. The motor section has a pressure-tight housing 0.72 m in length that contains an electronics package, consisting of a motor power supply (MPS) and basic bidirectional communication loop, to control the motor direction and send a signal back to the surface if the anti-torque is slipping rotationally while drilling. The motor power supply (MPS) is designed to handle an input voltage of up to 600 VDC. The output voltage varies with the

TABLE 6.2
Foro 1650 Component Dimensions

Hole Diameter, Dry	126.0 mm	Hollow shaft OD	30.0 mm
Hole Diameter, Wet	129.6 mm	Chips chamber ID	110.3 mm
Core Diameter	98.0 mm	Chips chamber OD	114.3 mm
Drill Head ID	99.0 mm	Chips chamber length	2.60 m
Core Barrel ID	100.0 mm	Motor section length	720.0 mm
Core Barrel Length	2.24 m	Anti-torque length	857.0 mm
Outer Barrel ID	113.0 mm	Total drill length	6.50 m
Outer Barrel OD	118.0 mm		
Outer Barrel Length	2.50 m		

input voltage, giving proportional speed control up to the motor's rated voltage of 220 VDC. At this point, any additional power supplied to the MPS is applied to keep the motor at constant speed, and the output voltage remains at 220 VDC. Over-voltage and over-current protection features are also built in to the MPS to protect the motor. This power supply design makes it possible to supply over 500 W of shaft power at the drill over the small diameter conductors of the winch cable. A Parvalux PM60 brushed DC motor is coupled to an 80:1 ratio harmonic drive gearhead. This gives a maximum output shaft speed of 63 rpm with 70 Nm of torque, assuming an 80% efficiency. The upper bulkhead and lower drive housing of the motor section are both removable from the central housing and kept leak tight with a pair of hydraulic piston seals on either end. Figure 6.11 shows section views of the anti-torque and motor section.

Connected to the bottom of the motor section are the chips chamber and the hollow shaft which drives the core barrel. Two configurations of the 304 stainless steel chips chamber have been made, one for dry drilling and one for wet drilling. They differ only by the wet chamber having 7200 1.5-mm-diameter holes drilled in it to aid in filtering the drilling fluid from the chips. The overall length of the chips chamber tube is 2.6 m with a usable length for chips of 2.3 m when configured with the pump, yielding a fill concentration of 56.5%. The output of the motor section attaches to the hollow shaft, which is 30 mm in diameter with three quarter-turn spring-loaded pins. Valve plates at the upper and lower ends of the hollow shaft open when the shaft is turned in reverse and close when the shaft is turned forward. The valves are set open when the drill is descending in the fluid-filled borehole to permit faster tripping speeds by reducing hydrostatic drag. The valves automatically close when the shaft is run forward, allowing the space between them to fill with cuttings and transport the cuttings back to the surface without being flushed out. Single-turn helical augers, often call boosters, can be placed on the hollow shaft at any position between the valves to facilitate the transport and packing of cuttings in the chips chamber. A 12-valve double piston pump can also be installed just above the lower valve for wet drilling (Johnsen et al., 2007). Alternatively, the pump can be replaced with a chips valve assembly designed by IDP, which prevents cuttings in the chips chamber from being able to be flushed out of the chips chamber when the drill is being tripped back to the surface. At the end of the hollow shaft is a bayonet housing that attaches to and drives the core barrel. The hollow shaft assembly is shown in Figure 6.12.

Following the chips chamber is the outer tube which contains the core barrel. IDP innovated the first use of filament-wound fiberglass tubing for the outer tube, shown in Figure 6.13. The tubing has 24 saw-tooth-like grooves running the length of the inside of the tube, which facilitate the transportation of cuttings up the helical flights on the core barrel. The tubes are very round and straight and cost a fraction of tubes made from metal. The outer tube attaches to the chips chamber with a slip fit connection that is held in place with three locking pieces that fit into a keyhole shape detail. The core barrel, which is sized to recover a 2-m-long core, is manufactured from 304 series stainless steel tubing. The top of the barrel has three channels machined into it that the bayonet housing pins run in. These details serve as a quick detach mechanism allowing the drill to be recovered, leaving the core barrel and cutter head behind, should the cutter head become irretrievably stuck while drilling. Also, using the mass of the upper portion of the drill, it can be used as a slide hammer to break core. Two configurations of this barrel have been made. The first has three flights with a 35° pitch angle, which are fabricated from ultra-high molecular weight polyethylene (UHMW). The second configuration, used for wet drilling with the pump, has three aluminum flights with a 48.5° pitch angle and at 2 mm high only partially fills the annulus between the core barrel tube and outer tube. These partial height flights provide a stirring action, and rely on the pump flow to provide the chip transport. Near either end of the core barrel, a short section of full height UHMW flights were added in for guidance to keep the core barrel centered in the outer tube.

The cutter heads follow the Hans Tausen drill design, having three cutters with a 42.5° rake angle and 15° relief angle, as shown in Figure 6.13. Two widths of cutters were made, both of which create

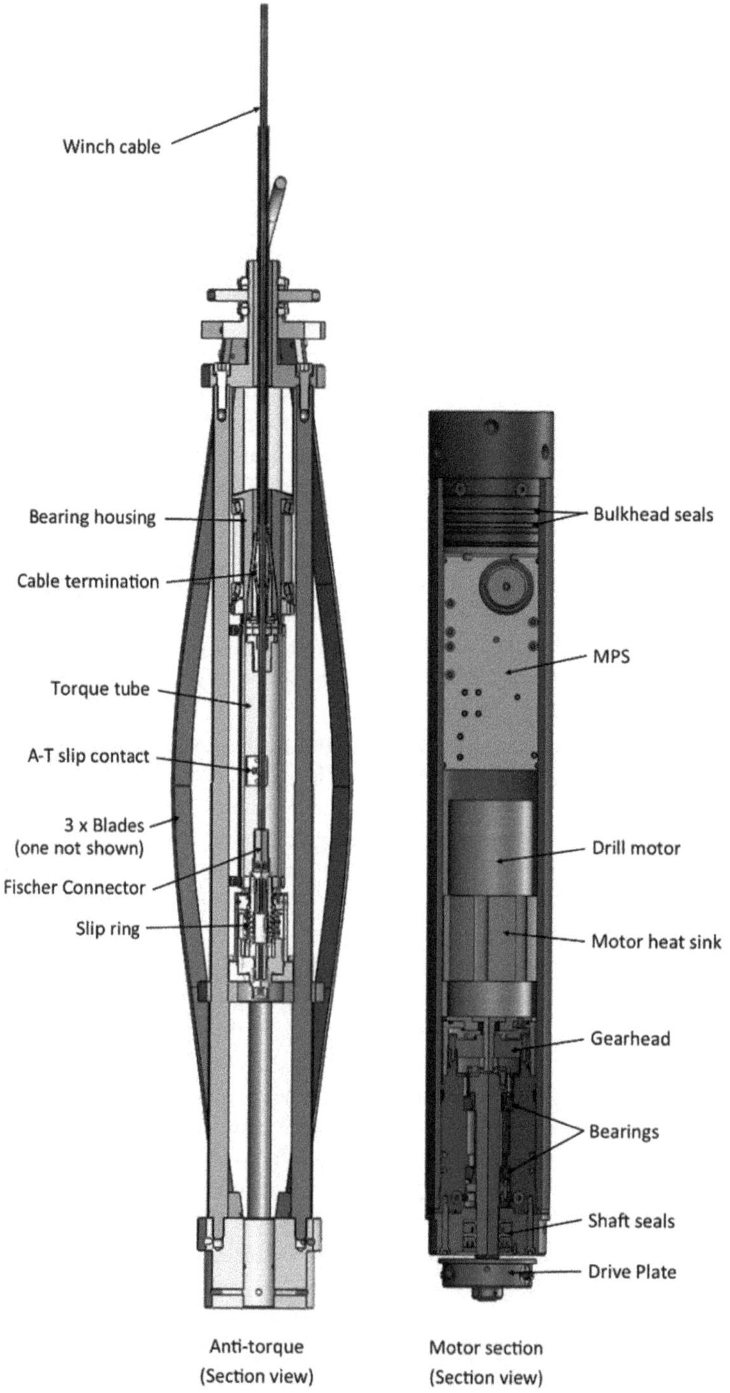

FIGURE 6.11 Foro 1650 anti-torque and motor sections.

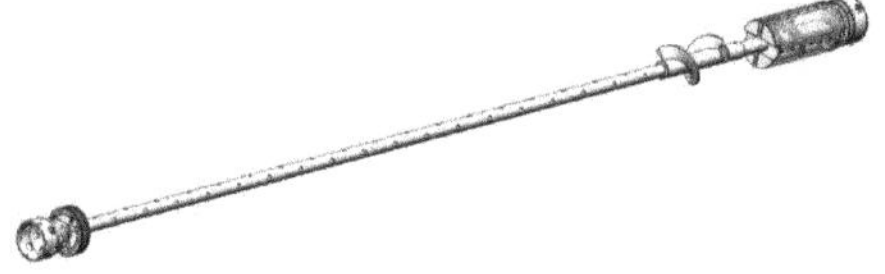

FIGURE 6.12 Foro 1650 hollow shaft assembly.

FIGURE 6.13 Filament-wound fiberglass outer tube and stainless steel cutter head.

98-mm-diameter cores. The narrower kerf cutter produces a 126-mm-diameter borehole while the wide kerf cutters make a 129.6-mm-diameter borehole and are used for wet drilling to provide more clearance around the drill for improved fluid flow. The three core dogs are double sprung, meaning a light spring is used to keep them retracted while a heavier spring is used to set a pre-load against the ice core. This method reduces damage to the ice core if the previous break was uneven. Two different cutter head configurations have been built. One is used for wet drilling and one for dry drilling. The only difference being the wet head has a smaller OD permitting improved fluid flow around it. Shoes mounted on the back of the cutting surface will provide cutting pitches between 0.5 mm and 3.0 mm per tooth. The head attaches to the core barrel using three eccentric headed bushings to pin it in place.

A tilting tower design was chosen for its ease to set up and ability to position the drill horizontally for servicing. The 6-m-long tower is fabricated from 1- and 2-m-long sections of 132-mm square aluminum tubing that break down for ease of transport and assembly. The aluminum top sheave is instrumented with a load pin and ring-type encoder for measuring cable payout. The ring encoder permits a very compact sheave design while providing 0.2-mm payout resolution. Up to 1700 m of 5.7-mm-diameter electromechanical cable can be spooled onto a removable all-aluminum winch drum with integral Lebus groove. The cable has three #24 AWG conductors and a breaking strength of 24.5 kN. The winch is driven by a 7.5 kW 460 VAC brushless servo motor with resolver feedback and built-in brake. It is coupled to the winch drum by a helical bevel gear reducer with a 56.38:1 reduction. This provides fully controlled line speed from 0.5 mm/s to 1.4 m/s and enough pulling power to do a 10-kN core break. A custom built level wind, using the similar control method the US Ice Drilling Program first designed for the DISC Drill, has been implemented. The level wind is a self-contained device (see Figure 6.14), requiring only 24 VDC, 170 W power, and operates fully independent of the winch.

The control system monitors the cable angle between the winch drum and fairlead roller and sets the required speed to keep this angle near zero. Further information on this novel method for cable level winding can found in a related article (Mortensen et al., 2014). The winch drum and drive are

FIGURE 6.14 Foro 1650 winch and level wind.

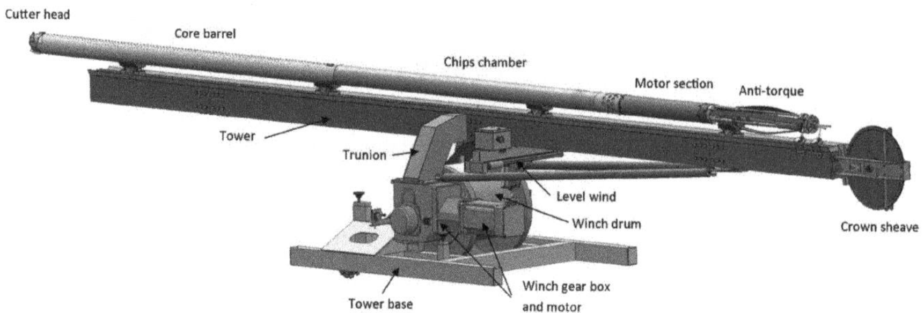

FIGURE 6.15 Foro 1650 tower and sonde.

mounted on a trunnion that pivots with the tower. There are two main advantages with this design. First, the drill does not move up or down the tower as it is tilted. Second, the winch drum translates vertically 15 cm as the trunnion tilts which makes it possible to install and remove the winch drum without additional rigging or lifting equipment. To remove the drum, two wood rails are first placed on the tower base. The trunnion is then tilted back until the drum rests on the rails. The drive flange is unbolted from one side and the bearing block from the other, freeing the drum so it can be rolled out and into its shipping case. The tower is tilted using an electric linear actuator with a built-in fail-safe brake and provides full variable speed control. The winch and tower assembly mount to an all-aluminum base frame, 1.7 m long by 1.2 m wide. A model of the complete winch and tower assembly is shown in Figure 6.15.

After the drill has been parked on the tower and tilted horizontal, the hollow shaft and core barrel are detached from the rest of the drill and pulled out either by hand or with a hand crank winch located at the end of the pull-out table. The cuttings are collected in a tub located between the end of the drill slot and the pull-out table for later processing. The core barrel, containing the core, is transferred by hand laterally to the core processing line. The ice core is pushed from the core barrel through a specially designed Fluid Evacuation Device (FED), which, using suction, removes most of the drilling fluid from the core surface. Ductile ice cores proceed down the line where they are measured and cut into 1-m lengths using a circular saw. At this point, they can be put into lay-flat tubing and packed for shipment. Cores that are recovered from the brittle ice zone are placed into polyethylene netting as they exit the FED, as seen in Figure 6.16. An initial length measurement is

FIGURE 6.16 Fluid Evacuation Device (FED). Polyethylene core netting is used for storage of brittle ice.

taken and the cores are then transported directly into a core storage trench that is excavated at the end of the drill trench. The cores will rest for one year before being cut and packed. A ladder lift is used to move packed ice core boxes from the trench to the surface where they are palletized for shipping.

The slurry of ice chips and drilling fluid removed from the chips chamber is transferred to a centrifuge where the slurry is spun to recover most of the drilling fluid. The cuttings are then carried out of the drill tent and discarded in a designated chips disposal area. Recovered fluid drains to a collection tank and passes through a final filter before flowing through a hose and back to the borehole. Drilling fluid is brought to the site in metal drums and staged next to the drill tent. A drum pump is used to transfer the fluid to a manifold in the drill trench. Fuel bladder material lines the slot end wall and floor to direct drilling fluid coming off the tower back to a catch pan around the casing and into the borehole. Fluid carried up on the winch cable is recovered using a vacuum connected to a custom built cable cleaner based on the one use by the Danish drilling program at the NEEM drill site in Greenland. A bailer was built for recovering cuttings that are not fully recovered by the drill. It consists of a perforated mesh tube, configurable to either 1 m or 2 m long, with a check valve assembly at the lower end. The bailer, shown in Figure 6.17, mounts to the motor section in place of the chips chamber. Chips are collected in a fabric mesh sleeve inside the tube as the tool is lowered in the fluid. The valve assembly retains the chips in the bailer while returning to the surface, yet permits easy removal of the chips when on the surface.

The drilling equipment is housed within a fabric covered tent. The floor level inside the tent is recessed 1.4 m below grade to reduce the required height of the tent and to help keep the interior of the tent cool. The floor is covered with 0.5 m × 1.0 m × 30 mm thick sections of polyethylene grid material to provide a reusable slip-free work surface. Stairs at either end of the trench provide access. The custom-made 19.5 m long × 4.9 m wide × 3.1 m high tent, which was fabricated by WeatherPort, features a steel frame with a fabric cover. Both end walls have personnel doors and a large double zipper entry for moving larger equipment in and out. The fabric is a PVC-impregnated polyester cotton blend with an acrylic top coat. The fabric has a cold crack resistance to lower than −80°C and it remains very pliable at −40°C for ease of setup. The structure has been engineered for a snow load rating of 269 kg/m^2 and 65 knot winds. The cover is white and non-insulated to provide good natural lighting. There is a red strip on either end to improve visibility in poor weather; this and other tent features can be seen in Figure 6.18. Four 305-mm-diameter wind directional vents were

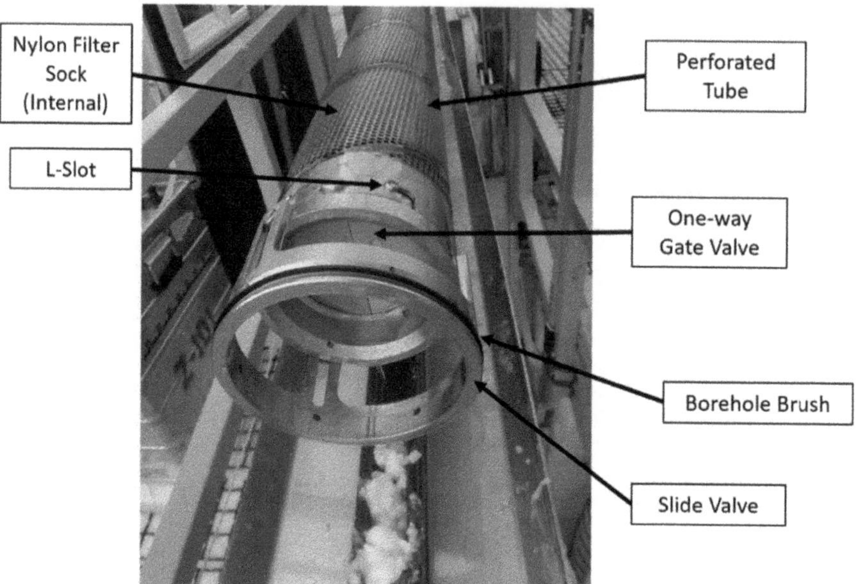

FIGURE 6.17 The bailer is a device used to collect chips in the borehole. It directs all fluid and chips through a perforated tube lined with a nylon filter sock while tripping down the borehole. When returning to the surface, the gate valve closes and the slide valve opens, allowing fluid to pass around the outside of the tube. At the surface, the slide valve assembly and nylon sock are removed using an L-slot and chips are emptied.

FIGURE 6.18 Interior view of the Foro 1650 tent.

installed on the ridge line to provide convection cooling. Inside the drill tent is a smaller, 1.2 m wide × 2.4 m long insulated tent which serves as the drill control room. Three windows were installed to provide the operators good viewing of all operations. The space is heated with an electric heater, and the brake resistor for the winch motor provides supplemental heat.

Ventilation in the Foro 1650 Drill tent is provided by two fan systems. An 1100 CFM centrifugal ventilator draws air from the bottom of the slot and exhausts outside the drill tent. The core dryer and cable cleaner vacuum systems are also vented into this system. The second ventilator is located next to the centrifuge and displaces 450 CFM. The combined air flow of the ventilators will provide seven air volume exchanges per hour.

6.2.2.2.1 SPICEcore Project

The first deployment of the Foro 1650 Drill (a.k.a. IDD) system was to the South Pole, Antarctica to recover ice cores for the South Pole Ice Core (SPICEcore) project. The Foro 1650 Drill system, with a shipping weight of 13,600 kg and volume of 63 m^3, was shipped to the South Pole and set up at the beginning of the 2014–2015 season. The entire project spanned three seasons, beginning in the 2014–2015 season and finishing mid-way through the 2016–2017 field season. A photograph of the SPICEcore drill site is shown in Figure 6.19. The first core was drilled on December 8, 2014. Dry drilling was completed to 160 m, and the borehole was reamed to 229 mm diameter to a depth of 130 m for installation of the casing. At this point, wet drilling began, and a final depth of 736 m was reached by the end of the season. The average length for the 98-mm-diameter cores was 1.82 m. The borehole diameter for both dry and wet drilling was 126 mm. All drilling, both dry and wet, was done using the "dry drilling" core barrel, which has full height UHMW flights. The dry hole chips chamber was used for dry drilling and the chips chamber with the 7200 vent holes was used for the wet drilling. Most of the drilling was done working in two 10-hr shifts. The first 600 m of ice was packed and shipped back to the US following the 2014–2015 season. The remaining ice was stored in the core storage trench for the winter.

While good drilling progress was made during the first season, there were several operational issues that had to be addressed. Chip collection by the drill was less than expected, resulting in 3–4 chip bailing runs having to be done each day. It was theorized this was due to fine chips being lost out of the drilled holes in the chips chamber as the drill was brought to the surface. If the borehole was not kept clean, then issues with penetration and difficulty drilling full-length cores was experienced. The −50°C ice was found to be very abrasive and would dull cutters quicker than experienced on past projects in warmer ice. The extreme cold temperatures also caused the cutters to chip while drilling. The ice also required higher motor torque than expected to drill. Standard cutters were modified into step cutters (Zagorodnov et al., 2005b) at the South Pole Machine Shop, where each cutter only cuts one-third of the kerf. This style of cutters required less torque and also reduced some of the penetration issues. Another complication in the first season arose from the drilling fluid used. Several drillers reported negative side effects after exposure to the Estisol 140 drilling fluid, despite measured vapor levels being well below published exposure limits. Side effects included mild (but prolonged) headaches, burning eyes, skin irritation, and temporary reduction in mental acuity.

During the second drilling season in 2015–2016, drilling operations were changed from two shifts to three to facilitate drilling operations 24 hrs each day and also to reduce the time drillers

FIGURE 6.19 SPICEcore drill site near South Pole Station, Antarctica.

spent exposed to the drilling fluid. In an effort to improve chip collection with the drill, the non-perforated chips chamber was used along with a modified hollow shaft. Modifications to the hollow shaft included drilling a pattern of 120, 12 mm diameter, holes through the tube and covering the shaft with an 80-mesh stainless steel filter sleeve, which allows drilling fluid to pass from the annular space in the chips chamber to the center of the hollow shaft and out back to the borehole. This configuration greatly reduced the amount of cuttings being lost to the borehole to the point where only three bailing sessions were required over the entire season. All cores drilled during the first part of the season, from 736 to 1078 m, were put directly into HDPE sleeve netting and into the core storage trench without cutting in anticipation that the ice would be brittle. This ice was stored in the trench to relax and was shipped off continent the following season. After 1078 m, the cores were again cut into 1-m lengths, bagged and packaged into insulated boxes for transport back to the US. Drilling continued to a final depth of 1751 m, as additional cable was available on the winch drum. A total of 1015 m were drilled during the second and final season over a 50-day period. Drilling progress for the entire SPICEcore project is shown in Figure 6.20.

Step cutters, visible in Figure 6.21, were used exclusively until drilling neared 900 m depth and the drill began experiencing penetration issues. Drill penetration was not an issue when the fullwidth kerf cutters were reinstalled. From approximately 1200 m onward, the remaining drilling was done with full-width kerf cutters. Motor current continued to be higher with this style of cutters, but it resolved penetration issues. IDP has yet to understand why the step cutters worked well for a period of drilling and then began to experience penetration issues. Prior to the second season, all cutters, which were made from A2 tool steel, were re-heat treated with an additional cryogenic processing step. This refined the materials' grain structure and improved the toughness and resolved the chipping issues experienced the first season. Negative side effects from exposure to the drilling fluid were still experienced the second season, but to a lesser extent. IDP attributes this to the shorter shift lengths and the greatly reduced amount of bailing that had to be done, which is an inherently messy process and exposes the operators to greater amounts of drilling fluid. The final days of the second season were spent starting disassembly and packing of the drill system. The casing was extended up to the trench floor level, and the slot was filled in.

The third and final field season, which took place from November 23, 2015 to December 13, 2015, focused on packing and shipping the remaining ice cores, conducting borehole logging,

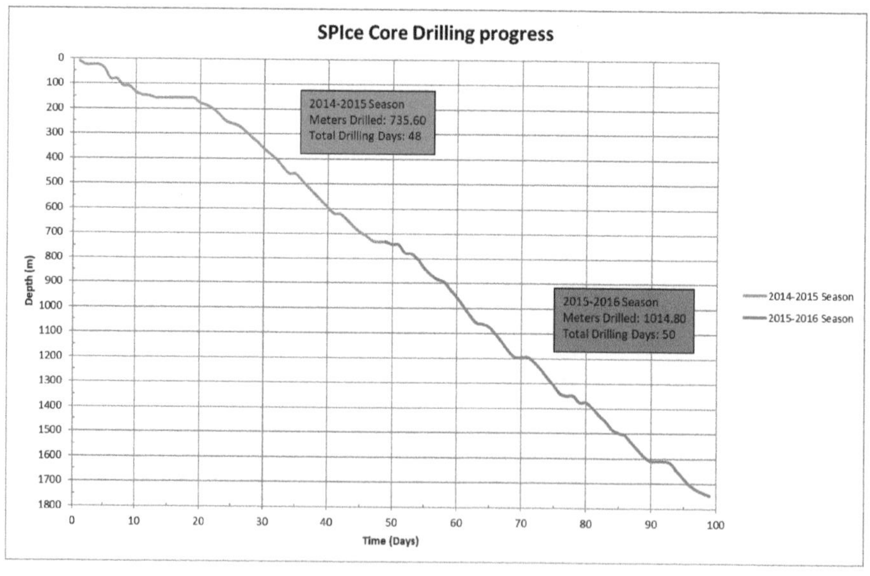

FIGURE 6.20 SPICEcore project drilling progress.

FIGURE 6.21 Step cutters on the Foro 1650 core barrel.

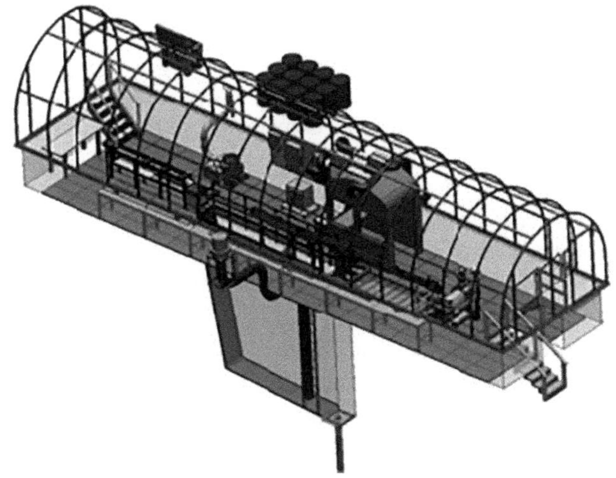

FIGURE 6.22 Foro 3000 Drill system layout.

packing the remaining drilling equipment and decommissioning the drill site. The remaining 614 m of ice were processed for shipment back to the US, which included cutting and boxing all of the brittle ice that had been relaxing over winter. A total of five logging runs were completed, which included one video log, two temperature logs, and two laser dust logs. Once the logging was complete, the remaining drill system components, including the core processing line and tent, were packed and all equipment was shipped off continent. The casing was extended to above the surface level and the drill trench was backfilled.

6.2.2.3 Foro 3000 Deep Ice Coring Drill

In order to recover ice cores from sites that are deeper than 1650 m and up to depths of 3000 m, the Foro 3000 Drill is being constructed; a model of the drill system layout is shown in Figure 6.22. This drill system features a larger winch, with 3100 m capacity of 7.2-mm-diameter four-conductor cable. The winch will be located beyond the end of the drill tower due to its larger size. A new tower base has been designed that pivots with a dual arm linkage so that the drill will remain stationary as

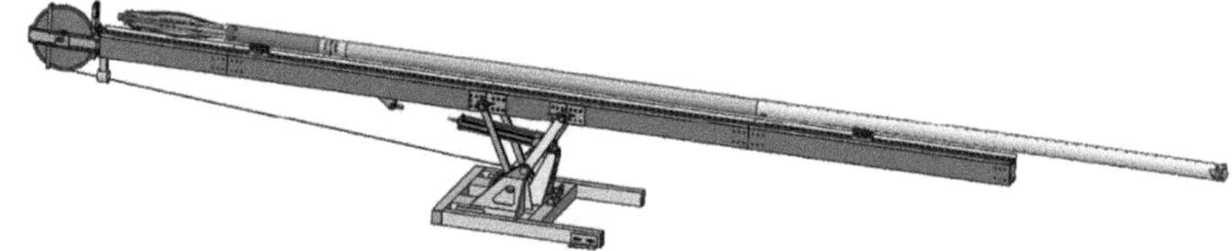

FIGURE 6.23 Foro 3000 tower design.

the tower tilts, as seen in Figure 6.23. The drill utilizes the Foro 1650 (a.k.a. IDD) anti-torque and motor sections mated to a longer chips chamber and core barrel for recovery of 3-m cores per run. A new instrument section electronics package is also in development that will include pressure and temperature sensors, inclination, and anti-torque slip detection using an accelerometer. The system will also include a custom-built DC-to-DC converter and integrate an off-the-shelf motor controller that will drive a brushless DC motor. The remaining components of the Foro 1650 system can be used or duplicated with little or no modifications. The entire Foro 3000 Drill and core processing line is designed to fit within the Foro 1650 tent dimensions, allowing the system to be configured for intermediate depth or deep drilling with minimal additional parts.

6.2.3 Large-Diameter Ice Coring

One of the major unsolved mysteries of Earth's climate system history is the question: why did the climate system change from a dominantly 41,000- to a 100,000-year glacial cycle approximately one million years ago? Numerous research endeavors have been and will be related to this transition and earlier through ice core records dating back to over a million years ago. There are two very different drilling approaches to retrieving ice this old; the first is to target a location where the ice has been subject to minimal ice flow, which requires coring through thousands of meters of ice. The second way is to find very old ice that, through glacial dynamics, has undergone flow and is now found at fairly shallow locations. Because the chemical evidence within the ice requires significant samples of ice, use of a large-diameter ice coring drill in the shallow locations facilitates retrieval of larger volumes of ice. In 2009, the US Ice Drilling Program developed a large-diameter drill for use in blue ice areas of Antarctica, and the large-diameter drill was called the Blue Ice Drill (BID).

6.2.3.1 Science Requirements of the Blue Ice Drill

The original scientific requirements for the Blue Ice Drill were:

- Ideal ice core diameter of 24.13 cm (acceptable range 22.86–24.77 cm) or 9.50 in (acceptable range 9.00–9.75 in).
- Minimum core length of 1.0 m.
- Maximum core length of 1.6 m.
- Core quality shall be such that the total surface area-volume ratio of 10 linear meters of consecutively drilled 24.13 cm diameter core does not exceed 21 m^2/m^3. This sets an upper bound on the number of breaks or fractures of the core.
- Samples must be free of contamination from oils, greases, exhaust fumes, and any carbon-containing lubricants or fluids.
- The drill system must be capable of reaching depths of at least 12 m.
- The core is to be collected with an angle of deviation less than 10° from vertical.
- The drill shall be capable of producing a minimum of seven sample cores per day. Note: for each hole, the top 5 m of ice will be augered or drilled and discarded. Sample core will then be drilled from this depth.

- Drill components shall be such that the entire drill system is transportable by one helicopter load. System design shall be based on the load capacity of a Bell 212 helicopter or similar (exact dimensions TBD).
- All components shall fit inside the helicopter.
- The optimum drill system weight is no more than 200 kg (440 lbs.), and the maximum weight should not exceed 500 kg (1100 lbs.).
- Individual drill components shall be of a size and weight that they are moveable by one–two people.

6.2.3.2 Blue Ice Drill Design

The Blue Ice Drill (BID) leverages design features of numerous existing electromechanical ice-coring drills, consisting of a double barrel coring assembly with a cutter head and a motor/reducer section (Kuhl and others, 2014). It currently has two configurations, one for shallow drilling to approximately 30 m depth and another version for core collection to 200 m. In the shallow version, an operator uses a T-handle to anti-torque the drill from the surface. The 200-m version is referred to as the BID-Deep. The drill assembly is supported by a 5.3-m aluminum tripod with dual sheaves on top. Ropes are used during shallow coring to raise and lower the drill string. The BID design incorporates an optional core recovery tool (CRT) to help break the core if needed. If cores are drilled without core dogs installed, or if the core dogs fail to successfully capture the core, the drill can be brought to the surface and the CRT barrel lowered over the core. The CRT induces a core break by tipping the core to the side, similar to a device designed and implemented by CRREL for an even larger 12-inch diameter drill in the 1980s (Rand and Mellor, 1985). Figure 6.24 shows the original BID system components.

6.2.3.3 Performance of the BID

Performance values for the original BID system are summarized in Table 6.3 as follows:

A custom stainless steel cutter head holds three cutters made of hardened tool steel. Cutters are spaced 120° apart, have a 30° rake angle and a 7° clearance angle from horizontal. The cutter head is shown in Figure 6.25. Both thin kerf (241-mm core, 288-mm hole) and wide kerf (240-mm core, 291-mm hole) cutters are available. Carbide cutter tips are available for drilling in dirty ice regions. Penetration shoes of various sizes may be implemented to control the depth per cut/penetration rate. Finally, the cutter head contains six spring-loaded core dogs also made from hardened tool steel. Three different lengths are available to adjust for varying ice conditions. During coring, the core dogs remain retracted into the cutter head, but engage the core when coring is completed and the operator pulls up on the drill string. This action fractures the core from the ice sheet and also holds the base of the core while the drill is brought to the surface.

Common to many ice coring drill designs, the BID incorporates a stationary outer barrel with a rotating inner barrel. The core barrel for this drill is made of fiberglass tubing, which is readily available and provides for a strong, round, straight, and lightweight core barrel. The outer barrel is painted white to reduce solar gain. Three helical flights are wrapped around the outside of the inner barrel, as shown in Figure 6.26. Chips are transported upward by the rotation of the helical flights inside the outer barrel, where they then enter the interior of the core barrel through three small windows. The chips then accumulate on top of the ice core. An optional plastic plug can be installed before drill descent to separate the core from the chips if desired.

The BID utilizes a custom 1.9 kW AC induction motor capable of 1730 rpm. A planetary gear reducer (28:1) provides for a 60-rpm cutter speed. An aluminum and stainless steel housing encloses the motor and reducer components as shown in Figure 6.27. Versions of the motor section are available both with and without a load-triggered slide hammer to assist in core break. Up to six neodymium magnets can be added or removed to adjust the actuation force of the slide hammer. The slide hammer can repeatedly produce clean core breaks. Two versions of the motor section exist. The version of the motor section without a slide hammer is shorter and lighter, and is used with the CRT.

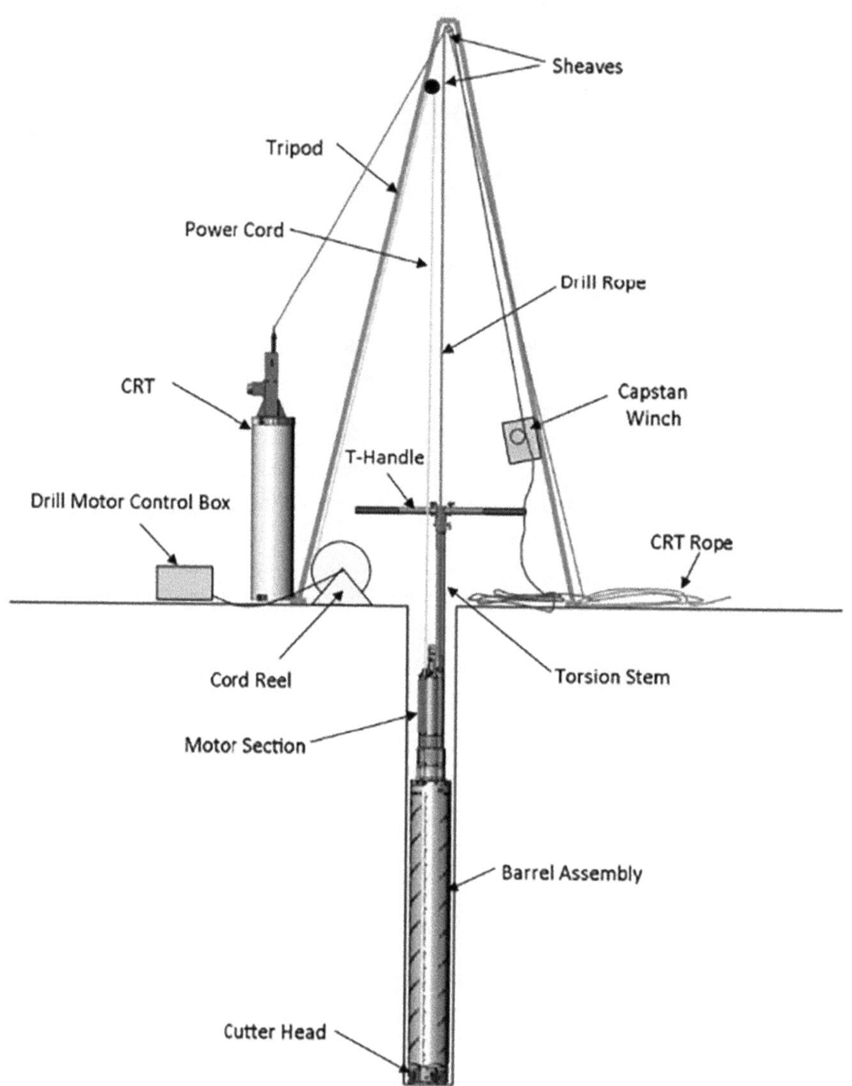

FIGURE 6.24 Original Blue Ice Drill components.

The CRT consists of a 1-m-long fiberglass tube with three core dogs at the bottom. The top of the CRT is illustrated in Figure 6.28. As described in Kuhl et al. (2014) a tilt mechanism, illustrated in Figure 6.29, is attached to the top of the core barrel with three bayonet-style lock-pins. The tilt mechanism transforms a vertical force from the suspending rope to a horizontal force via a cam profile/follower actuating a spring-loaded piston. A vertical force of 700 N is sufficient to actuate the tilt mechanism, breaking the core at the base. The core dogs engage in the core to resist the vertical actuation force and hold the core for retrieval to the surface. Use of the CRT can save time during operations by being deployed to retrieve a previously drilled core while the drill sonde is being cleaned and readied for the next drill run. Use of the CRT prevents operators from having to separate the inner and outer core barrels of the primary drill sonde after each coring run. This method provides efficiencies when drilling to approximately 30 m or less.

Power for the BID system is provided by a single 6.5-kW generator (240 V, 20 A continuous). A custom control box and variable frequency drive convert the 240-V, single-phase generator output

TABLE 6.3
Performance of the BID System

Cutting Pitch (depth per revolution)	15 mm (5 mm depth of cut per cutter)
Rotational Speed	60 rpm (cutter head)
Weight on Bit (WOB)	Minimum possible (negative WOB ideal)
Drill Motor Power	0.5 kW typical (1.5 kW max) while coring
Core Length (max repeatable)	1.15 m
Core Quality	1 piece cores, excellent surface finish in crack-free ice
Coring Rate	1 m per minute
Core Production Rate (maximum)	60 m per 10-hr shift (2 holes in close proximity)
Hole Depth (maximum)	30 m
Hole Inclination (measured)	<0.5°
Contamination	None identified
System Weight (including complete spares and tent)	1580 kg (500 kg with minimal equipment)

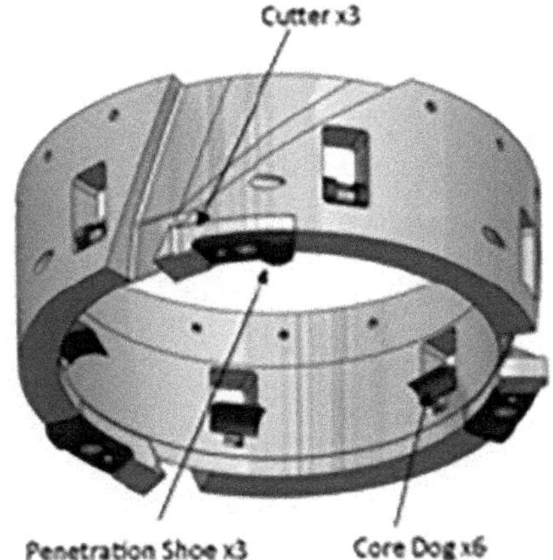

FIGURE 6.25 BID cutter head.

to three-phase power to operate the drill motor. 120-V single-phase power is supplied to two outlets on the control box to power the winch motor and other components. A spring-loaded cord reel sends power down to the drill motor. The control box allows the operator to turn the drill on and off, control its speed and direction, control the variable frequency drive (VFD) settings and has an integrated emergency stop button. A small screen provides the operator with motor load data. The drill can be controlled via either the control box or via two interlock switches on the torsion stem handles during shallow coring. A schematic of the BID power and control system is shown in Figure 6.30.

The BID incorporates a 5.3-m tripod made of aluminum pipe that can be broken down for transport. The tripod can be seen in Figure 6.31. Two of the three legs can be adjusted to ensure the drill hangs plumb when set up on uneven surfaces. Two sheaves mounted on top of the tripod allow for both the drill and CRT ropes to remain installed and ready for use. A capstan winch attaches to one of the tripod legs and is operated by a foot-switch. A custom HDPE sled is available for projects

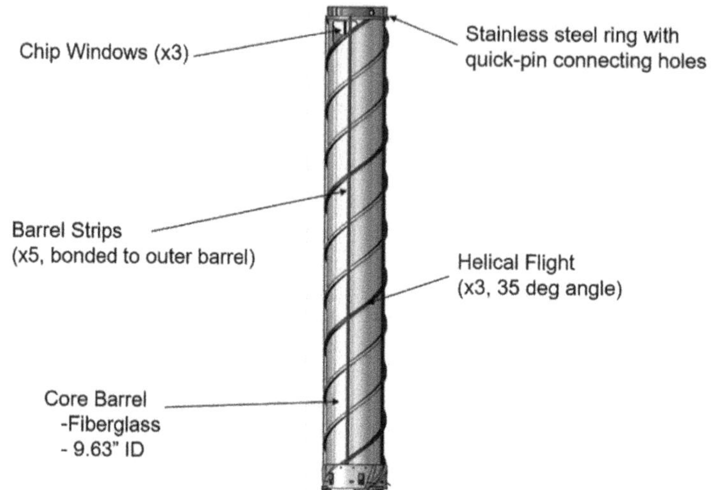

FIGURE 6.26 BID core barrel.

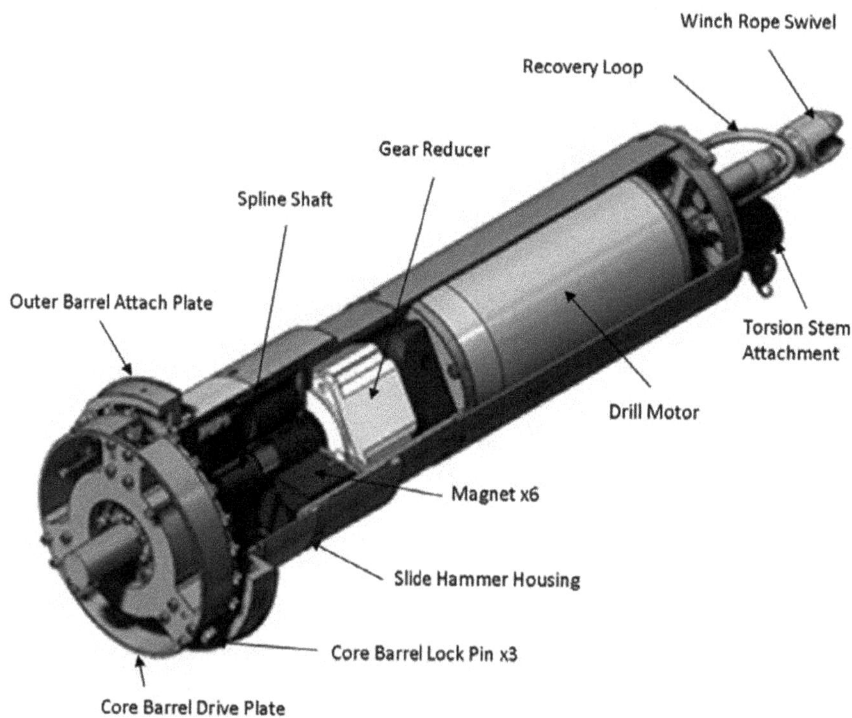

FIGURE 6.27 BID motor section.

requiring numerous drill moves between holes. Due to the height of the tripod, the sled may only be used on relatively flat terrain.

Operation of the BID requires at least two operators. One operator controls the capstan winch for lowering and raising the drill and CRT. This operator also monitors and operates the control box. A second operator operates anti-torque handle (T-handle) attached to the core barrel or rigid extensions. Both operators perform core handling and drill cleaning tasks.

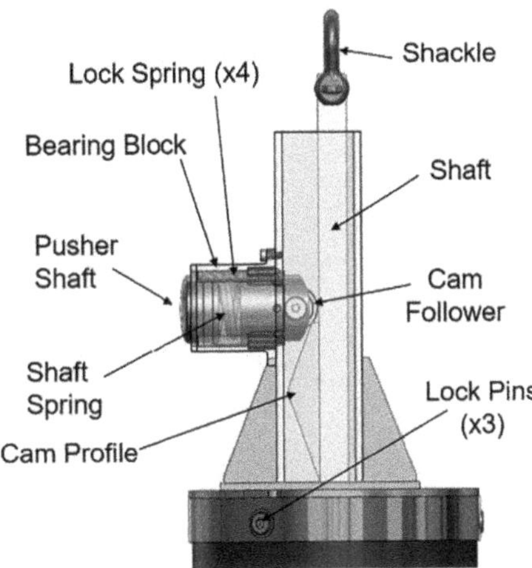

FIGURE 6.28 Top section of the BID core recovery tool.

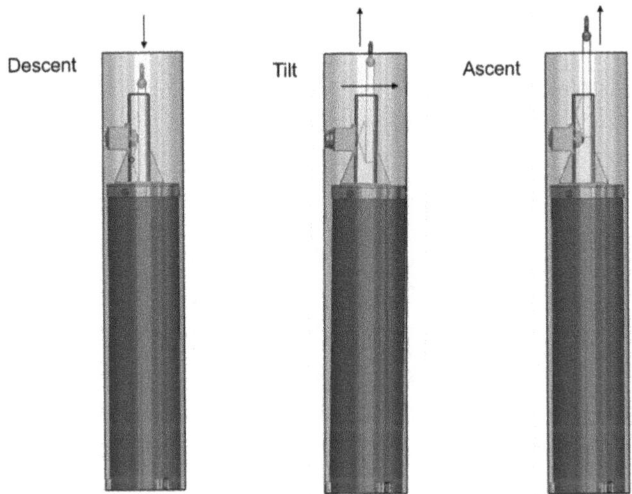

FIGURE 6.29 Core recovery tool tilt mechanism actuation.

6.2.3.3.1 BID-Deep

In light of the success of the original BID design and a desire to reach deeper targets of scientific interest, the US Ice Drilling Program designed and implemented a number of modifications and upgrades to extend the system's depth capability to 200 m. A cable winch was integrated with one of the tripod legs along with a steel electromechanical cable (9.6 mm diameter). The winch utilizes a 4-kW motor with a 119:1 gear reducer. Anti-torque blades were added to the motor section, to provide downhole anti-torquing capability. To accommodate the bend radius of the new steel cable, a larger sheave was implemented on top of the tripod. The BID-Deep is powered by a single 6.5-kW generator. Tie rods are used to join the three tripod legs to increase its load capacity. Improvements to the magnetic slide hammer were also made. Figure 6.32 shows a schematic of the BID-Deep components.

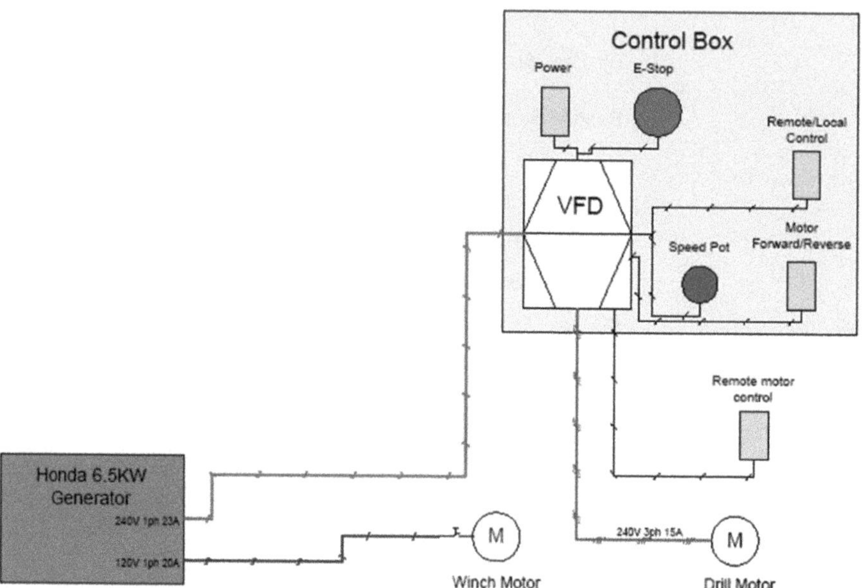

FIGURE 6.30 BID power and control system diagram. The VFD is a variable frequency drive.

FIGURE 6.31 BID tripod deployed in Taylor Valley, Antarctica.

A weather-tight hand-held operator's pendant, connected to the winch via a power cable, allows the winch operator to control both speed and direction of the winch. A hand wheel is also available for finer BID-Deep drill position control. Two emergency stop buttons are incorporated, with one on the pendant and one on the winch itself.

6.2.3.3.2 BID Drill Tent

Due to the height of the BID system, an existing tent design was not found that could accommodate the BID. In an effort to maintain operations during inclement weather, IDP worked with Fabricon

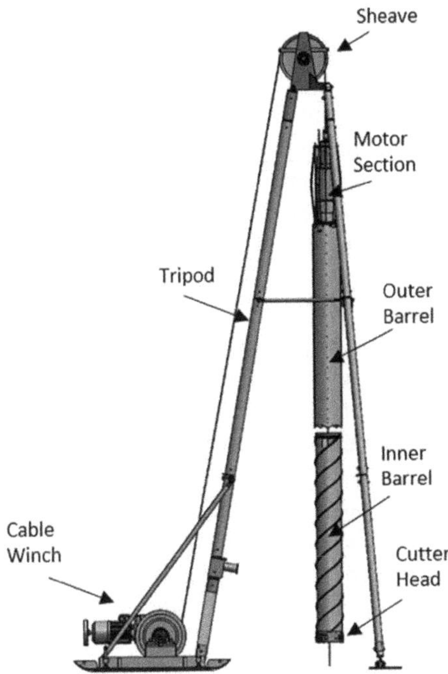

FIGURE 6.32 Model of the BID-Deep system.

FIGURE 6.33 BID drill tent in use at Law Dome, Antarctica.

LLC in Missoula, MT; the custom tent is shown in Figure 6.33. The tent was designed to sustain wind gusts up to 50 knots or 20 psf of snow load. The tent attaches to the BID tripod via a ring mounted near the top of the drill tripod. A pipe frame forms two door arches on either end of the tent as well as provides support for the main arch. Guy ropes are used to secure the tent. White WeatherMax80 nylon fabric helps reduce solar gain inside the tent during drilling operations. The drill tent proved invaluable during its first field use at Law Dome during the 2018–2019 Antarctic field season. A similar tent was also constructed by Fabricon LLC for use with IDP's Foro 400 Drill.

6.2.3.3.3 Cleanliness Requirements

Because the large-diameter ice cores collected by the BID are traditionally melted immediately in the field to capture gases contained in the ice, projects utilizing the BID thus far have required a cleanliness protocol be implemented to ensure the drill is free of contaminants and, historically, to ensure carbon-free sampling. The following describes the pre-shipping cleaning protocol implemented by IDP, as outlined in the BID Operations and Maintenance Manual.

- Metal parts that come in direct contact with the core (coring heads, cutters, dogs, CRT dogs and holders, hardware for all, mango carabiners and rings, etc.) undergo ultrasonic cleaning in acetone, then 190–200 proof ethanol, then DI water.
- Fiberglass and plastic parts (core barrels, outer barrels, CRT Barrel, mangoes, CRT unloading ring, etc.) are scrubbed with ethanol, rinsed with DI water, air dried overnight, and then wrapped in large lay-flat tubing.
- Metal parts are baked at 50°C overnight in a Tenney environmental chamber. Non-stainless steel parts are rinsed with DI water just before insertion into the Tenney chamber to minimize rust.
- Clean small parts are packed in clean plastic parts organizers or new plastic Ziploc bags. Coring heads and other large clean parts can be wrapped in large lay-flat tubing.
- Plastic organizers must be cleaned before putting in clean parts if they are new or were contaminated during a previous season. Ethanol and DI water are used for this application.
- The remainder of the sonde(s) is wiped down with ethanol to degrease it before being packed. Anything that goes downhole should be degreased.
- Clean coring heads are reassembled and the heads, barrel sets and CRT barrels and core dog mounts are then wrapped in lay-flat tubing and packed into cases.
- Approximately 4 gallons each of acetone and 190–200 proof ethanol are needed for the cleaning process.

6.2.3.3.4 Performance Experience

The BID can quickly produce large amounts of core during a single field season. The BID was first used in Taylor Valley, Antarctica during the 2010–2011 field season where approximately 575 m of ice were drilled. The drill has deployed regularly for several Arctic and Antarctic field seasons. Production rates as high as 1200–1400 m have been achieved during a 6–8 week long field season.

The BID-Deep system was first tested in Greenland during the 2014 Arctic field season, reaching a depth of 187 m (approximately 80 m of firn and the rest ice). The BID-Deep capabilities were further tested in Taylor Valley in Antarctica during the 2014–2015 field season down to a depth of 70 m in a blue ice area with almost no firn. The drill system has since been used in both the Arctic and Antarctic in blue ice areas and in areas with overlying firn. In the majority of projects, deterioration of core quality was observed approximately 70 m below the firn-ice transition, or 70 m below the surface in blue ice areas. Core quality was very poor with severe fracturing resulting in cores made up of 3 to 10 pieces per meter. Numerous combinations of drill parts and drilling techniques were tried with no noticeable improvement. Depth capability is largely influenced by site/ice characteristics. The current equipment is likely reaching its design limits; ensuring good quality core collection to depths of 200 m would require a redesign of the system.

6.2.3.3.5 Modifications

In an attempt to improve core quality at depth, IDP engineers surmise that the fiberglass core barrel may not be sufficiently rigid for deeper drilling, because the fiberglass may be inducing flex and breakage in the cores. A carbon fiber core barrel has been designed and the barrel will be tested in Antarctica during the 2019–2020 field season.

During the first use of the drill tent with the BID during the 2018–2019 Antarctic season, IDP engineers determined that use of the tent requires a stronger support structure than the BID tripod, as well as a simpler, safer erection method. To this end, IDP engineers have designed a new truss-style tower, floating base, and new sheaves for use with the BID-Deep. The new designs will be tested during the Antarctic 2019–2020 field season. The tripod may still be used for shallow coring configurations without the tent.

6.3 CORING ICE-ROCK COMPOSITES

The demise of glaciers and ice sheets under current climate change, with accompanying rise in sea level, has created urgent scientific questions about the extent of ice sheets during the previous interglacial period. Cosmogenic nuclides in bedrock beneath ice sheets reveal clues to the former extent of ice sheets and the timing and duration of past exposure periods. Several innovative technologies for retrieving bedrock core from beneath hundreds of meters of glacial ice using nimble methods have been developed and used by the US Ice Drilling Program. Designed for reconnaissance recovery of short rock cores for cosmogenic nuclide techniques to quantify periods of exposure (ice free) and burial (ice cover), the Agile Sub-Ice Geological Drill (ASIG) and the ice-adapted Winkie Drill are useful for retrieving meters of bedrock core near glacial rock outcrops and near the ice sheet margins.

6.3.1 Agile Sub-Ice Geological Drill

Under ice less than 700 m thick, nimble methods for reconnaissance recovery of small rock cores are needed for use near outcrops and near the ice margins. The US Ice Drilling Program organized iterative discussions between scientists and the IDP-Wisconsin engineering staff in order to create the following IDP Science Requirements for the Agile Sub-Ice Geological Drill:

- Produce 700-m borehole to base of ice with drilling and retrieval of 10 m of bedrock core and or unconsolidated frozen sediment core.
- Ice drilling will include the possibility that the ice is entrained with rocks.
- Ice drilling will be to dry, frozen-bed conditions, and will not be done in areas where there is subglacial water.
- Retrieve several short ice cores (~50 cm long) at up to 700 m depth.
- Ice drilling may be in ice that is within 2.0°C of the pressure melting point.
- Required ability to drill at ice borehole temperatures as low as −40°C, and surface temperatures as low as −30°C.
- Retrieve 10 m of bedrock cores of maximum 33 mm (1.3″) diameter beneath the ice sheet.
- Maximum site altitude for the design should be 2500 m.
- Maximum time at a site, including setup and core retrieval, should be six days.
- Stand-alone capability is needed for operation at small field camps at remote sites.
- Minimal staff (four) for drilling operations in the field; other field camp staff in support of drilling operations to be provided separately.
- Drilling fluid or a fluid "system" (to be determined) will be immiscible with water.
- Drilling fluid should not be a boron-rich fluid.
- Drill system must be transportable by Twin Otter aircraft, or helicopter with sling load.
- Drilling depth of each core collected should be determined and recorded.
- Drilling and core handling history should be recorded.

The Agile Sub-Ice Geological Drill (ASIG) Drill is the first drill with the capability of retrieving meters of rock from under hundreds of meters of ice. The drill system design is based on a commercially available conventional drilling rig with rod extensions, as commonly used for minerals exploration, which IDP has adapted for drilling through ice and for ice coring. The system is designed to drill access holes through ice less than 700 m thick and subsequently collect meters of bedrock cores from beneath glaciers.

6.3.1.1 System Overview

The ASIG Drill system uses a modified version of the Discovery MP1000-Man Portable Core Drill Rig from Multipower Products Ltd.; the schematic is shown in Figure 6.34. The system is designed to be field-portable by Twin Otter aircraft for example, as illustrated in Figure 6.35. Specifications for the system are provided in Table 6.4. Permeable layers are cased and sealed to impermeable ice

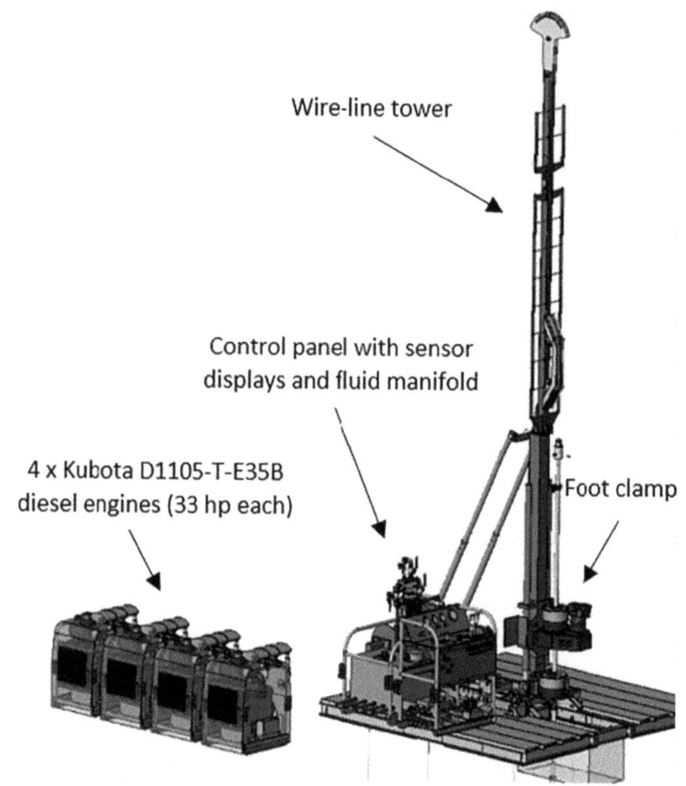

FIGURE 6.34 As-built CAD illustration of the ASIG Drill rig.

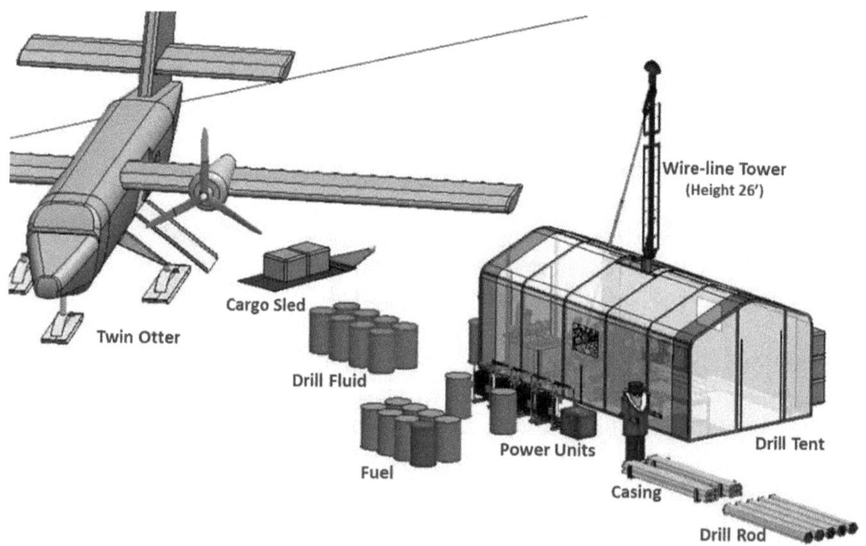

FIGURE 6.35 Schematic of a typical ASIG Drill site. *Source:* Courtesy of Gibson et al. (2015).

TABLE 6.4
IDP ASIG Drill System Specifications

Drill Type	Surface driven rock coring rig
Power Unit	Four Kubota D1105-T-E35B diesel engines (33 hp each)
Drill String	Rigid, single wall drill rod (Sandvik WL56)
Rod Tripping Mechanism	Rig mast hydraulics/chuck
Drill Fluid	Isopar K (Exxon-Mobil)
Fluid Filtration	Continuous-shaker table, secondary filter, chip melter
Rod/Core Barrel Configuration	Sandvik WL56 thin-kerf metric
Core Size [mm]	39 (larger core possible with different drill rod)
Maximum Core Length [m]	1.5 or 3.0
Available Bit Configuration	Hardened steel (ice), diamond-impregnated, Geoset, PDC
Depth Capacity [m]	700 (~1500 m max with modifications, needs testing)
Drill Rod Material	Steel
Rod Weight [kg/m]	3.8 (for 39-mm core)

with an inflatable packer. The system uses industry-standard downhole tools with minimal modification. Ice is drilled in a continuous manner with a full-hole bit to create an access hole. Traditional rock coring equipment and techniques are used for sub-glacial rock sampling. Custom and off-the-shelf rock coring bits complement the ice drilling bits to core ice, rock, and transitional ice layers containing sediment; the drill bits, which require continuously circulated drilling fluid, are depicted in Figure 6.36.

Rigid pipe is used when cutting chips, creating an access hole and when drilling ice and rock core. For ice and rock core collection, a wireline recovery system using the Sandvik WL56 core barrel assembly is deployed down the center of the drill pipe. This method greatly reduces cycle-time as compared to removing the rigid pipe drill string to recover core. After the casing is set, continuous drill fluid circulation and chip filtration is maintained by a positive displacement piston pump and a custom filtration system. Drilling fluid is continuously recycled during drilling and the vast majority of the drill fluid can be recovered at project completion and reused. Circulation pressures can be more than 100 psi (0.69 MPa), so it is critical that drilling be performed in competent ice in an environment with a frozen bed. Reverse circulation is used for cutting chips and creating an access hole and normal or forward circulation is used when a core has been collected. Adaptation of the system for use in environments with an aqueous ice-rock interface is conceptually possible and would require additional engineering development.

6.3.1.2 Logistics

To address the need to deploy the system to remote sites, the system is designed to be useful with minimal deep field logistics to be transported via light fixed-wing aircraft, helicopter, or tractor traverse. Heavy equipment is not required for assembly at the field site. When heavy equipment is available, however, it does speed operations in setup, teardown, and transport between holes. All of the ASIG Drill system equipment is limited to a 600 lbs. max single-piece weight. Total weight of the system is highly dependent upon project requirements including the number of holes, the depth of required casing, and total depth to bedrock; weights for two example applications are listed in

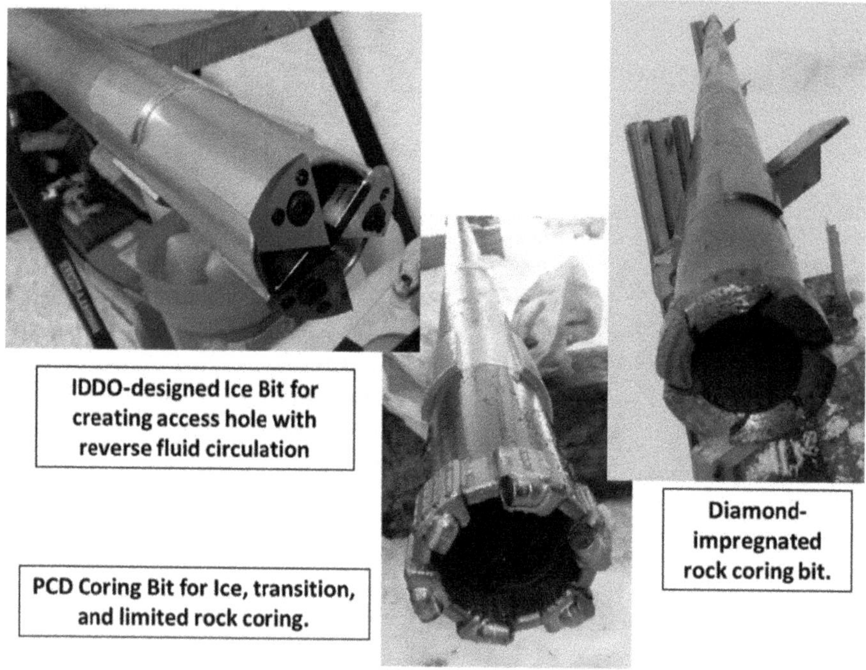

FIGURE 6.36 Drill bits for ice, rock, and mixed media in transition zones.

Table 6.5. Spare components and extra drilling fluid are recommended, but may increase system weight significantly.

The approximate required time on-site for drilling operations is estimated as follows:

- 200-m hole with 10-m core recovery = 100 working hours (four–five people)
- 700-m hole with 10-m core recovery = 150 working hours (four–five people)

These are approximate values. In practice, large casing depths, drilling problems, mechanical issues, adverse weather, etc. may significantly increase hours to completion and would need to be included in planning estimates. Also useful in planning estimates is a breakdown of drilling times for specific activities, as provided in Table 6.6.

6.3.1.3 Field Deployment

The ASIG Drill system was tested in a 50-ft. ice test well at the University of Wisconsin-Madison in February, 2016. From November 2016 to January 2017, the drill system was successfully deployed to remote west Antarctica near the Pirrit Hills in support of the Ex-Probe science project. Drillers used the ASIG Drill system to drill through approximately 150 m of ice and then collected 8 m of 39-mm diameter rock core of excellent quality. Nearly 5 m of ice core of poor quality was also collected near the ice-bedrock transition.

Two holes were attempted in this initial field season. In the first hole, a casing was set with the inflatable packer and drilling continued to approximately 90 m of ice. At that depth, a fracture of the ice formation occurred and drill fluid pressure was lost, stopping circulation and forcing a halt to drilling. The drill was disassembled and transported to the second site.

At the second site, the drill was reassembled, casing was set and drilling continued to the target ice depth of approximately 150 m. By pulling the drill string and changing to appropriate coring bits,

TABLE 6.5
Sample ASIG Drill System Weights for Shallow (200 m) and Deep (700 m) Projects. Weights are Given in lbs. (kg)

Drill Equipment	200 m	700 m
Drill Rig lbs. (kg)	4565 (2070)	4565 (2070)
Drill Rod lbs. (kg)	1900 (861)	6971 (3162)
Tools/Equipment lbs. (kg)	6533 (2963)	6608 (2997)
Total Equipment Weight lbs. (kg)	13,088 (5936)	18,144 (8230)
Twin Otter Flights[a]	6–8	8–10
Consumables	**200 m**	**700 m**
Casing lbs. (kg)	310 (140)	310 (140)
Drill Fluid lbs. (kg)	2749 (1247)	7588 (3442)
Fuel lbs. (kg)	2649 (1201)	4013 (1820)
Total Consumables Weight lbs. (kg)	5708 (2589)	11,911 (5403)
# Twin Otter Flights[a]	3–4	6–8
Total		
Total System Weight lbs. (kg)	18,796 (8526)	30,055 (13,633)
Total # Twin Otter Flights[a]	9–12	14–18

[a] Number of flights based on a 250 nautical mile flight and the standard fuel capacity with no optional cabin auxiliary tank. Actual cargo capacity will vary with specific conditions and the amount of fuel required.

TABLE 6.6
ASIG Drill System Performance and Operation Values

Number of Operators	3 drillers, 1–2 core handlers
Initial System Assembly (hours)	30
Time-to-Depth (200 m, 10-m core, hours)	50
Time-to-Depth (700 m, 10-m core, hours)	100 (estimated)
Pilot Hole (auger, casing, m/h)	10
Access Hole Drilling, total (m/h)	8
Coring, Total (m/h)	1
Auger Max. ROP (firn, m/min)	1
Ice Max. ROP (full hole, m/min)	1
Rock Max. ROP (coring, m/min)	0.15
System Disassembly/Packing (hours)	20

FIGURE 6.37 Drilling with the ASIG Drill in the drill tent at Pirrit Hills, Antarctica.

FIGURE 6.38 The first subglacial rock core drilled using the ASIG Drill system. Core breaks are performed using a collet inside of the drill head.

5 m of basal ice and 8 m of granite core were recovered. Core breaks in the granite were achieved through use of a collet and pulling power of the drill rig. Figure 6.37 is a photograph of the drill rig in operation within the drill tent. The first successful retrieval of meter-scale rock core from beneath glacial ice is shown in Figure 6.38.

6.3.1.4 Future Work

Based on experiences of the first field season of drilling with the ASIG Drill, a number of system modifications have been identified to simplify operations and improve performance. Several modifications may significantly reduce the time required to set casing. Casing must be set at a depth beneath the firn-ice transition depth where impermeable ice is adequate to support the packer and fluid pressures. In order to determine the appropriate packer depth, it is helpful to collect ice cores and measure their density. To this end, new augers with a central clearance hole will be implemented to rapidly create the casing pilot hole and facilitate ice core sampling. Coring tools will be developed to quickly collect the ice core through the auger using the existing wireline core recovery equipment. As a complement to the pilot hole augers, a wireline bailer system will be developed to remove cuttings left behind after the augers are removed from the pilot hole.

By improving cycle time for drilling the main ice access hole and retrieving ice core, the potential exists to significantly decrease time in the field. By modifying the foot clamp, a wider range of tools can be accommodated without disassembling the equipment. A piston-driven foot clamp will be installed, accommodating tools up to nearly six inches in diameter. This will accommodate drill rod, casing, augers, reamers, and all drill bits.

The existing filtration system used a screw press to remove ice chips from the drill fluid. This required significant maintenance, demanding the attention of a third person during drilling. To reduce the need for this effort, a shaker table will be installed that is potentially maintenance free. After the bulk of fluid is removed, a melter tank using waste heat from the hydraulic power packs proved very useful in recovering the remaining fluid in the first field season. This melter tank will be enhanced for efficiency and ease of operation. Fluid recovery will be further improved by use of additional splash guards during all drill operations. These not only minimize losses but also improve cleanliness of the drill site.

In addition to improvements to filtration, the drill fluid circulation system will be enhanced to improve drilling fluid pressure monitoring and control. A pressure accumulator, Flexicraft MHY1650500, will be added to the outlet of the pump to dampen pressure spikes and improve stability of the gauge reading. A high precision pressure relief valve, Sun RPGE-LEN, will replace the existing pop-off valve that exhibited an excessively large activation range of over 25 psi. A simple cartridge sieve will also be added to protect the piston pump from any foreign material.

These modifications make the ASIG Drill a capable system, although potential for further improvements remains. Continued efforts in managing drilling fluid pressures will be of particular benefit going forward. Two areas of interest in managing fluid pressures are the quantification of drill fluid pressures and acceptable limits during all stages of drilling operations, and the design of equipment to allow drilling through a wet ice-rock interface or less competent ice.

6.3.2 Ice-Adapted Winkie Drill

Many scientifically interesting areas of Greenland and Antarctica are in locations that challenge the logistical capabilities of any nation. Hence there is a large need for small, light, agile drills that can be easily transported with the drilling completed with only several drilling staff. The US Ice Drilling Program has adapted a commercially available Winkie rock drilling system from Minex, and has modified and upgraded it to add ice augering and ice coring capability for a total drilling depth of 120 m. The resulting ice-adapted Winkie system can be transported by a Twin Otter or similar-sized aircraft.

The as-purchased Winkie system is capable of rock coring to a maximum depth of 120 m with AW34 drill rod, producing core with a diameter 33.5 mm. If smaller EW drill rods are used, the drill can reach a maximum depth of 145 m. The system uses an 8 horsepower, 2-stroke, gasoline engine to power the drive head. Power from the engine is transferred through the two speed transmission before being applied to the drill rods. The system is top driven; the drill rods are screwed to the drive shaft of the drive head rather than clamped into a chuck. The drive head is lifted and pushed with a hand wheel that allows for a maximum 1.53-m stroke length. The published weight of the drive head and frame is 84 kg. However, this does not include drill rod, downhole tooling, circulation components, or spares. The drill can easily be disassembled into pieces manageable by teams of two people. The drill was designed to be brought to extremely remote locations as a probing or exploration rig so man-portability is paramount to its success.

The simplicity and minimal weight of the Winkie Drill makes it an ideal solution for rock coring in the logistically difficult polar regions of the world. Several key modifications needed to be made to adapt the Winkie Drill into a useful tool for collecting subglacial samples; the gasoline engine was replaced with a more reliable electric motor, a base was built to anchor and support the rig on ice or firn, oversized drill rods and core barrels were procured, an access borehole system was developed, specialized bits were procured, and a closed circulation system was designed.

The ice-adapted Winkie Drill was first deployed to Antarctica in 2016. The rig was sent to the field with the gasoline engine that came installed from Minex. The engine performed well, but the reliability of the engine was a weak point in the system. Due to the extreme isolation of many projects proposed for the ice-adapted Winkie Drill, it was decided that the engine should be upgraded to a more reliable alternative. Winkie systems have been retrofitted with hydraulic motors by several independent operators. This option was explored but eventually ruled out due to the increased system weight and complexity. An electric motor proved to be ideal because it uses already-available electric power produced by a generator, is reliable and requires minimal auxiliary equipment, and it can be safely operated inside a drill tent.

Many motor options were explored but few could closely match the power and speed produced by the US820 gasoline engine, which is visible in Figure 6.39. The closest option available is the Evo3 brushless DC motor by Sonceboz. The motor included a fully integrated controller in a sealed housing. Additional equipment required to operate the motor includes a 3-kW AC-DC rectifier and simple drive head-mounted control box. As a direct replacement for the US820 motor, the Evo3 was less powerful; however, its reliability, simplicity, and improved operator comfort justified any reduction in penetration rate. Another added benefit to the Evo3 motor is the ability to add modular gear reducers between the motor and powerhead, optimizing speed and torque for specific conditions. For example, the 2017–2018 Ong Valley project requested oversized cores consisting of a mixture of ice and rock debris. To achieve the required torque while also slowing down the surface speed, a 3:1 gear reducer was added to the assembly. The graph in Figure 6.40 compares the raw torque of several gasoline engines and electric motors.

The conventional Winkie Drill uses a small base to support the drill during operation. The base is bolted directly to the rock formation being sampled with concrete anchors. This is not an option

FIGURE 6.39 The Winkie Drill as operated in the Ohio Range, Antarctica. Note the gasoline engine driving the powerhead and aluminum pallet base.

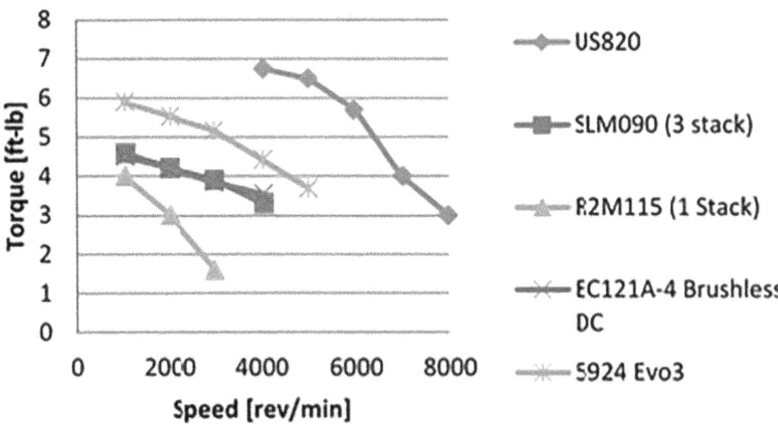

FIGURE 6.40 Comparison of the US820 gasoline engine with various electric motors.

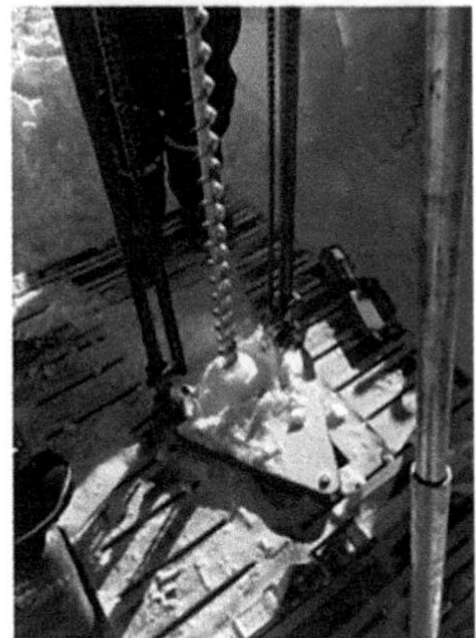

FIGURE 6.41 At the Ohio Range, access holes were drilled in ice using Kovacs 2-inch augers.

when operating on an ice or firn surface. A custom aluminum pallet with HDPE base was fitted to the drill to distribute core break loads onto the glacial surface as well as provide a solid and safe work platform, as seen in Figure 6.39.

The most significant capability added to the system was the ability to drill access holes through glacial ice to reach the desired subglacial samples. First implemented and tested was the use of a continuous string of augers to quickly drill through blue ice and make contact with the bedrock below. A test of the augering concept was conducted during the 2015–2016 Antarctic season at a site near Crater Hill outside of McMurdo Station. Two auger systems were tested, the 2-inch Kovacs augers proposed for use with the Winkie Drill and custom 5.75-inch augers proposed for use with the ASIG Drill. The test was beneficial in determining the power requirements for augering as well as the efficiency of the auger string at clearing the hole of chips. The 2-inch augers required much lower torque than expected; a 30-m string could be rotated by hand even when full of chips. Even the 5.75-inch augers could be operated with the ice-enabled Winkie Drill although it has a fraction of the power of the ASIG Drill. The most significant finding of the test was the inefficiency of the augers to clear chips; chips filled between 46% and 27% of the access borehole.

The Kovacs augers were again used during the 2016–2017 field season when the ice-enabled Winkie Drill was deployed to the Ohio Range, as shown in Figure 6.41. This site was not covered

TABLE 6.7
Ice Chip Depth that Remained in the Borehole after Augering at the Ohio Range was Measured to Determine the Efficiency of the Augers at Each Borehole

Borehole	Borehole Depth (m)	Depth of Chips in Borehole (m)	Fill Ratio (%)
#1	26.5	0.99	3.7
#2	12.1	N/A	N/A
#3	12.9	0.55	4.3
#4	27.0	0.80	3.0
#5	28.3	1.23	4.3
#7	25.5	0.5	2.0

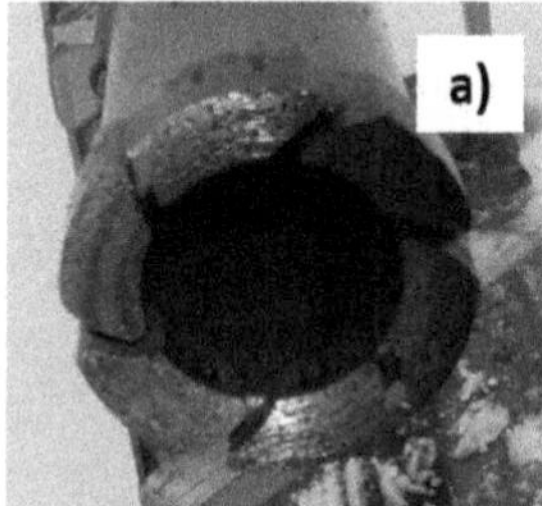

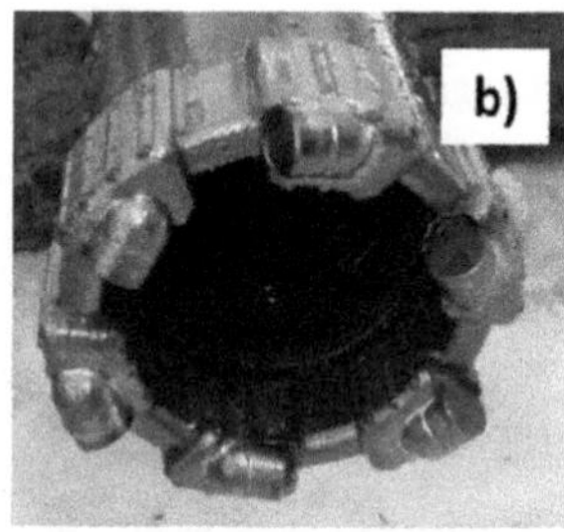

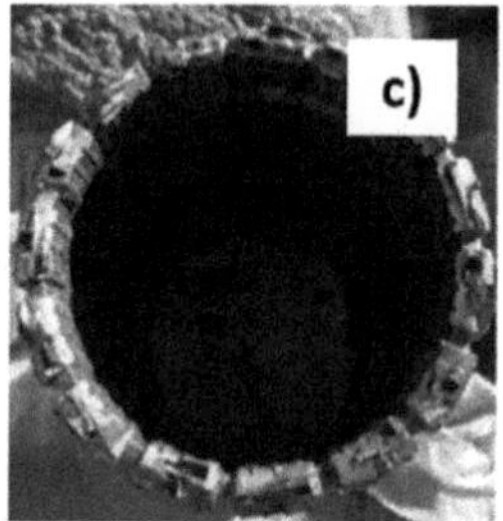

FIGURE 6.42 Photographs of the drill bits deployed for drilling mixed-media ice-rock mixture cores in Ong Valley: (a) diamond-impregnated bit, (b) PDC bit, (c) Geoset bit.

in firn, rather the solid glacial ice was at the surface at this location. At this bare ice site, the augers were much more effective, leaving less than 5% of the borehole filled with chips in boreholes as deep as 28 m. A comparison of the situations is given in Table 6.7. The huge reduction in ice chips was the result of drilling through solid ice, rather than drilling through porous firn, before reaching the underlying rock. The solid ice borehole allowed the drill string to be spun as fast as 1500 rpm to clear the chips. At sites that are covered in firn, the borehole could not support the augers at that speed; the augers would become less stable and the borehole would be enlarged as a result. Once bedrock is reached, conventional downhole tooling is utilized. A double barrel, AW34 core barrel collected rock samples at the Ohio Range.

The next project, which was conducted during the 2017–2018 season in Ong Valley, Antarctica, involved drilling through ice-rock mixed media and required retrieval of larger cores, so an oversized 86T2 core barrel was adapted to the drive head. Neither this nor the system used at the Ohio Range system utilized a wireline retrieval system due to the relatively shallow coring depths and limited coring runs required. For both coring systems, several bit options were deployed with the drill: diamond-impregnated, PDC, and Geoset, as depicted in Figures 6.42 to 6.44. The most effective bit for solid rock is a diamond impregnated bit. Soft matrix bits were chosen in part because of the hard rocks predicted at the sites but also because limited core is required and generally a softer matrix produces a higher penetration rate. Maximizing the penetration rate through rock is beneficial in the case of total fluid loss drilling.

For mixed-media drilling (see rock-ice mixed-media cores in Figure 6.43), it has been found that a Geoset-style bit (shown in Figure 6.44) is most effective. Penetration rates differed greatly depending on the media being collected but varied from 0.4 to 3.0 cm/min. The fastest penetration rates

FIGURE 6.43 Mixed-media cores recovered from Ong Valley. The most effective bit for drilling this type of media was found to be a Geoset-style cutter.

FIGURE 6.44 This Geoset-style coring bit was used exclusively in Ong Valley where only mixed media was cored.

occurred in almost dry rubble and the slowest through large rocks where the full kerf was engaged. An impregnated bit is ineffective whenever ice is present at the cutting surface because the ice does not fracture with the same mechanism as rock so an ice glaze is formed over the bit and no penetration is possible. PDC bits were also tested during the Ong Valley project. A PDC bit more closely resembles a traditional ice cutting head with individual teeth that shave ice rather than crush. The PDC bit was not effective during that project, always plugging the water-ways and stopping fluid circulation. This may have been the result of insufficient fluid flow rate or insufficient annular space for chip removal.

The fluid circulation scheme is designed as a closed system, filtering Isopar K drilling fluid to remove particulates before pumping it back down the borehole. Ideally, there is very limited fluid loss; however, the cumulative effect of handling wet rods and evaporation leads to fluid loss. The mud pump recommended by Minex is a portable, gasoline powered unit. To reduce weight and minimize the number of engines running, an electric pump was instead procured. The pump was fitted with a motor drive to allow for flow rate control. The assembly uses a triplex piston pump and can produce 300 psi and a 3-gpm flow rate. A complete, spare pump is sent with the system to the field and can be added to the system to increase the flow rate if needed. The rig can operate only in forward circulation. The Ohio Range project utilized a cavity dug into the blue ice at the surface to

FIGURE 6.45 Sump pit cut from the solid blue ice at the surface. The drill fluid is pumped through the drill rod and returns to the surface where it is collected in the sump.

act as a sump, collecting the returning drill fluid where it could be pumped back into the filter tank, as shown in Figure 6.45. The filter tank is a 66-gal HDPE tank that uses gravity to filter the dirty fluid through 50 or 100 micron filter socks. When drilling is complete, a fluid bailer is used to collect the drill fluid that is left in the hole after the drill rods are removed. The fluid bailer is simply a drill rod with a one-way valve at the bottom, allowing fluid to fill the tube but sealing the gate while the tube is brought to the surface.

When solid ice is not present at the surface but instead the surface is covered by meters of porous firn, the firn borehole must be cased. The Winkie Drill uses thin wall BTW drill rod as casing. The BTW inner diameter is nearly identical to the AW34 borehole diameter, 1.909 inches and 1.895 inches respectively, and is 10% lighter than standard AW casing. The casing is sealed to the borehole with an inflatable packer designed by QSP Packers, LLC. The packer has a 2.0 inch through diameter and 3.5 inch resting diameter. The inflatable element length is 34.5 inches and can be inflated to a maximum diameter of 6.5 inches to a pressure of 700 psi. The element has been wrapped in nitrile rubber for chemical compatibility with the Isopar K drilling fluid. The packer is inflated by air, compressed with a portable air compressor and dried through a desiccant. This packer has not yet been used in the field but is scheduled for deployment with the system during the 2019–2020 Antarctic season. Similar inflatable packers have been successfully implemented in ice with the ASIG Drill.

The ice-enabled Winkie system has successfully completed two Antarctic campaigns, and it is instructive to compare the operations at both. The first campaign to the Ohio Range resulted in the retrieval of six sub-glacial rock cores. Borehole and sample depths can be found in Table 6.8. The five-person crew and all the camp and drill gear was flown to the drill site in six Twin Otter flights. The camp was in place for 25 days, with 15 of those days dedicated to drill operations. All boreholes were through solid ice at the surface, without the presence of firn. One borehole intersected a crevasse, but sufficient drill fluid was available at the site to collect the core with total fluid loss. A total of 160 gallons of Isopar was lost during the season.

The second deployment of the system was to the debris-covered Ong Valley, at the site shown in Figure 6.46. A seven-member crew was deployed for 28 days, including 17 days of drill operation. Two boreholes were drilled, each collecting samples from the surface to the final depth. The first

TABLE 6.8
Ohio Range Sample Depths and Core Lengths

Borehole	Depth (m)	Rock Core Length (cm)
#1	26.53	57
#2	12.08	38
#3	12.90	67
#4	27.00	Gravel
#5	28.33	60
#6	30.00	n/a
#7	25.50	28
#8	54.86	n/a

FIGURE 6.46 The ice-enabled Winkie Drill as assembled at borehole #1 at Ong Valley.

borehole reached a depth of 9.45 m and the second, 12.36 m. Unlike the bedrock samples collected at Ohio Range that were solid rock, the mixed-media cores from Ong Valley were a mixture of ice and rock particles. The mixture varied from clear ice veins to almost dry sand, as can be seen in Figure 6.43. Only 20 gallons of Isopar K was lost over the entire season.

Deployment of the ice-enabled Winkie Drill during the 2019–2020 Antarctic season will be at a site that has a significant amount of firn covering the glacial ice. This project will leverage the use of the existing IDP Badger-Eclipse Drill to create the access boreholes through the firn and ice to bedrock. This decision was made in part because using augers would result in a significant amount of chips left in the borehole, and also for weight and ice sampling considerations. The method used to create an access hole through firn and ice to the underlying bedrock depends on site conditions, available logistics, and nature of cores required by the science project.

6.4 DRILLING ENGLACIAL ACCESS HOLES

On topics ranging from glacial dynamics, sedimentary systems, and polar biology, scientists often seek to measure characteristics of the environment within or beneath the ice sheet, and hence they need a hole for access into or through the ice, but not an ice core. Similarly, for seismic studies, holes are needed down to depths of approximately 100 m to serve as shot holes for explosives. Drilling access holes in the ice can often be achieved at a much faster rate than retrieving ice cores, although for deep drilling, the amount of equipment and fuel becomes large in either case. The following sections describe various methods of drilling holes in glacial ice, and the resulting performance of the drills.

6.4.1 Rapid Access Isotope Drill

A novel method of drilling has been developed that creates a borehole but also permits low-resolution isotopic sampling of the chips retrieved from drilling the hole. The Rapid Access Isotope Drill (RAID) is a novel, field-proven technique for rapid ice drilling to a theoretical depth of 600 m that has been developed by the British Antarctic Survey (BAS) (Rix et al., 2019). It has successfully drilled to 461.58 m in approximately 104 hrs at Little Dome C in Antarctica, creating a borehole and also retrieving the chips from the drilling that facilitated 20-cm resolution isotopic sampling of the ice to 25 kyrs.

The RAID was designed to help in the search for suitable sites for a deep ice core through kilometers of ice where ice older than one million years may exist. For this search one of the most important unknowns is the Geothermal Heat Flux (GHF). Ice sheet modeling suggested that only a short temperature profile of the upper 20% of the ice sheet is required to constrain a model that would allow reasonable estimates of GHF. The accuracy of the GHF and the basal temperature estimates is greatly improved when ice samples are also collected to give a paleo-accumulation record, and vertical advection measurements are made using phase sensitive radar (pRES). This modeling and operational requirements in the areas where one-million-year-old ice is likely to be found were used to create the following science requirements for the RAID:

- Produce 600-m borehole as quickly as possible to allow the deployment of temperature sensing system to measure the borehole temperature profile.
- Provide ice samples for stable water isotope analysis.
- Total system to fit in and be light enough (<1000 kg) for a single Twin Otter aircraft to facilitate easy access to remote sites.
- Drill to work in ice at temperatures close to −55°C.
- System to be easily set up and operated by two persons.
- Operation to be simple as operators may be working outside at high altitude.

In order to reach the system weight requirement, it was decided that the RAID would be a fast cable-suspended electromechanical ice drill. The drilling time for a cable-suspended drill becomes dominated by the winching time as the depth increases. By carrying out full-face drilling, no core breaking is required and the winch can be geared solely for tripping up and down the borehole as fast as possible. Ice chips are collected, not an ice core, so core quality is not a consideration, allowing aggressive cutting of the ice. These chips can then be ejected from the drill sonde on the surface by reversing the motor, in the vertical position, minimizing time on surface.

The complete drill sonde is shown in Figure 6.47, which was developed over two test seasons and two field science seasons. A 3-inch (76.2 mm) diameter was chosen for most of the sonde as this allowed for stock off-the-shelf tube to be used for many components. This diameter is a compromise, as a smaller diameter hole requires less power to cut through the ice; space constraints become a problem when smaller than this. The anti-torque section is a reduced size version of the BAS shallow drill which uses both springs and a cam to press six blades against the borehole wall. The motor section houses a 400-W brushless motor with an epicyclic reduction gearbox along with power conversion electronics and a motor controller. An unusual feature in the design is that the outer barrel is attached to the cutters and rotates while a central auger spiral is stationary. An off-the-shelf 10 SWG (3.125 mm) thick barrel was chosen to form the chip collection chamber, at nearly the full allowable length to fit into a Twin Otter aircraft. The barrel section with auger spiral is shown in

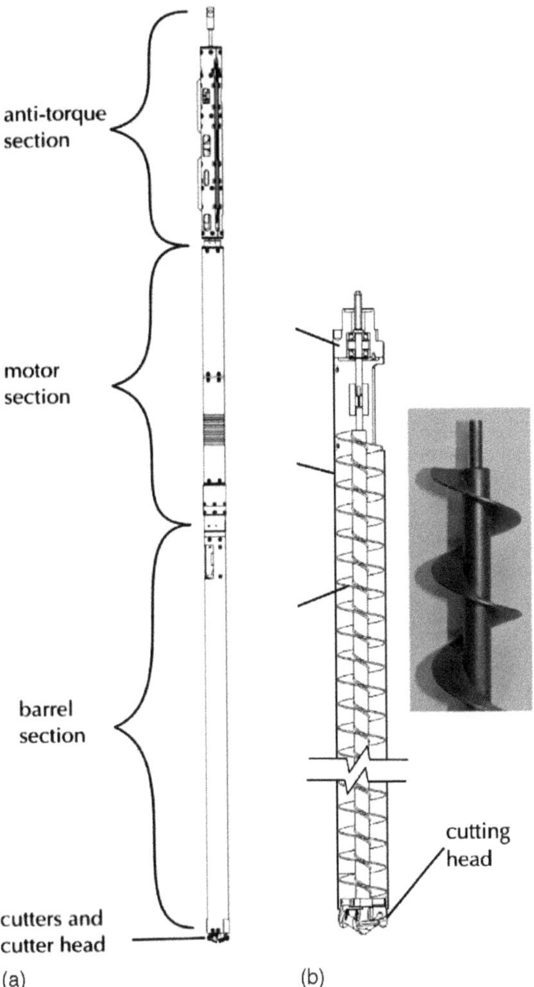

FIGURE 6.47 (a) RAID drill sonde showing four main sections of the drill (with short barrel attached). (b) shortened sectional view of barrel showing left-handed inner stationary auger spiral for chip transportation. A photo of the small section of the Xylan coated auger spiral is shown inset. *Source:* Courtesy of Rix et al. (2019).

more detail in Figure 6.48a. Scoops at the bottom of the barrel, visible in Figure 6.48b, push chips onto the spiral and the spiral directs them up the barrel in roughly the order that they were drilled. This design allows the outer barrel to be the higher torque carrying component, rather than the auger spiral, though this necessitates a short barrel to start the drilling process until the anti-torque section is below the surface with the long barrel. Cutters are scaled down laser-style ice auger cutters more commonly used for opening holes in frozen lakes for fishing.

The drill sonde is suspended from a 650-m, 4.72-mm-diameter cable with four 24 AWG (0.2 mm^2) conductors. Two conductors are used to carry the nominal 385 VDC for power. The other two conductors and the armor are used for CAN Bus communications from the surface controller to the motor controller in the drill sonde. The winch uses a 2.2-kW single phase motor with encoders on both the cable drum and the motor. This allows both low-speed drilling and high-speed tripping to be well controlled. Drill speeds as low as 0.05 m/min at the bottom end and higher than >80 m/min when tripping can be achieved; however the winch is only capable of a pull force of 175 kg.

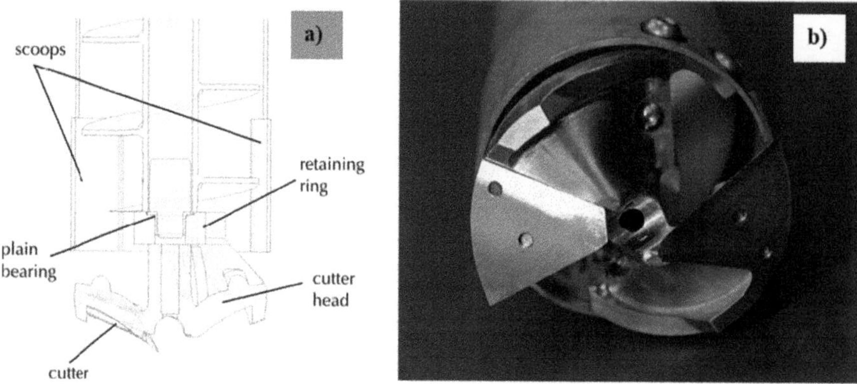

FIGURE 6.48 (a) Schematic of the cutter head and scoops; (b) photograph of the cutters with 82.5-mm outer tip to outer tip dimension and cutter head. *Source:* Courtesy of Rix et al. (2019).

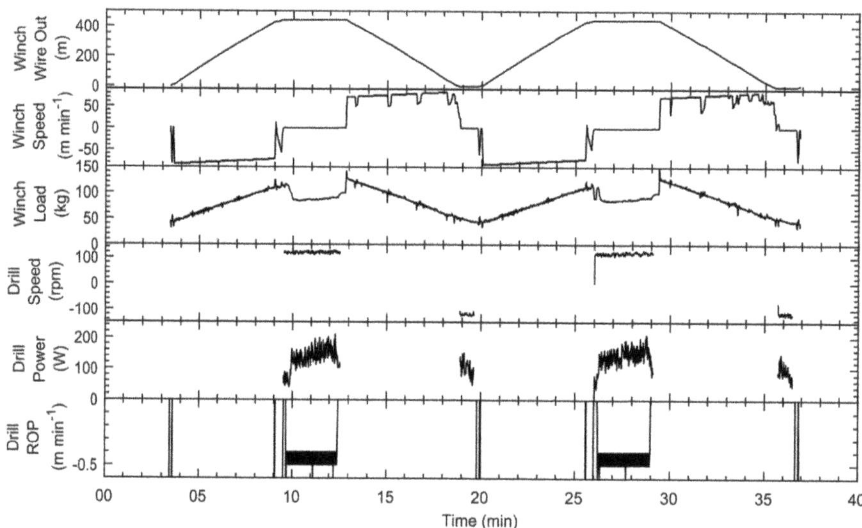

FIGURE 6.49 Typical data showing information logged by the drill controller. Drill current has been converted to drill power. This plot shows that winching speeds >80 m/min are achieved and drilling rate of penetration is ~0.45 m/min. Where the drill speed is negative, the chips are being unloaded on the surface which can be achieved in less than a minute. *Source:* Courtesy of Rix et al. (2019).

The winch controller uses a single axis joystick with a mechanical zero position interlock for manual control but also has controls for automatic deployment, drilling, and recovery of the drill sonde. The controller can be interfaced to another controller or computer to allow fully automated drilling but this has not been implemented yet. An RS-232 port also outputs all winch data (cable speed, wire out, deepest depth reached, and load) which is logged every second by the drill controller.

The drill controller provides the high voltage supply to the drill sonde and communicates with the motor controller in the motor section of the drill. Drill speed and motor current data is returned digitally from the motor controller and logged by the drill controller. Typical data logged for a couple of drilling runs are plotted and shown in Figure 6.49.

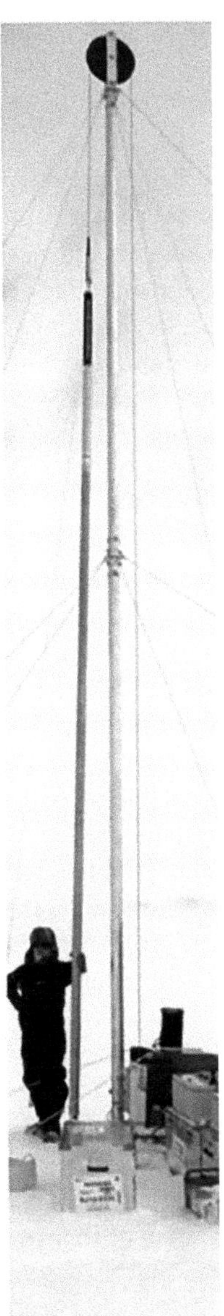

FIGURE 6.50 Photo of RAID system during test season at Sky Blu.

A mast that splits into two parts attaches to the base of the winch. The highest point of the mast when raised is 9.9 m above the snow surface. The system can be ready to drill about 4 hrs after arrival on site. Full system weight is approximately 650 kg including tools, generator, and spares. A photograph of the Rapid Access Isotope Drill in the field at the Sky Blu site is shown in Figure 6.50.

Having successfully proved the concept of this drill, scientists have suggested new science applications. Drilling to bedrock to obtain a coarse resolution climate record at Sherman Island in Antarctica will be attempted in the 2019–2020 Antarctic season. A percussive rock drilling head has also been designed and built for collecting rock samples once the RAID has drilled to bedrock. The percussive head, P-RAID, attaches to the RAID anti-torque section and reuses much of the RAID system. A rock core of ~20 mm in diameter and up to 30 cm long will be collected for cosmogenic dating. Full automation of the drilling utilizing the remote control capability of the winch controller should allow unattended drilling to occur. A large-diameter (230 mm) version of the drill, BigRAID, for deploying instrumentation to a depth of about 200 m is also being designed.

6.4.2 Hot Water Drilling

Access to depths within or beneath a glacier or ice sheet that does not require a core can be rapidly achieved through use of hot water drills. These are often used to access subglacial lakes, or the ocean beneath an ice shelf, or to provide access for sampling subglacial sediments. The use of hot water to melt a hole through the ice does not require a drilling fluid, thus is also amenable to sterilization for clean access in the case of biological sampling within or beneath the ice. The amount of equipment and fuel needed for hot water drilling rises as the hole diameter and depth requirements rise. Description of several recent systems and their operating characteristics are described next.

6.4.2.1 Scalable Hot Water Drilling Systems

Since the advent of hot water drilling in the 1970s, drill systems have generally been designed for specific projects with a narrow range of borehole depth and diameter requirements. The majority of subglacial access boreholes are less than 1000 m, encompassing many grounded ice areas, grounding line regions where the ice sheet goes afloat on the ocean, and all but the thickest parts of the floating ice shelves. Subglacial access holes provide safe passage for oceanographic or glaciological probes, samplers, and permanently deployed instrumentation strings into the underlying ocean or basal sediments.

Recently, new drills have been made scalable by using multiple compatible modules, allowing easy expansion or reduction of the drill system size to meet borehole and logistics requirements of specific field projects in remote locations. In 2009, the British Antarctic Survey (BAS) started building a new scalable hot water drill, primarily for use on the ice shelves around Antarctica, to enable the study of ice-ocean interactions, ocean properties, ocean circulation, and collection of water samples and sediment cores. The original drill requirements were:

- Provide access holes through ice shelves between approximately 100–1000 m thick.
- Operate in ice temperatures down to −30° C.
- Drill and maintain 30-cm diameter access holes.
- Deep 1000-m holes to be drilled in less than 15 hrs.
- Be transportable by Twin Otter aircraft and skidoo traverse.
- Be movable and operable by a maximum of four people.
- Facilitate safe recovery of probes and samplers from ocean into borehole.
- Maximize fuel efficiency.
- Minimize drill system weight but maximize system reliability.
- Minimize drill setup and teardown times.

The scaling concept was to optimize the system components to allow additional components to be added as energy needed for drilling requirements increased. The BAS ice-shelf hot water drill (HWD) is an example of a scalable drill, using a standardized range of modular units to build 500-m and 1000-m versions; the 500-m-depth version is shown in Figure 6.51. Initially configured for depths up to 500 m, the drill was first used during the 2011–2012 Antarctic field season to access the ocean cavities and sea floor sediments beneath sites on Larsen C and George VI ice shelves. Over three field seasons, beginning in 2014–2015, the scaled-up 1000-m system provided access holes, up to 891 m deep, at ten locations on Filchner-Ronne Ice Shelf, Antarctica; as seen in Figure 6.52.

Schematics of both systems are shown in Figure 6.53. The scalability allows for redundancy in components, and also can result in logistical savings in shallower borehole cases when a single larger system would be oversized.

FIGURE 6.51 Ice-shelf HWD, 500 m configuration. (a) Primary drill pump 5.5 kW submersible pump in water tank, (b) 250 kW water heaters without heat recovery units (center) and 15 kVA generator with one on standby (right and left of water heaters), and (c) the hose winch reel, capstan, and tower with instrumented sheave.

FIGURE 6.52 Ice-shelf HWD, 1000-m configuration. (a) Mounted on a plastic sled (left to right), hose reel, four generators, three water heaters with heat recovery units, four primary surface pumps with one as standby, (b) two water tanks, (c) hose winch system and plastic sled with drill units, and (d) drill transportation between sites.

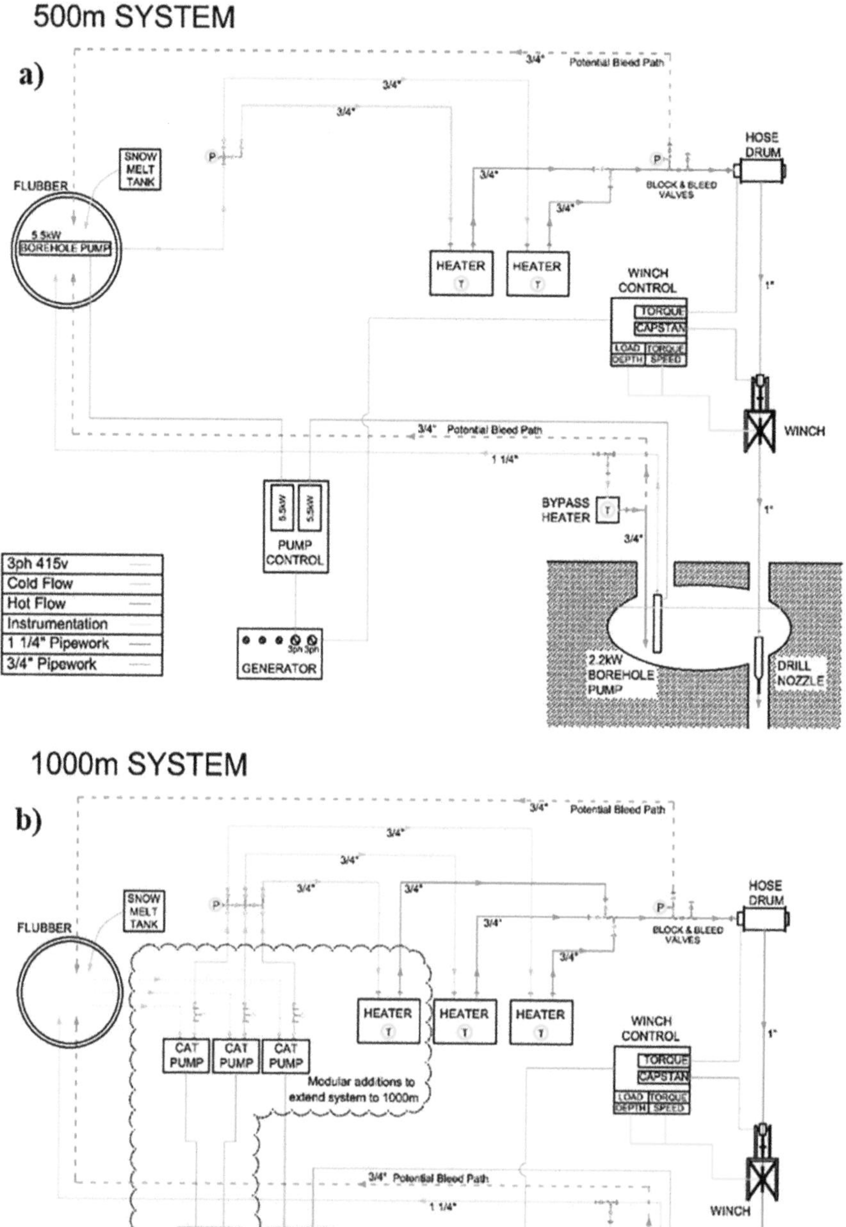

FIGURE 6.53 Schematic of the hot water drill system for (a) up to 500 m ice depth and (b) up to 1000 m ice depth. The additional equipment required for the upgrade to 1000 m is inside the scalloped outline, and the 5.5 kW borehole pump is reused in the cavity. *Source:* Courtesy of Makinson and Anker (2014).

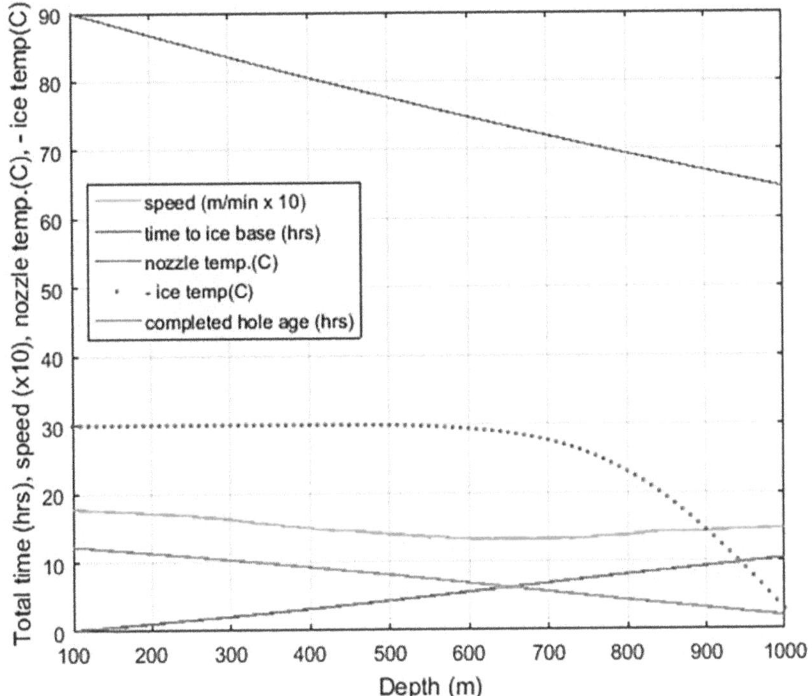

FIGURE 6.54 Plot showing the variation of nozzle water temperature (°C), an estimated (negative) ice temperature (°C), x10 for clarity drilling speed (m/min), the time taken (hours), and the age of the 30-cm-dimeter hole at completion (hours), with depth. Drilling commences from the cavity depth.

6.4.2.1.1 Thermal and Electrical Requirements

The diameter, depth, and time requirements to drill a 1000-m hole, for example, defines the minimum upper thermal requirements. To melt a 1000-m column of ice, 30 cm in diameter at −30°C requires almost 26 GJ to melt, hence a hot water drill with a 750-kW thermal capacity, for example, would require almost 10 hrs to deliver that energy. In addition, further energy and drilling time is needed to account for heat conduction into the surrounding −30°C ice.

Simple thermal modeling of the drilling process accounting for the thermal losses along the drill hose, the energy delivered to the drill nozzle and the time dependent refreezing of the hole, indicates that the 750-kW thermal input is sufficient for the 1000 m, as shown in Figure 6.54. The heat loss along the thermally leaky drill hose is not wasted; rather it is essential in reducing or preventing refreezing above the drill nozzle, reducing the risk of entrapment when the drill is recovered to the surface.

HWD electrical power requirements are largely defined by drill water pumping and the recovery of the return water to the surface, which are a function of pressure and flow. With the flow rate defined by the thermal requirements and assuming a maximum operating temperature of 90°C, the drill operating pressure is defined by the hose diameter, as the length is fixed. The operating pressure of the primary pump in the water tank on the surface pressure can be calculated using forms of the Darcy-Weisbach equation,

$$Pressure = \frac{f_d L \rho v^2}{2D} = \frac{8 f_d L \rho Q^2}{\pi^2 D^5} \tag{6.1}$$

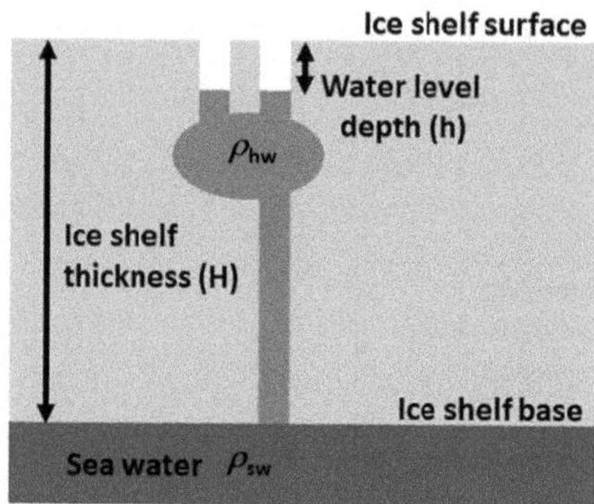

FIGURE 6.55 Schematic of borehole water level depth on an ice shelf.

where the dimensionless friction coefficient (f_d) is <0.02 for hoses and 0.03 for heat exchanger coils, L and D are the hose length (m) and internal diameter (m), v and Q are the mean flow velocity (m/s) and volumetric flow rate (m^3/s), and ρ is water density (kg/m^3), and pressure is in units of Pa.

The pressure at the submersible pump in the borehole, which is used to recover cold return drill water to the surface of the ice shelf, is the sum of the pressure from the hose friction and the elevation gain from the water level in the borehole to the surface. A schematic of the situation is shown in Figure 6.55. On ice shelves, the water level is determined from the ice shelf thickness (H (m)). The water level depth (h (m)) can be estimated using the following equation:

$$h = H - \frac{\rho_{sw}\left(0.892H - 17\right)}{\rho_{hw}} \tag{6.2}$$

where ρ_{sw} is the seawater density (1027 kg/m^3) and ρ_{hw} the hole water mean density which is typically fresh melt water (1000 kg/m^3). For ice shelf thicknesses of 100 m to 1000 m, the water level depth ranges from approximately 26 m to 102 m or 28 m to 125 m if dense seawater fills the borehole. On ice shelves where surface melt occurs, surface firn densities will increase, reducing h by several meters, to a maximum of around 17 m when only solid ice is present. Note: 10-m water column exerts a pressure of approximately 100 kPa.

For the drill water pressure, the height between the surface and water level effectively provides additional pressure, though this is usually offset by the pressure across the drill nozzle. With operating temperature, flow rate, thermal power, hose diameter, pressure, and electrical power defined, optimization of scalable drilling equipment modules and procedures to meet the remaining drill requirements can be calculated.

6.4.2.1.2 *Drill Equipment*

A unique feature of hot water drilling in polar environments is that even brief stoppages can result in rapid freezing of water in surface equipment, hoses, and the drill hole, which can lead to further delays and even loss of the drill, the hole, and damage drilling equipment. Drill equipment and procedures need to be sufficiently robust to accommodate any system failure to ensure consistent drilling operations. All drill equipment should include some level of redundancy and flexibility, and ideally should be repairable in the field. A wide range of spares, common to as many units as possible, should also be available. The components detailed next form the modular and scalable BAS ice-shelf hot water drill.

6.4.2.1.3 Variable-Frequency Drives (VFD)

With increased reliability, cost effectiveness, power efficiency, and dynamic flexibility, three phase motors with Variable-Frequency Drives (VFD) are fitted to all pumps and winches. Consequently, much smaller, lighter three-phase 50-Hz generators, with petrol engines can be used. Large electric motors were then sized to take approximately all, half, or quarter of the power delivered by each generator for maximum efficiency. Furthermore, as most motors can easily run at 60 Hz, motor speeds can be increased by 20% if required, offering greater drill system flexibility.

6.4.2.1.4 Electrical Generator

Three-phase (400 V) and single-phase (230 V) power at 50 Hz is delivered using Europower super-silenced EPS15000TE 12.5 kVA generators. Minimizing weight and volume were key logistic constraints; hence, petrol engines at approximately half the weight of equivalent diesel units were selected. The 500-m system operates on one unit, with one on standby, and the 1000-m system uses three, with one on standby.

6.4.2.1.5 Surface Water Storage

To ensure a supply of water for the drilling process, surface storage in the form of robust flexible coated fabric tanks with a 10,000-L capacity are used; see Figures 6.51 and 6.52. Water held in the tank is usually maintained at 5°C to 15°C to mitigate against freezing in hoses and pumps; therefore insulation under the tank is needed to prevent it from melting into the snow surface. When not in use, these tanks are lightweight and pack down into a small volume.

6.4.2.1.6 Primary Drill Pump on the Surface

With relatively low operating pressure, the 500-m system uses a Caprari E6X25-4/24 multi-stage centrifugal submersible pump with a 5.5-kW motor, capable of delivering 90 L/min at 2100 kPa, and is located in the surface water tank. A key safety feature of centrifugal pumps is that they can never over pressure and damage the drill system. Operating at higher pressure, the 1000-m system uses three positive displacement CAT1531 plunger pumps, each with a 5.5-kW motor, and capable of delivering ~40 L/min at 6900 kPa. These pumps require pulsation dampers and pressure relief valves with backups fitted on each unit.

6.4.2.1.7 Borehole Pump and Umbilical

Cylindrical multi-stage centrifugal submersible pumps are used to return water to the surface via a plaited umbilical consisting of a 32-mm bore thermoplastic return hose, and a 19-mm bore thermoplastic hose to deliver hot water to the umbilical and pump to prevent freezing, and a three-phase power cable. A current loop water level sensor with a shielded cable also runs alongside the umbilical. The 500-m system uses a Caprari E4XP35/20 multi-stage centrifugal submersible pump with a 2.2-kW motor, capable of delivering over 120 L/min from 70 m depth. The 1000-m system uses a Caprari E6X25-4/24 multistage centrifugal submersible pump from the 500-m system, which is capable of delivering over 140 L/min from 125 m depth. The return flow to the surface is regulated to balance the flow of water in and out of the subsurface cavity.

6.4.2.1.8 Water Heaters

Commercial high-pressure 250-kW water heating units have been modified for field use and can be transported by Twin Otter aircraft. The units use kerosene or Jet-A1 fuel and are fitted with adjustable high-temperature and low-flow cut-off switches. Secondary heat exchangers are fitted to the exhausts, and the recovered heat warms the water storage tank. Two heaters are used for the 500-m system, and three for the 1000-m system. A smaller 60-kW water heater maintains the downhole pump and umbilical against freezing.

6.4.2.1.9 Drill Hose Winch System

The winching system is comprised of a powered hose reel, a capstan drive mechanism, an instrumented sheave tower, and control panel, all of which can be disassembled and transported by Twin Otter aircraft. The hose reel holds up to 1000 m of drill hose, and a three-phase AC motor in torque mode provides the continuous tensioning in the hose leading to the capstan unit, irrespective of rotation rate or rotation direction. The AC motor is fitted with an encoder to ensure its smooth operation in all modes of operation. In the event of power loss, a motor brake is applied immediately to prevent freewheeling of the reel. During recovery of the drill, the hose is winched to the surface and the hose reel level winding is done manually.

The capstan consists of a grooved wheel with a circumference of 2.5 m and rubber beading at its base to increase friction with the drill hose. The groove is matched to the drill hose diameter and the three-quarter wrap around the capstan provides sufficient grip to power the drill hose up and down the hole, provided 5%–10% back tension is applied by the hose reel unit. The speed range of 0–9 m/min can be controlled in increments of <0.05 m min^{-1} which is needed during drilling and recovery. The capstan is powered by a three-phase AC motor, fitted with an encoder, via self-locking worm gearbox that prevents overhauling of the motor by the hose tension.

The winch tower has an instrumented sheave wheel, identical to the capstan wheel, which gives a clearance of 3 m above the hole for easy deployment of the drill nozzle as well as probes and samplers. The sheave and capstan wheels are both oversized for the hose bend radius; however, if the hose has couplings, the large diameter helps prevent bending damage to the hose at the coupling ends. The tower can be used with or without the capstan for the deployment of coring and sampling equipment, oceanographic instruments, and moorings.

6.4.2.1.10 Drilling Hose

To eliminate or minimize hose couplings in the drill hose, 500-m lengths of standard 1PDN25 Kutting thermoplastic hose (26-mm bore) consisting of a polyester elastomer lining, a single polyester braid and a polyurethane outer jacket are used. 500 m is the maximum length that will fit in a Twin Otter aircraft. The hose has a maximum working pressure of 6900 kPa with a 4:1 safety factor for dynamic applications, an operating range of −40°C to 100°C, and weighs 0.52 kg/m in air. It is slightly negatively buoyant in fresh water, countering the 0.02 kg/m buoyancy of the hot water in the drill hose.

6.4.2.1.11 Drill Monitoring

Sensors and display units output the key parameters of drill water temperature, pressure, and flow, as well as drill hose load, drilling speed, drill depth, and borehole water level, which are logged and displayed graphically. Numerous other sensors associated with individual units can also provide useful diagnostics during drilling operations.

6.4.2.1.12 Antifreeze System

When parts or all of the hot water drill are not in use, water must be removed from the system. A tank of 200–400 L of colored propylene glycol, axillary pumps, and an air compressor are used to flush items with antifreeze, which is removed with compressed air. For large drill hose reels, it is necessary to clear the hose by rotating the reel until no further water or antifreeze remains.

6.4.2.1.13 Drill Nozzle

The main body of the drill nozzle assembly is a 50-kg brass pipe, 1.5 m long and 75 mm in diameter. Drilling at a predefined rate ensures the drill hangs freely within the hole, with gravity providing the steering mechanism for a straight and vertical hole. Typically, the pressure drop across the nozzle would be in the range of 500–1000 kPa to give an exit velocity of 30–40 m/s which is essential for rapid hole formation in advance of the drill nozzle. Operating in low ice temperatures with a single forward pointing water spray, a potential problem exists. In drill hoses where a loss in pumping

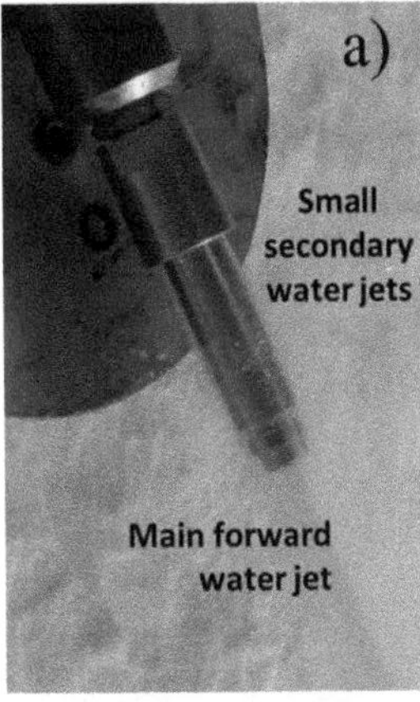

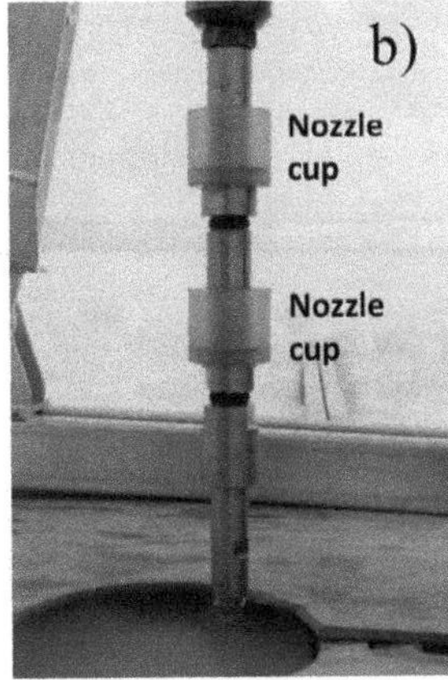

FIGURE 6.56 Drill nozzle attachments. (a) Point jet spray with six small secondary water jets behind, (b) nozzle cups for sediment recovery, and (c) bidirectional reamer at rear of the drill nozzle.

pressure results in hose elongation, a system failure that also includes loss of winching power can quickly result in a spray tip frozen to the hole bottom and blocked with ice. With no heat flow to the nozzle, no melting can take place, resulting in the nozzle, hose, and hole being lost. By adding six small forward-pointing water jets a short distance behind the main spray tip, this problem is avoided and once flow is re-established, melting out of the tip will occur and full drilling can be re-established. Photographs of the drill nozzle are shown in Figure 6.56.

6.4.2.1.14 Drill Tools

Tools that are used with this hot water drill include sediment cups, reamer, and brush reamer. Each is described as follows.

6.4.2.1.14.1 Sediment Cups

On ice streams and even ice shelves, substantial amounts of rock debris have been found within the ice, make drilling difficult. To remove or sample such material from drill holes, several collection cups can be added in series to the drill nozzle, so that in the highly turbulent environment of the nozzle tip, sediments are lifted into these simple sediment traps and then recovered with the drill nozzle.

6.4.2.1.14.2 Reamer

The scalable reamer ranges in diameter from 150 mm to 350 mm in steps of 50 mm and is built up from a series of aluminum plates containing holes and channels that are stacked together. It is used to guarantee a minimum hole diameter, particularly before the deployment of larger instruments and is activated when contact is made with a narrowing in the hole when traveling either down or up the hole. When activated, hot water flows through the network of holes and channels and sprays laterally to enlarge the hole.

6.4.2.1.14.3 Brush Reamer

This reamer enlarges the access hole specifically at the ice shelf base only, without needing to enlarge the entire access hole as is the case with the standard reamer. It consists of eight horizontal fan sprays and a flexible brush with 1-mm-thick plastic bristles that can traverse a 30-cm borehole and unfold to 1 m diameter once beneath the ice shelf base, as seen in Figure 6.57. Pulled up against the ice shelf, the brush separates the hot drill water from the cold underlying seawater, preventing the loss of heat into the ocean and therefore widening the hole at the ice shelf base. By creating a wide bell shaped hole at the ice shelf base, this greatly assists in the recovery of instruments into the hole or the deployment of free-fall probes measuring turbulence at the ice-ocean boundary.

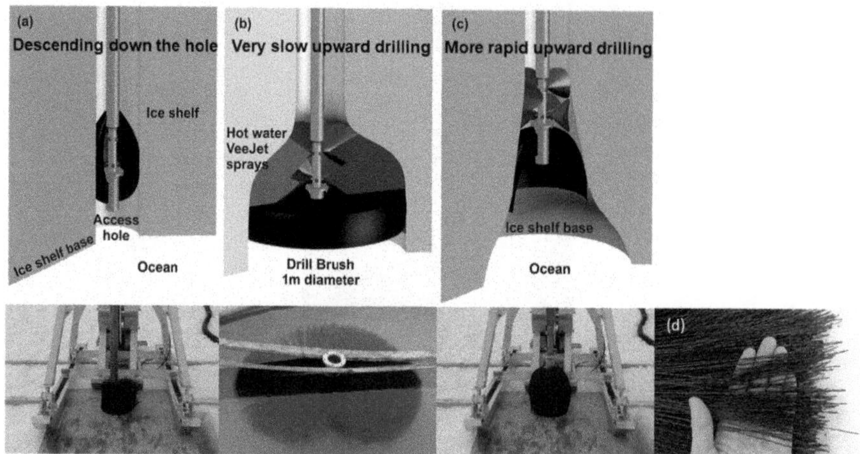

FIGURE 6.57 Schematic of the brush reamer at the ice shelf base (from Makinson and Anker, 2014) and illustrative surface photographs: (a) brush going down the hole, (b) brush at base and drilling slowly upward, (c) brush near base and drilling quickly upwards, and (d) close up of 1-mm dimeter brush bristles.

6.4.2.1.15 Drilling Method

Hot water drills and their drilling process are fundamentally the same using generators, water pumps, water heaters, and a hose with a weighted drill nozzle that delivers a high-speed jet of hot water to melt ice ahead of the drill. Water then flows up the melted hole toward the surface while providing support pressure to the ice walls.

The drilling process initially uses the 10,000-L water supply to melt the first hole to a depth below either sea level on an ice shelf or the local hydrological level on grounded ice. The drill is paused to widen the hole and form a cavity before recovery to the surface. A submersible borehole pump is deployed to recover water while a second parallel hole, offset by approximately 0.6 m, is drilled and interconnected with the first hole, establishing the water recirculation system cavity. To minimize the period of continuous drilling, the reliance on longer periods of good weather, and the need to operate two drilling shifts, drilling operations cease at this stage. The cavity is pumped dry and the pump recovered. Excess water is stored in a second tank. This temporary stop in drilling and creation of a dry borehole allows water seepage from the porous firn to freeze, therefore avoiding the pump and umbilical potentially becoming frozen to the hole walls.

A day or more later, drilling recommences from the cavity and the borehole pump is redeployed. Once the cavity is refilled, water recirculation is re-established. In the final stages of drilling, a sudden change in water level provides clear confirmation the base of the ice shelf has been reached; after that point the borehole will always contain some water that is susceptible to freezing. Even for a 1000-m access hole, drilling and drill recovery should be complete within 15 hrs to avoid shift work. If required, additional reaming at the ice base using a nozzle brush attachment could also be undertaken. Hole availability for deployment probes and samplers is limited to about 12–16 hrs as refreezing reduces the hole diameter by at least 5 mm/hr. If required, periodic reaming maintains the hole diameter before deploying a permanent sub-ice shelf instrument mooring.

Typically, full-site turnarounds can be as little as four days for shallow holes and seven days for the deepest holes. New tractor traverse and plastic sled developments now enable transport of the 1000-m drill system almost fully assembled between sites accompanied by a 1350-L insulated bladder holding warm seed water to prime the drill system at the next site. Drill equipment is mounted on several 1.83 × 2.44 m steel pallets, attached to a 21.9 × 2.44 m tractor-towed plastic sled. These logistical changes result in highly efficient multi-site hot water drilling operations during a single Antarctic field season.

6.4.2.1.16 Recent Developments

Building on the scalability and modularity of the 1000-m system by adding further generators, a heater, a high pressure pump, and a borehole pump and umbilical, the depth capability has increased to over 2000 m. However, this has only been possible by increasing the drill hose diameter from 26 mm to 32 mm and providing a much larger hose winching system for the single length hose. At different sites on Rutford Ice Stream, West Antarctica, this drill was successfully used in early 2019 when three access holes up to 2152 m deep were drilled to the base of the ice stream.

6.4.2.2 Clean Deep Hot Water Drilling

Subglacial lakes, rivers, and non-oceanic aqueous environments below ice sheets in Antarctica and Greenland are sensitive environments that have been biologically disconnected from the rest of the planet for potentially millions of years. The Antarctic Treaty Code of Conduct requires that exploration of the subglacial aqueous environments be accessed in a "clean" fashion to minimize disturbance and contamination.

The University of Nebraska-Lincoln successfully built and deployed a clean hot water drill system (CHWDS) for deep hot water drilling at Subglacial Lake Whillans; it includes a filtration unit and UV-treatment system to decrease contaminants in the drilling water and provide clean access to the subglacial environment (Priscu et al., 2013). The filtration technology was successful at reducing microbial bioload in the drilling fluid as per the Antarctic Treaty Code of Conduct. At the Subglacial Lake Whillans site, the CHWDS created an access hole through 800 m of ice; details are further described in Rack et al. (2014) and Blythe et al. (2014). After repair and upgrade, the drill was recently used again with success for the SALSA project at Subglacial Lake Mercer, Antarctica in 2019. A schematic of the drilling at Subglacial Lake Mercer is shown in Figure 6.58. Figure 6.59 is a photo showing the layout of the drilling and science operations at the site.

6.4.2.3 Shallow Hot Water Drills

Shallow hot water drills are in use by many nations for drilling shallow holes in glacial ice. The IDP Small Hot Water Drill (SHWD) is a non-coring drill used most often for shot holes for seismic work but has also been used for access holes through a thin ice shelf. The drill is transportable by light aircraft and helicopter.

IDP has two SHWD systems; in 2015 and 2016, they were modified to improve performance and lessen logistical demands. The heaters were refurbished, the system was upgraded to include modern controls, and a new nozzle kit was created. Lightweight Siglin sleds were purchased to replace the aluminum sled and covers were fabricated for protection from the elements. One of the two IDP systems has a 30-m-depth capability and the second has a depth capability of 60 m. Each can produce a 2.5-inch-diameter hole 25–30 m deep in about 12 minutes. A photograph of the system in the field is shown in Figure 6.60, and the system specifications are provided in Table 6.9.

The SHWD is designed to drill quickly through firn. During drilling, no water is recirculated back from the hole. The drill's thermal energy is used both to melt the hole and to melt snow for makeup water through a hot water recirculation loop. The main system components consist of a gasoline-powered pump/generator module and two 60-kW fuel-fired water heaters, a water tank and two fuel tanks (AN8 for heaters and gasoline for generator and pump module), as shown in the schematic in Figure 6.61.

The water is supplied from the water tank and is drawn into the high-pressure pump through a coarse screen filter on the suction line. The high-pressure pump pushes the water through the system and is driven by an 18 HP Honda GX630 gasoline engine. The Honda engine also drives an AC generator. The heaters can be powered from this generator; however, a separate 5-kW electric generator can be substituted for independent power. The engine throttle is adjusted to a 60-Hz generator output. At this speed, the pump provides 7 gpm of flow. The maximum discharge pressure of the pump is 1000 psi. A pressure relief valve on the discharge line protects the system from overpressure.

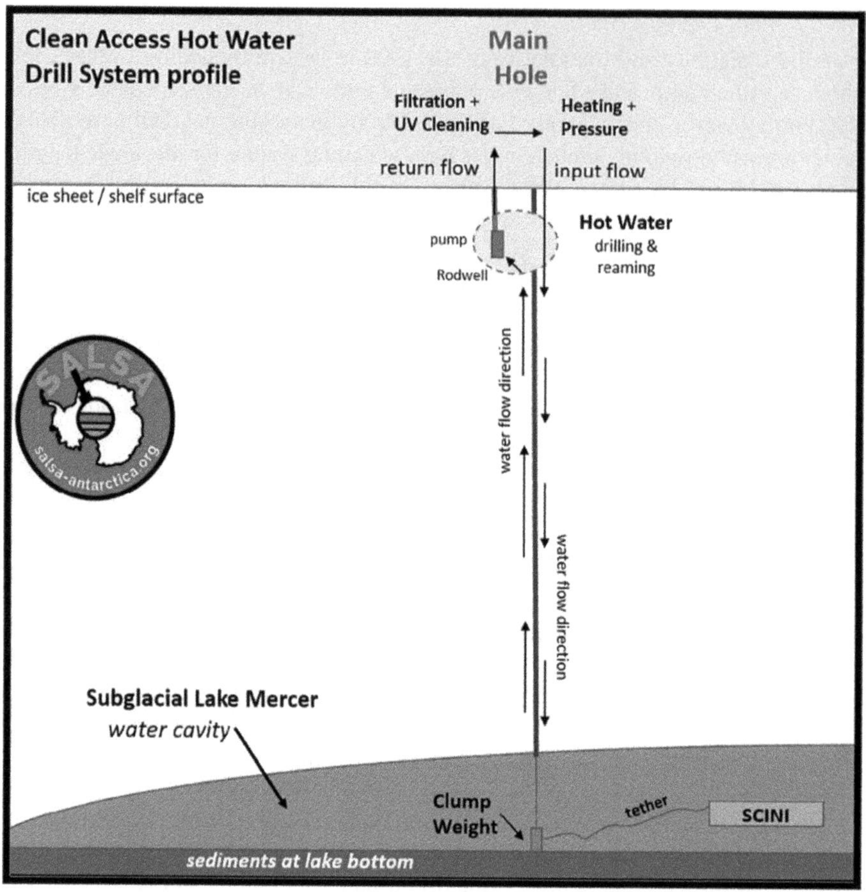

FIGURE 6.58 Schematic of the clean access hot water drill system profile at Subglacial Lake Mercer.

FIGURE 6.59 Photograph of the CHWDS operations at Subglacial Lake Mercer in 2018–2019.

Cold water exits the pump and is directed to the two water heaters through the cold water manifold which is instrumented with pressure and temperature gauges. Flow balancing between the heaters is accomplished by adjusting two brass needle valves, also on the cold manifold.

The customized Whitco Stinger Heaters are diesel-fired, rated to 2000 psi, and instrumented with a temperature gauge. A series of safety interlocks including thermostat-controlled temperature,

FIGURE 6.60 Drilling seismic shot holes with the Small Hot Water Drill on Beardmore Glacier, Antarctica, during the 2012–2013 summer field season. *Source:* Courtesy of Maurice Conway (icedrill.org).

TABLE 6.9
IDP Small Hot Water Drill System Specifications

Dry Weight	2200 lbs.
Max Piecewise Weight	280 lbs. (heater unit)
Nominal Hole Diameter	Variable (10 cm nominal)
Max Practical Hole Depth	Dependent on conditions; typically reliable and efficient to a depth of 25–30 m; max depth 60 m
Fuel Types	Heaters: AN8 Generator/Pump: Gasoline
Crew Size	2
Drill Time	12 min
Cycle Time[1]	30 min

[1] Cycle time for a typical 30 m × 10 cm diameter hole; includes 15 min for setup and teardown but does not include time to travel between holes.

low-flow switch, and over-temperature switch provide reliable operation. Combustion is provided by a standard forced-air burner, comprised of a blower fan, fuel pump, igniter, igniter coil, and burner control module. The output from each heater has its own shutoff ball valve.

Flows from both heaters are recombined in the hot manifold, which is also instrumented with pressure and temperature gauges. Hot water is then split again into the two primary system loops: local recirculation back to the water tank, and flow to the drill head. Local recirculation water is hot water sent back to the water tank to aid in snow melting. The drill head flow is sent through the hose reel and used for creating the borehole. To adjust the flow between the two loops, a flow control needle valve is provided on the recirculation loop. Partially closing this valve will send more flow to the drill nozzle. A vacuum break is also provided on the drill flow circuit at the hot manifold to drain the hose after drilling to prevent freezing.

For the 30-m configuration, a hand-powered hose reel is provided. An electric motor assist is provided for the 60-m hose reel configuration shown in Figure 6.62. The motor is controlled by a foot-switch and a small speed control box. This reel is powered by a separate 5-kW electric generator.

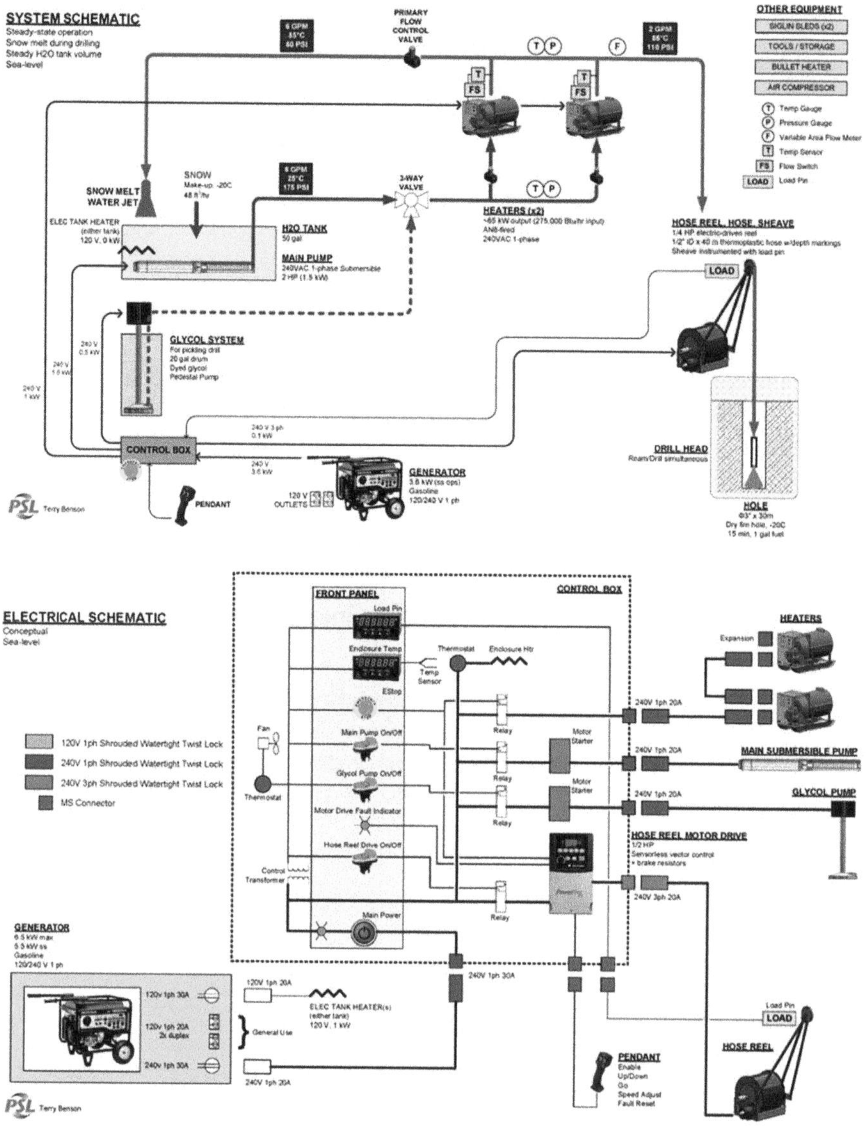

FIGURE 6.61 Schematic of the Small Hot Water Drill System.

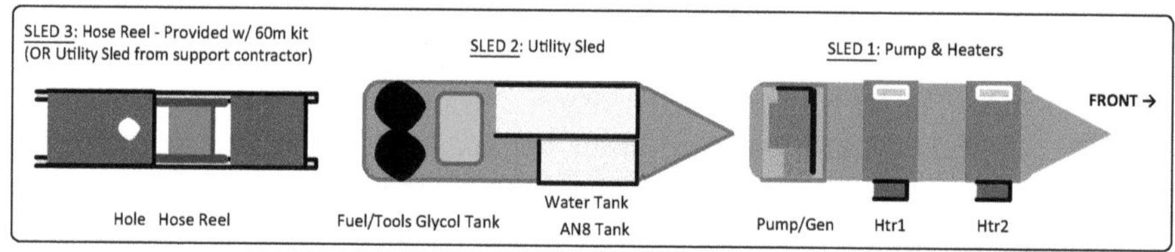

FIGURE 6.62 Layout of the IDP Small Hot Water Drill System.

The drill nozzle stem provides weight so that drilling can be steered straight by gravity, and also integrates the drill nozzle. A nozzle kit is provided with instructions for nozzle selection. When melting snow in the reservoir tank while drilling, a large amount of water is diverted from the drill head for this purpose. With a relatively low flow at the drill head, a nozzle with a small orifice is needed. In this case, about 2 gpm is used for drilling while 5 gpm is used for snow melting. If water reserves are adequate, nearly all the flow can be directed downhole and a nozzle with a larger orifice is required to keep system pressures from becoming too high.

A simple fuel system supplies the heaters. An elevated marine fuel tank (about 27 gallons) provides slight gravity assistance for the fuel system. The tank is refueled in place using a hurdy-gurdy pump or other delivery system. Burn rate is about 4 gal per hour of AN8 fuel. Heaters are set up with a supply hose and a return hose to help avoid having to bleed air from the fuel lines. With many years of field history and recent refurbishments, the IDP SHWD systems are expected to continue to be in demand for numerous applications in upcoming Arctic and Antarctic field seasons.

6.5 CONCLUSION

New technologies spawn new scientific discoveries, and this has been and will continue to be true for scientific drilling in glacial ice. Glaciers, ice sheets, and the subglacial environment contain natural records of past climate, providing insights and clues to understanding future climate. Current climate change is the most pressing environmental issue of our time, affecting every nation on the planet. Scientific discoveries from ice core science have impactful messages, for example the discovery that dramatic changes in climate can occur in less than ten years, and direct evidence from atmospheric gases trapped in polar ice providing the important context that current greenhouse gas levels in the atmosphere are higher now than they have been in over 800,000 years. Rising seas under current climate change threaten major coastal cities around the world; drilling access holes in glacial ice enables scientific discoveries in ice dynamics and subglacial conditions that are important for prediction of future sea level rise.

Innovations in ice drilling in the past decade have yielded both great improvement over earlier drilling techniques, as well as innovative new drilling technologies that foster new avenues of science. Completely inaccessible for cosmogenic dating, bedrock sampling under many meters of glacial ice was impossible before the development of the Agile Sub-Ice Geological Drill; this drill, and its smaller comrade the Ice-Enabled Winkie Drill, are the first of their kind, opening doors to new scientific geological discoveries from previously inaccessible realms. Similarly, scientific demand for evidence of past abrupt climate change from ice cores placed very high demand on ice from certain depths; while drilling additional full ice cores would have been both time- and cost-prohibitive, the new replicate coring capability development enables recovery of additional specific bands of ice without the need to drill an entire new core.

Science and engineering go hand in hand. Working closely with scientists in the ice-science community, the engineering community continues to rise to meet the need. Major goals in the development of new equipment includes the aim to reduce the footprint of equipment, the amount of fuel required, and where possible adopt renewable energy technologies and environmentally friendly processes in order to minimize environmental impacts of drilling operations. As we look to the future, further innovations in ice drilling technologies will continue to foster new scientific discoveries of importance to all people.

ACKNOWLEDGMENTS

The authors express our sincere thanks to editors Yoseph Bar-Cohen and Kris Zacny for inviting us to write this chapter, and for their patience while we worked through the task. The authors would like to thank Victor Zagorodnov, Ohio State University, Columbus, OH; Matthias Huether and Jan Tell, Alfred Wegener Institute, Bremerhaven, Germany; and Boleslaw Mellerowicz, Honeybee Robotics, Pasadena, CA, for reviewing this chapter and providing valuable technical comments and suggestions.

Contributions by Keith Makinson and Julius Rix were supported by the British Antarctic Society. The US Ice Drilling Program is a National Science Foundation Cooperative Agreement with Dartmouth including subawards to the University of Wisconsin and the University of New Hampshire; efforts of the US Ice Drilling Program were made possible through funding from the US National Science Foundation through NSF Cooperative Agreements 1327315 and 1836328.

REFERENCES

Bentley, C.R., B.R. Koci, L. J.-J. Augustin, R.J. Bolsey, J.A. Green, J.D. Kyne, D.A. Lebar, W.P. Mason, A.J. Shturmakov, H.F. Engelhardt, W.D. Harrison, M.H. Hecht, V. Zagorodnov (2006). "Ice Drilling and Coring," *Chapter 4 in Drilling in Extreme Environments*, Y. Bar-Cohen and K. Zacny, eds., Wiley-VCH Verlag GmbH & Co., KGaA, Weinheim. ISBN: 78-3-527-40852-8.

Blythe, D., Duling, D., & Gibson, D. (2014). "Developing a hot-water drill system for the WISSARD project: 2. In situ water production," *Annals of Glaciology*, 55(68), 298–302. doi:10.3189/2014AoG68A037

Gibson, C.J., J.A. Johnson, A.J. Shturmakov, N.B. Mortensen, J.J. Goetz (2014). "Replicate ice coring system architecture: mechanical design," *Annals of Glaciology*, 55(68), 165–172.

Gibson, C.J., T.W. Kuhl, J.A. Johnson, G.V. Boeckmann, J.J. Goetz (2015). Presentation from the May 28, 2015 Design Review of the Agile Sub-Ice Geological Drill System; Icedrill.org

Johnsen, S., S.B. Hansen, S.G. Sheldon, D. Dahl-Jensen, J.P. Steffensen, L. Augustin, P. Journe, O. Alemany, H. Rufli, J. Schwander, N. Azuma, H. Motoyama, T. Popp, P. Talalay, T. Thorsteinsson, F. Wilhelms, V. Zagorodnov (2007). "The Hans Tausen drill: design, performance, further developments, and some lessons learned," *Annals of Glaciology*, 47, 89–98.

Johnson, J.A., N.B. Mortensen, C.J. Gibson, J.J. Goetz (2014a). "Replicate ice coring system testing," *Annals of Glaciology*, 55(68), 331–338.

Johnson, J.A., A.J. Shturmakov, T.W. Kuhl, N.B. Mortensen, C.J. Gibson. (2014b). "Next generation of an intermediate depth drill," *Annals of Glaciology*, 55(68), 27–33.

Kuhl, T. et al. (2014). "A new large-diameter ice-core drill: the Blue Ice Drill," *Annals of Glaciology*, 68, 1–6.

Langway, C.C. (2008). "The early polar ice cores," *Cold Regions Science and Technology* 52, 101–117.

Makinson, K., & Anker, P. (2014). "The BAS ice-shelf hot-water drill: Design, methods and tools," *Annals of Glaciology*, 55(68), 44–52. doi:10.3189/2014AoG68A030

Mortensen, N.B., J.J. Goetz, C.J. Gibson, J.A. Johnson, A.J. Shturmakov (2014) "Replicate ice coring system architecture: electrical, electronic and software design," *Annals of Glaciology*, 55(68), 156–164.

Mulvaney, R., O. Alemany, P. Possenti (2007). "The Berkner Island (Antarctica) ice-core drilling project," *Annals of Glaciology*, 47, 115–124.

Priscu, J.C., A.M. Achberger, J.E. Cahoon, B.C. Christner, R.L. Edwards, W.L. Jones, A.B. Michaud, M.R. Siegfried, M.L. Skidmore, R.H. Spigel, G.W. Switzer, S. Tulaczyik, T.J. Vick-Majors (2013). "A microbiologically clean strategy for access to the Whillans Ice Stream subglacial environment," *Antarctic Science*, 25(5), 637–647. doi:10.1017/S0954102013000035

Rack, F., Duling, D., Blythe, D., Burnett, J., Gibson, D., Roberts, G. Fischbein, S. (2014). "Developing a hot-water drill system for the WISSARD project: 1. Basic drill system components and design," *Annals of Glaciology*, 55(68), 285–297. doi:10.3189/2014AoG68A031

Rand, J. and M. Mellor (1985). "Ice-coring augers for shallow depth sampling," CRREL Rep. 85–21.

Rix, J., R. Mulvaney, J. Hong, D. Ashurst (2019). "Development of the British Antarctic Survey Rapid Access Isotope Drill," *Journal of Glaciology,* 65(250), 288–298. doi: 10.1017/jog.2019.9.

Shturmakov, A.J., D.A. Lebar, C.R. Bentley (2014). "DISC drill and replicate coring system: a new era in deep ice drilling engineering," *Annals of Glaciology*, 55(68), 189–198.

Slawny, K.R., J.A. Johnson, N.B. Mortensen, C.J. Gibson, J.J. Goetz, A.J. Shturmakov, D.A. Lebar, T.W. Wendricks (2014). "Production drilling at WAIS divide" *Annals of Glaciology*, 55(68), 147–185.

Talalay, P.S., (2016). *Mechanical Ice Drilling Technology*, Geological Publishing House, Beijing, China, ISBN 978-981-10-0559-6.

Zagorodnov, V.S. (1988). *"Antifreeze-Thermodrilling of Cores in Arctic Sheet Glaciers," Ice core Drilling. Proceedings of the Third International Workshop on Ice Drilling Technology* (eds C. Rado and D. Beaudoing), 97–109.

Zagorodnov, V.S. (2019). Personal communication.

Zagorodnov, V., L.G. Thompson, P. Ginot, V. Mikhalenko (2005a). "Intermediate-depth ice coring of high-altitude and polar glaciers with a lightweight drilling system," *Journal of Glaciology*, 51(174), 491–501, doi: 10.3189/172756505781829269

Zagorodnov, V. and L. Thompson (2014). "Thermal electric ice-core drills: history and new design options for intermediate-depth drilling," *Annals of Glaciology*, 55(68), 322–330.

Zagorodnov, V., L.G. Thompson, P. Ginot, V. Mikhalenko (2005b). "Intermediate-depth ice coring of high-altitude and polar glaciers with a lightweight drilling system," *Journal of Glaciology*, 51(174), 491–501.

7 Environmental Drilling/ Sampling and Offshore Modeling Systems

Roy Long,
DOE/FE National Energy Technology Laboratory, Houston TX
Peter Lucon,
Montana Technological University, Butte, MT
Ernie Majer, and
Lawrence Berkeley National Laboratory, Berkeley, CA
Kelly Rose
DOE/FE National Energy Technology Laboratory, Albany, OR

CONTENTS

7.1 INTRODUCTION AND BACKGROUND

Environmental sampling can involve a wide variety of methods. This chapter focuses on those sampling systems related more toward "drilling engineering" than, for example, simple push core sampling systems for acquiring shallow surface samples. Such systems are typically associated with those systems requiring a drilling "rig" or surface system controlling the down-hole tools.

A drilling rig can be thought of as an engineered system designed to achieve a specific subsurface access or sampling purpose. Regarding environmental drilling, the initial development of that purpose was established on September 27, 1962 with the publication of Rachel Carson's book, *Silent Spring*, which is acknowledged for giving rise to the modern environmentalist movement (Piddock, 2009). The book referenced the contamination being caused by uncontrolled use of pesticides. The enabling sensor technologies that would advance the ensuing basis of control by allowing field measurements/monitoring did not gain momentum until solid state electronics allowed commercialization of much smaller, more rugged instruments (Kennedy, 2013). It was at this point that all forms of field monitoring began to develop that drove innovation in environmental drilling and subsurface monitoring/sampling.

Some sampling technologies were already starting to develop as oil and gas and mining exploration began to have similar sampling needs. An additional boost to deep "non-contaminating" sampling occurred in the late eighties at the Department of Energy's Yucca Mountain Site Characterization Project with the initiation of Unsaturated Zone sampling.

7.2 ONSHORE CHALLENGES AND TECHNOLOGIES

For the purpose of the onshore discussion in this chapter, conventional versus environmental drilling can be differentiated by the intended emphasis of each system. Drilling, per se, mainly consists of a focus on rapidly and economically gaining reliable access to some subsurface resource or geologic interval of interest. The primary focus of environmental drilling is that of obtaining subsurface samples, typically in situ fluids for environmental studies, or mitigation of subsurface contamination. Some of the first drilling systems designed to allow subsurface sampling were associated with acquiring oil and gas industry cores from exploratory boreholes and needs for in situ samples of ore bodies as part of mining industry evaluations (National Driller-1, 2019).

7.2.1 Commercial Environmental Monitoring Drilling Systems

There are several specific goals of onshore environmental drilling/sampling that form the drivers for the systems that will be discussed in this chapter: (1) evaluation of the impact of a process/activity (typically surface based) on the environment, referred to as environmental monitoring; (2) minimizing the risk of a subsurface process from contaminating the environment, such as "big hole" drilling developed for underground nuclear testing, and; (3) mitigating the impact of the drilling process itself on potential environmental contamination, as in the Department of Energy's site characterization of Yucca Mountain for potential high-level nuclear waste storage.

7.2.1.1 Auger Drilling

One of the older concepts of drilling systems to achieve the first goal (monitoring) is the auger drill. Auger drills are used extensively in construction for the same reason they are used for environmental monitoring: they are cost effective, have simple operations, and the ability to drill/sample relatively extensive areas quickly. While typically used for less consolidated soils, a specialized version called a hollow stem auger can be used for applications such as that shown in Figure 7.1 for acquiring deep ice cores. Similar systems can be utilized to acquire shallow cores in loosely consolidated soils for monitoring ground water or soil contamination. One of the primary advantages of this type of system is that circulation of drilling/coring fluid that would alter the in situ condition typically is not required. Alternatively, the auger drill is limited to rather shallow depths on the order of 30 m ([NAS], Figure 7.2).

FIGURE 7.1 Ice core mechanical drill head. *Source:* Courtesy of Profaizer (2006).

FIGURE 7.2 Auger drilling of monitoring wells. *Source:* Public domain, courtesy of Wikimedia Commons, the free media repository.

7.2.1.2 ODEX/TUBEX/STRATEX Drilling

These are all trade names of rotary percussive systems that advance a protective casing while the drilling assembly simultaneously advances the borehole (Fortin, 2019). The primary difference between the models is how the drilling assembly engages the casing drive shoe. For example, the air hammer bit strikes the formation while the body of the drilling assembly strikes an upset machined into the inner diameter of the drive shoe attached to the bottom of the casing. When it is time to change the bottom-hole assembly (BHA) in the ODEX "Overburden Drilling System," the drill rod is rotated counterclockwise and an eccentric lobe of the bit rotates inward allowing the bit to be pulled back into the casing and removed from the hole (Figure 7.3).

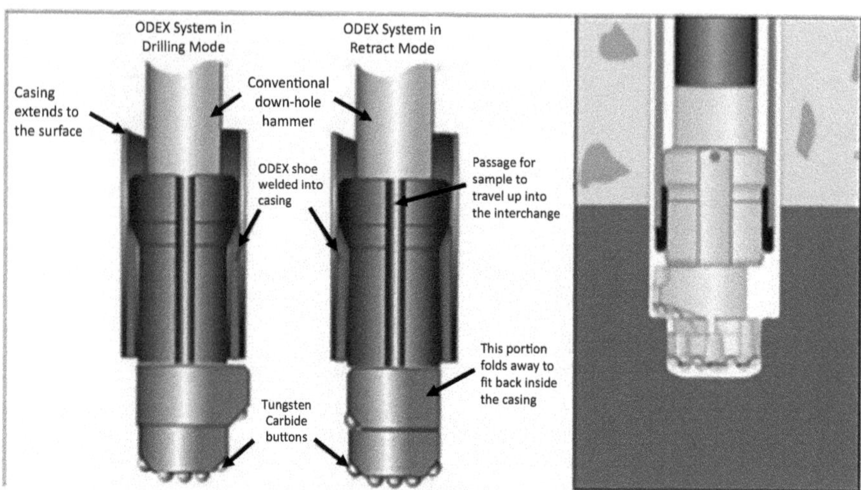

FIGURE 7.3 Drilling system features/concept. *Source:* Courtesy of ODEX (2019).

This type of system is advantageous for sampling when the earth gets too hard or boulder laden for auger drilling. However, one of the weaknesses of the system is that, if the drive shoe/casing connection assembly is not modified in some manner, the continuous impact on the drive shoe sometimes causes the thread connecting it to the casing to fail leaving a segment (fish) in the hole that can be very difficult to retrieve. This typically occurs when the casing friction gets too great due to excessive drilling depth or type of material being drilled. These systems typically drill on the order of 150 m.

7.2.1.3 Sonic Drilling

7.2.1.3.1 Background

Today, there are five major sonic drill companies worldwide. The major original equipment manufacturers (OEMs) in sonic drilling are Boart Longyear, GeoGroup, Terra Sonic International, Geoprobe, and SonicSampDrill. Sonic drills are being utilized more in the mining industry but are typically limited to use in the drilling of shallow wells, frequently for water, or to take more representative samples, which approach undisturbed samples, of the underground strata. The sampling recovery rates by this method are commonly approaching 100% (Barratt, 2012). In recent years, the technology has also proved well-suited for sampling of various soil types and conditions (Barratt, 2012).

Because sonically drilled shallow holes are typically drilled much more rapidly than conventional drilling, it is becoming the method of choice for quick, shallow jobs. However, sonic drilling is often more expensive than traditional drilling per hour of operation because of the requirement for highly trained operators as well as the greater consequences of drilling failures. It is relatively easy to allow the resonant sonic drill to exceed its pipe stress limits. In order for operators to attempt to control the resonance system within safe operating conditions, they require knowledge from experience.

The sonic drill is an advanced, hydraulically driven system. Through the use of a sonic drill head, shown in Figure 7.4, a series of high-frequency, sinusoidal wave vibrations are imparted to a steel drill pipe to create a cutting action at the bit face. In its resonant condition, each energy pulse imparted to the drill pipe is exactly superimposed on each reflected energy pulse wave. In this condition, the direction and magnitude of movement of each molecule in relation to another stays the

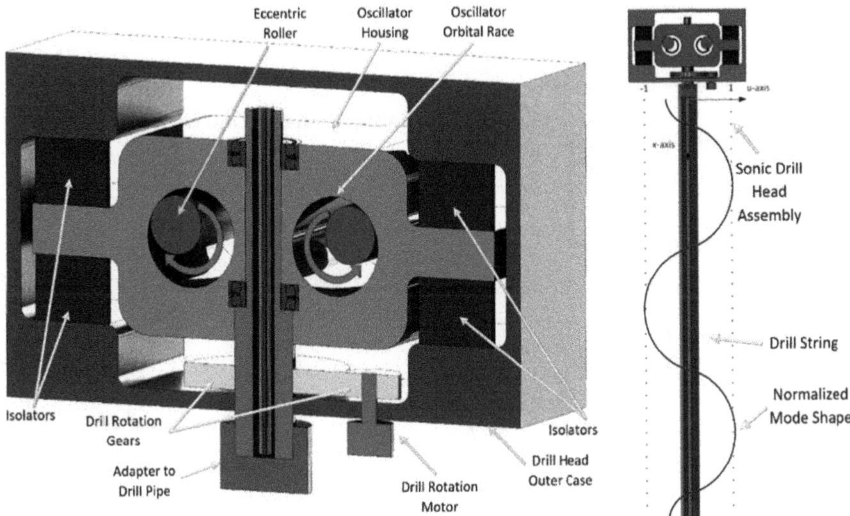

FIGURE 7.4 Sonic head section view diagram and diagram of the sonic drill. *Source:* Courtesy Lucon Engineering, Inc.

same and creates a situation where the energy stored in the pipe can greatly exceed the energy being dissipated in the form of "work" on the medium being drilled. This leads to efficient drilling as long as the energy is directed into the drilled medium. However, the lack of this energy sink can lead to excessive energy building quickly within the drilling string, leading to catastrophic tubular and/or surface failure.

In a non-resonating state (as occurs with traditional vibratory equipment), the energy waves are not superimposed in a reinforcing pattern and tend to cancel each other out as they move up and down the pipe. Consequently, the pipe is unable to utilize higher horsepower inputs from the drill head and the drilling rate is greatly reduced.

Presently, sonic drill rigs are operated primarily by "feel" and by "ear." Although provided with numerous gages, successful sonic drilling is accomplished through the expertise of the operator; less practiced drillers do not perform well on sonic rigs. Drilling with resonance is unlike any other drilling method.

7.2.1.3.2 Sonic Drill Models

A sonic drill typically operates between 20 and 250 Hz. The sonic drill head is where the input force is applied to the drill string and is always located at the top of the drill string as displayed in Figure 7.5. Sonic drill systems are coupled to a drill rig by a compliant member (spring or spring/damper). A spring could be mechanical springs or air springs. A spring/damper is typically designed as rubber isolators.

A sonic drill model can be developed by deriving the governing differential equation (GDE) of motion for the axial direction (along the length of the drill string), which can be by either force balance or energy balance. If the force balance method is used, then the free body diagram (FBD) in Figure 7.5 is used to derive the GDE, given in Equation 7.1 where ρ is the density of the drill string material (kg·m^{-3}), **A** is the cross-sectional area of the drill pip (m^2), **u(x, t)** is the deflection along the length of the drill string relative to time (m), **b** is the drill pipe damping constant (s^{-1}), **a** is the drill pipe shaft elasticity constant (s^{-2}), **E** is the drill pipe elastic constant (Pa), and **q(x, t)** is the body force along the length of the drill pipe (N·m^{-3}). The damping constant '**b**' and the spring

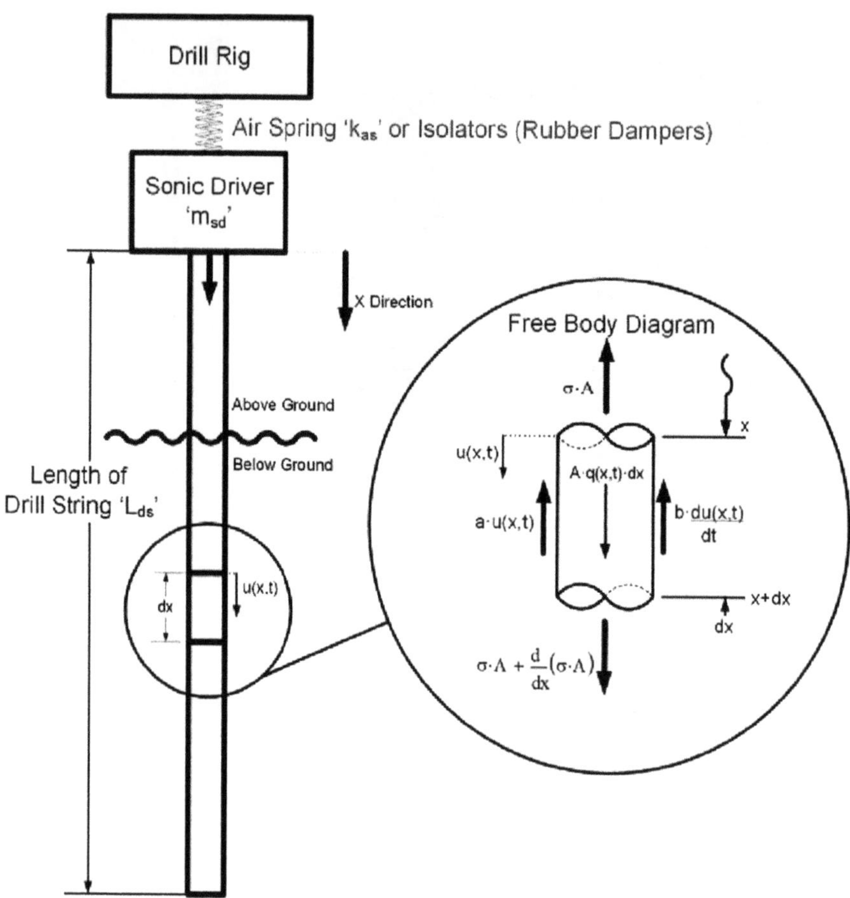

FIGURE 7.5 Sonic drill free body diagram. *Source:* Courtesy Lucon Engineering, Inc.

constant '**a**' are derived and explained further in the work by Don C. Warrington (Warrington, 1997).

$$\rho A dx \frac{\partial^2 u(x,\ t)}{\partial t^2} + 2bA dx \frac{\partial u(x,t)}{\partial t} + \left(aA dx\, u(x,t) - EA dx \frac{\partial^2 u(x,t)}{\partial x^2} \right) = q(x,t) A dx \tag{7.1}$$

However, by manipulating and transforming the force balance-derived GDE (Equation 7.1) for the sonic drill from the time domain into the Fourier and then the Laplace domains, the previous equation is transformed into the simpler Equation 7.2.

$$s^2 + 2bs + c^2\theta^2 + a = F_o \tag{7.2}$$

The damping ratio 'γ' is defined for the sonic drill system, as shown in Equation 7.3.

$$\gamma = \frac{b}{\sqrt{c^2\theta^2 + a}} \tag{7.3}$$

The assumed solution of the governing differential equation of the sonic drill, displayed in Equation 7.4, is a function of both the length along the drill string '**x**' and time '**t**', while also being in the same form as the sinusoidal input force, shown in Equation 7.5, which is applied at the sonic drill head.

$$u(x,t) = (A_o \sin(\theta_f x) + B_o \cos((\theta_f x))(C_o \sin(\omega_f t) + D_o \cos((\omega_f t)) \tag{7.4}$$

$$F(t) = F_{ecc} \sin(\omega_f t) \tag{7.5}$$

The '**x**' radial frequency 'θ_f' with respect to the length along the drill string '**x**' is related to the forcing angular frequency 'ω_f' by the relation depicted by Equation 7.6, that shows that the square of the ratio between 'ω_f' and 'θ_f' is equal to the Young's Modulus '$\mathbf{E_{ds}}$' of the drill string divided by the drill string density 'ρ_{ds}'.

$$\frac{\omega_f^2}{\theta_f^2} = c^2 = \frac{E_{ds}}{\rho_{ds}} \tag{7.6}$$

In order to solve this problem and find the unknown coefficients $\mathbf{A_0}$, $\mathbf{B_0}$, $\mathbf{C_0}$, and $\mathbf{D_0}$ for the assumed solution for the model dynamics of the sonic drill system, the boundary conditions for the sonic drill system must be defined, as displayed in Figure 7.6.

The sonic driver mass, the input force from the sonic driver, and the air spring all reside where '**x**' is equal to zero. The sonic driver mass and the air spring are always boundary conditions; however the input force can either be a boundary condition, or an input into the sonic GDE as **q(0, t)**.

At the drill tip of the string, where '**x**' is equal to length of the drill string '$\mathbf{L_{ds}}$', a boundary condition caused by coupling of the sonic drill tip to the material being drilled through exists. All boundary conditions are located on the ends of the drill string and because of this, all the conditions have to equal the apparent forces at the end conditions. The forces for the ends are found by taking the drill string's elastic constant '$\mathbf{E_{ds}}$' multiplied by the cross-sectional area of the drill string '$\mathbf{A_{ds}}$' and also multiplied by the partial derivative of the local deflection '**u**' with respect to the location in space '**x**' and setting this equal to the boundary condition, displayed in Equation 7.7.

$$E_{ds} A_{ds} \frac{\partial u(x,\ t)}{\partial x} = Boundary\ Condition \tag{7.7}$$

If the drill string tip was held in place on the ends, then the end condition would be fixed, making the local displacement always equal to zero, displayed in Equation 7.9.

$$u(L_{ds}, t) = 0 \tag{7.8}$$

The boundary conditions for both the driver (top) and bit (tip) ends of the drill string form two separate independent equations, displayed in Equations 7.9 and 7.10.

$$E_{ds} A_{ds} \frac{\partial u(0,\ t)}{\partial x} = m_{sd} \frac{\partial^2 u(0,\ t)}{\partial t^2} - F_o \sin(\omega_f t) + k_{as} u(0,t) \quad Top\ Boundary\ Condition \tag{7.9}$$

$$E_{ds} A_{ds} \frac{\partial u(L_{ds}, t)}{\partial x} = -m_{em} \frac{\partial^2 u(L_{ds}, t)}{\partial t^2} - c_{em} \frac{\partial u(L_{ds}, t)}{\partial t} - k_{em} u(L_{ds}, t) \quad Tip\ Boundary\ Condition \tag{7.10}$$

the sonic drill system must be defined, as displayed in **Figure 6**.

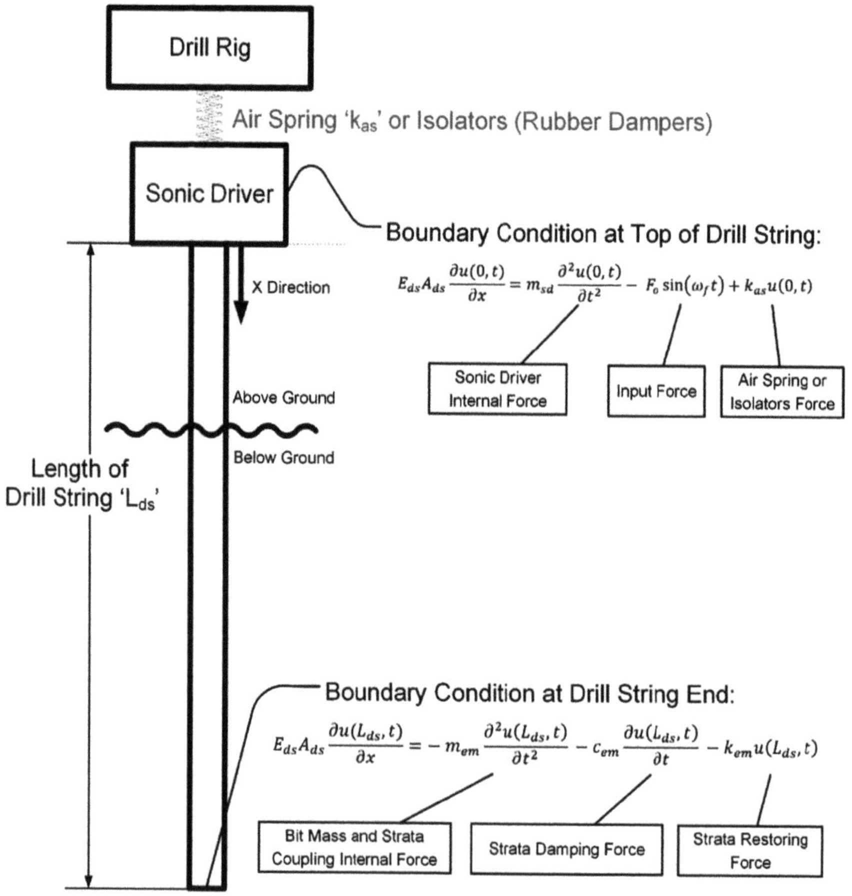

Drill Rig

Air Spring 'k_{as}' or Isolators (Rubber Dampers)

Sonic Driver

Boundary Condition at Top of Drill String:

$$E_{ds}A_{ds}\frac{\partial u(0,t)}{\partial x} = m_{sd}\frac{\partial^2 u(0,t)}{\partial t^2} - F_o \sin(\omega_f t) + k_{as}u(0,t)$$

Sonic Driver Internal Force

Input Force

Air Spring or Isolators Force

X Direction

Above Ground

Below Ground

Length of Drill String 'L_{ds}'

Boundary Condition at Drill String End:

$$E_{ds}A_{ds}\frac{\partial u(L_{ds},t)}{\partial x} = -m_{em}\frac{\partial^2 u(L_{ds},t)}{\partial t^2} - c_{em}\frac{\partial u(L_{ds},t)}{\partial t} - k_{em}u(L_{ds},t)$$

Bit Mass and Strata Coupling Internal Force

Strata Damping Force

Strata Restoring Force

FIGURE 7.6 Sonic drill boundary conditions. *Source:* Courtesy Lucon Engineering, Inc.

By placing the assumed solution to the governing differential equation, displayed in Equation 7.4, into these two independent boundary condition equations, the unknown solution coefficients, $\mathbf{A}_0$, $\mathbf{B}_0$, $\mathbf{C}_0$, and $\mathbf{D}_0$, can then be solved. These coefficients can be solved by the two independent equations because each equation can also split into two more independent equations, where one includes the sine terms and the other the cosine terms, creating four independent equations for the four unknown coefficients. Now that model has been developed, the perturbation parameters used to determine the significant factors of the sonic drill system are defined.

The sonic drill models are explained further in the PhD dissertation by Dr. Peter Lucon (Lucon, 2013) and in Lucon (2009). An example of the resultant mode shapes and system response from 50 Hz to 150 Hz for a 130-m long drill string is shown in Figure 7.7.

7.2.1.3.3 Select State-of-the-Art Sonic Drill Application

TI Geosciences Ltd. (TIG) is an Offshore Geotechnical Contractor focused on the deep-water environments owned by Royal IHC. The deep-water geotechnical drill SWORD was developed by the Royal IHC, specifically to ensure speed of execution and safety of operations. Sonic drilling was selected as the most efficient technology for this purpose. Sonic drilling parameters, as well as

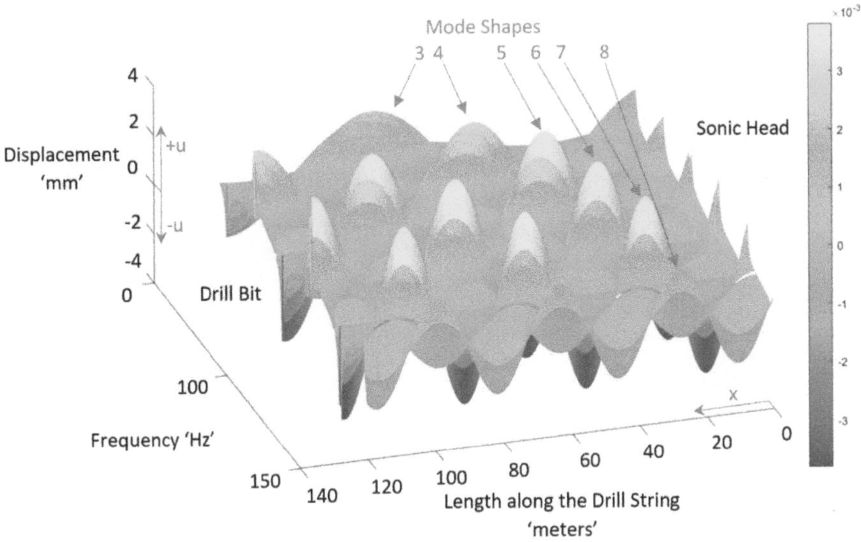

FIGURE 7.7 Sonic drill system response between driving frequencies of 50 Hz and 150 Hz. The mode shapes can be seen and are labeled from 3 through 9. *Source:* Courtesy of Lucon Engineering, Inc.

all other relevant operational and scientific information, are collected electronically and reported online, thus enabling fast decisions and further analyses.

Royal IHC enables its customers to execute complex projects from sea level to the ocean floor in the most challenging of maritime environments. Royal IHC is a reliable supplier of innovative and efficient equipment, vessels, and services for the offshore dredging and wet mining markets.

Additional links to industry applications of the Sonic Drill can be found in the references (Hafner, 2017a, b, c).

7.2.2 Specialized Environmental Purpose Drilling Systems

While similar systems to the ones discussed in this section are now commercial, the initial specialized systems were developed to accommodate specific testing requirements of the US government, hence the term "specialized" since much of the technology and methods used had not been invented. The first is focused on an example of a specialized-type drilling system designed specifically for minimizing the risk of a subsurface process from contaminating the environment (Figure 7.8).

7.2.2.1 Nevada Test Site "Big Hole" Drilling

Development of "Big Hole" drilling on the Nevada Test Site (NTS) was undertaken in earnest with the initiation of underground nuclear testing at the Nevada Test Site starting in July 1962 as part of the ratification of the Limited Test Ban Treaty [LTBT] of 1963. This type of drilling became known as "Blind Shaft Drilling" (BSD) and was some of the first drilling of its kind at that time. One technical report published in 1993 refers to the level of development as "The State of the Art" of Blind Shaft Drilling (Rowe, 2019). The report notes that BSD was an evolutionary process starting as back in 1957. It was initiated in response to the nuclear test programs of Los Alamos (LANL) and Lawrence Livermore National Laboratories (LLNL). The hole size was driven by the nuclear test package designs developed at these labs. Diameters ranged from 48 in (122 cm) to 144 in (366 cm) with holes ranging in depth up to 5000 ft (1524 m) (Rowe, 2019). And there was an occasional need to increase the diameter to 252 in (640 cm) using a reaming system (Figure 7.9).

FIGURE 7.8 NTS big hole drilling bit assembly. *Source:* Courtesy of US Department of Energy.

FIGURE 7.9 NTS big hole rig and BHA. *Source:* Courtesy of US Department of Energy.

In order to ensure the best possible sealing of the package within the shaft, there was an additional requirement to make the shaft as vertical as possible. At the time these holes were drilled, directional tools had not been developed for such large equipment; instead, very large weights were installed on the drilling assembly just above the plate holding the drill bits so that the system would act somewhat like a plumb bob. For this approach to be successful, the amount of weight used to drill had to be controlled. It was found that using only 40% of the weight of the drilling assembly to drill resulted in acceptable performance. The success was also a function of the relatively low compressive strength of the formations drilled. Despite the low compressive strength, the requirement for a near vertical shaft resulted in bottom-hole assembly weights of up to 300,000 lb (136,078 kg) for drilling a 96 in (244 cm) diameter shaft. While such a system allowed drilling up to 100 ft (30 m) per day at a rotation rate of 30 rpm, the torque required to turn 18 drill cones (bit/base plate consisted of multiple drill cones) in this scenario resulted in excessive stress on the surface systems. Initially conventional drill rigs equipped with heavy duty rotary tables were used. However, the makeup torque requirement associated with the thread makeup of 13-3/8 in (340 cm) pipe initially resulted in the use of casing running equipment. Subsequently it was recognized that specially designed surface equipment was going to be needed instead of conventional drilling rigs.

The geological situation at the NTS precluded the circulation of a full column of drilling liquids due to low rock strengths. "The dual string circulation system was developed at the NTS because formations would not support a hole full of fluid" (Rowe, 2019). The method used to mitigate this situation was reverse circulation; that is to say that cuttings and drilling fluid were returned up the center pipe of the dual-wall drill string. Air was pumped down to the bit assembly in similar fashion to that shown in Figure 7.10.

The NTS system configuration was relatively simple initially with the dual circulation paths created by an inner pipe run within the casing. Figure 7.11 shows the integral dual-wall pipe now available for use in circulation systems used today.

As noted previously, the goals of the Big Hole Drilling were to drill a shaft large enough to allow emplacement of the packages and to keep the alignment as vertical as possible; this was intended as an aid in securely sealing the annulus to minimize the probability of gas leaks to surface, in the case where the sealants were put in place using separate "tremie" pipes run down the annulus of the completion package. Drilling problems associated with lost circulation and poorly cemented, sloughing formations occasionally made achieving this goal even more difficult. The drilling fluid used was water which had enough velocity to provide effective cuttings carrying capability when combined with air and lifted to surface via the inner smaller diameter pipe. The flow channels under the drill bit assembly were optimized to provide effective sweep of cutting across the face of the assembly toward the center of the system where they were picked up and carried to surface. The bits initially used on the drilling assembly were conventional oilfield bits; however, there was some early experimentation using disc cutters, the results of which appear to have been undocumented.

7.2.2.2 DOE Yucca Mountain Project Dual-Wall Reverse Drilling

Yucca Mountain was the site of one of the DOE's most comprehensive site characterization studies. The Yucca Mountain Site Characterization Program was developed to investigate the potential of the mountain to serve as a high level nuclear waste storage facility. Although high level nuclear waste was being temporarily stored at a number of sites across the US, Yucca Mountain was to be the only major permanent storage facility that could eventually be sealed. As a result, the area and geology had to be thoroughly investigated to ensure a level of knowledge of both the geologic and hydrologic characteristics of the subsurface that would support a license application for long-term storage of high-level radioactive waste. This section is not intended to discuss the site characterization requirements outlined in the USGS Yucca Mountain Site Characterization Plan other than to relate how these requirements affected the design of the drilling/coring/sampling system used for Unsaturated Zone investigations focusing on detection of potential fast transport pathways to the groundwater related to faulting in the area. Contamination of the groundwater was of particular concern since the regional flow had been determined to potentially enable flow out of the area reserved for the repository site (Figure 7.12).

Perhaps one of the better descriptions of the rigorousness of investigations involved in this effort is the Yucca Mountain Progress Report Number 12, published in 1995 ([SCPR-12]; US Department of Energy, 1995). By this time, the tunnel boring machine had competed creation of the drift through the mountain for testing within the proposed repository and much of the Unsaturated Zone testing was complete.

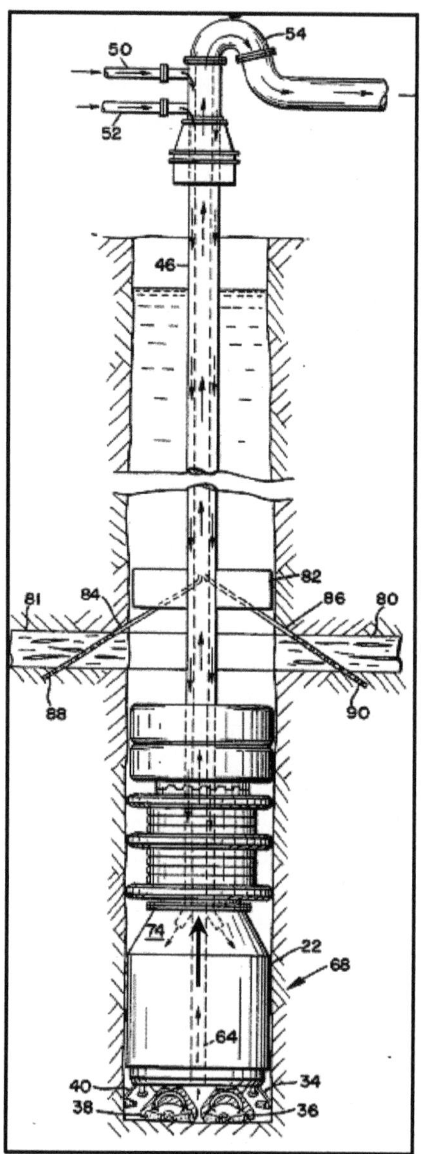

FIGURE 7.10 Blind shaft drill concept. *Source:* Courtesy of US Department of Energy.

One of the basic testing techniques was investigation for the existence of Chlorine-36 (36 Cl).

> Chloride and 36C1 concentrations in ream cuttings and sampled water from the subsurface were measured. Chloride concentrations provide an indirect measure of evapotranspiration rates and hence infiltration rates. Chlorine-36 is used to identify recent water that was generated during surface nuclear bomb tests.
>
> (Thamir et al., 1998)

A tracer gas, sulfur hexafluoride, was added to the air used for circulation to enable researchers to be able to tell when the borehole had returned to in situ conditions. Following drilling, a vacuum was placed on the borehole and air withdrawn until no tracer gas was detected.

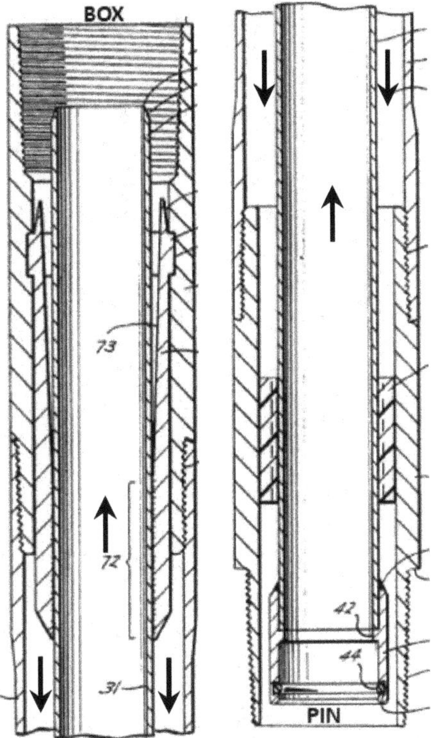

FIGURE 7.11 Dual-wall pipe concept. *Source:* Courtesy of US Department of Energy.

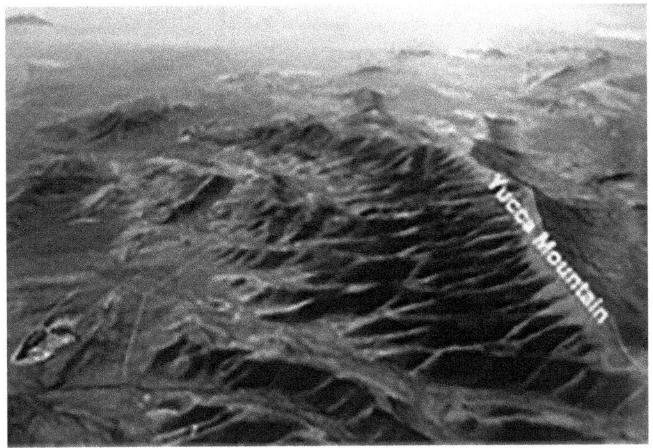

FIGURE 7.12 Yucca mountain. *Source:* Courtesy of US Department of Energy.

An early method used to minimize contamination of in situ gases with atmospheric air was a vacuum drilling unit such as that shown in Figure 7.13 for the drilling of the Unsaturated Zone borehole USW UZ-1 (Whitfield et al., 2020). The method was relatively effective for minimizing contamination due to drilling; however, a number of disadvantages were recognized when drilling deeper. Primarily there was not adequate energy being produced at the bit for both effective cuttings

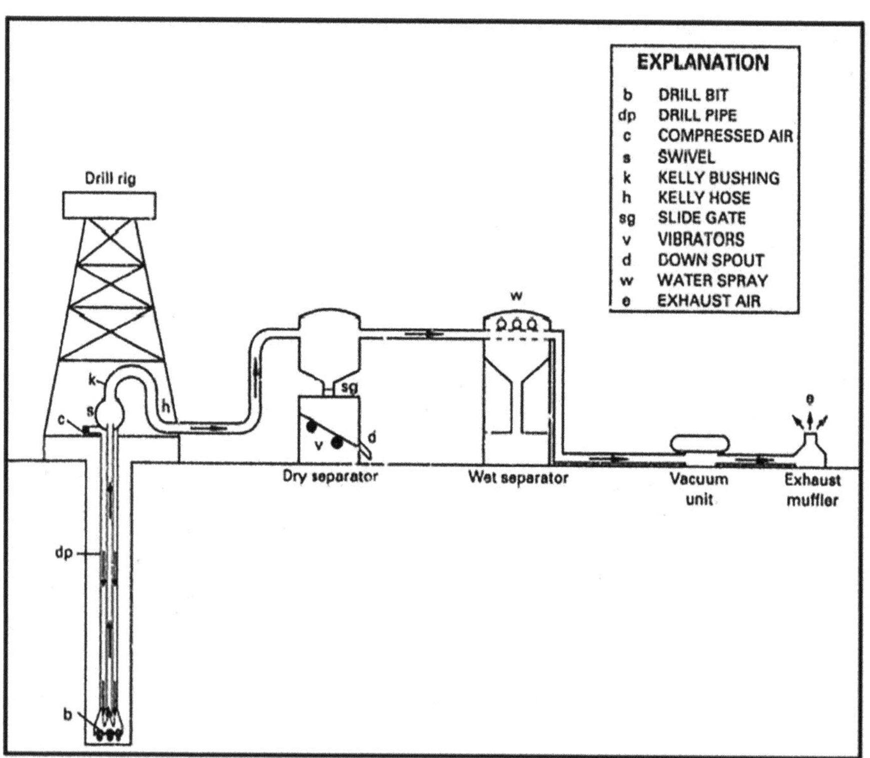

FIGURE 7.13 Vacuum drilling system. *Source:* Courtesy of US Department of Energy.

removal and handling a small amount of water, such as the small perched water zones above the water table that occasionally entered the borehole. As noted in the USGS report, "When the rock-moisture content exceeds 5 percent by weight, plugging of the system occurs and drilling needs to be stopped, or the drill bit will be buried in drill cuttings and will become stuck." Another drawback of the system was that the drill string had to be removed from the borehole whenever it was necessary to obtain a core. It was subsequently determined that continuous coring would be required in addition to drilling a borehole of sufficient diameter to install permanent monitoring instruments and seal them in the borehole.

In order to resolve the drilling/sampling needs for deeper, approximately 2000 ft (610 m) investigation, the following dual-wall drilling and coring system was developed.

The system shown in Figure 7.14 utilized dual-wall drill pipe to achieve the following Unsaturated Zone testing needs:

1. Provided high pressure air to the drill bit for effective hole cleaning and circulation of cuttings up the inner tube of the pipe during drilling, resulting in a relatively undisturbed borehole wall.
2. Provided a protective string for running core rod to depth without the need to pull the drill pipe.
3. Provided an inner return for cuttings and air such that a vacuum could be applied to both coring and drilling return lines.

Circulation issues were resolved by metering the injected and return air rates. This allowed localized high pressure air for coring/drilling while keeping the net air charge at the bottom of the hole

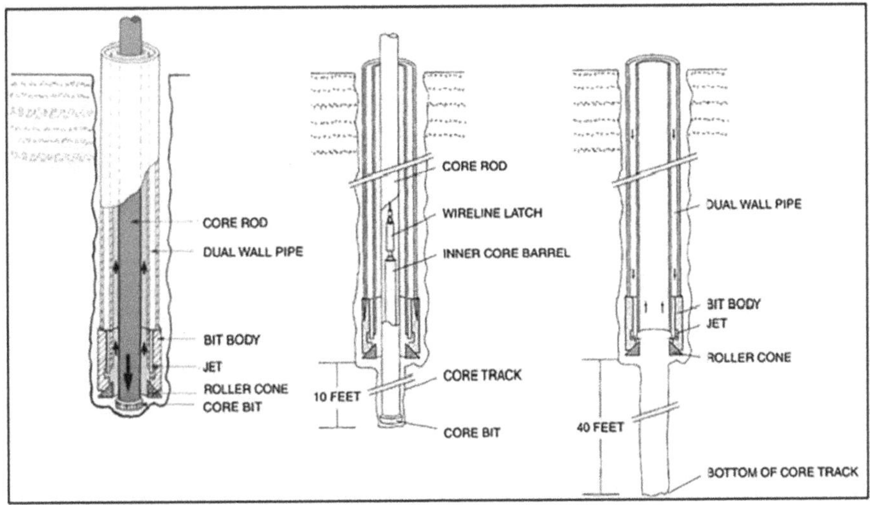

FIGURE 7.14 Dual-Wall Pipe and Coring Sequence. *Source:* Courtesy of US Department of Energy.

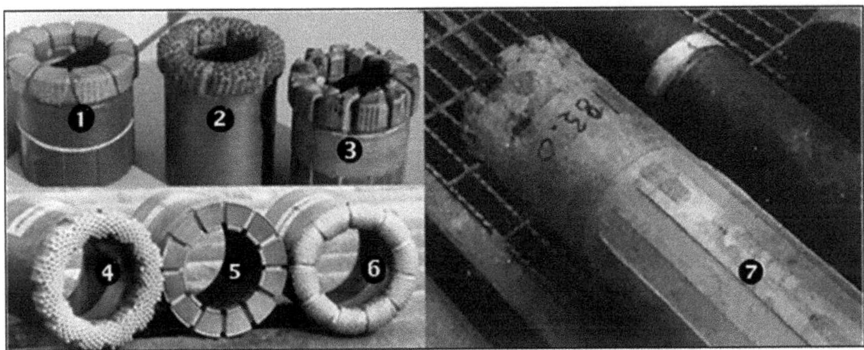

FIGURE 7.15 Core Bits and Core Barrel Reaming Shell. *Source:* Courtesy of US Department of Energy.

slightly below atmospheric pressure. This system allowed relatively rapid drilling/coring despite the need for continuous core to planned depth. The wireline core was pulled every 10 ft (3 m) and continuous coring continued until a maximum core ahead interval of 40 ft (12 m) was reached. It was found that if core runs were kept to this interval or less, there was no need to include an insert in the center of the reaming bit as was initially considered. Coring beyond this interval before ream down resulted in enough difference between the core track departure compared to the reaming track being offset enough to result in bit plugging with cuttings too large to pass through.

The use of the coring assembly in air proved to be a challenge. Typical coring operations were accomplished in a drilling mud environment using diamond impregnated core bits such as number 5 shown in Figure 7.15. Use of these bits in an air coring environment resulted in low penetration rates and localized heating of the samples such that there was concern regarding potential impact to analysis of the cores. The use of Polycrystalline Diamond Cutter (PDC) core bits was relatively new at the time, and, there was some debate regarding their effectiveness for coring hard rock. As it turned out, new sintering and bonding methods were just starting to be used in the fabrication of the PDC core bits such that one (a Christensen Stratapac 20, CS20, similar to number 3 shown in

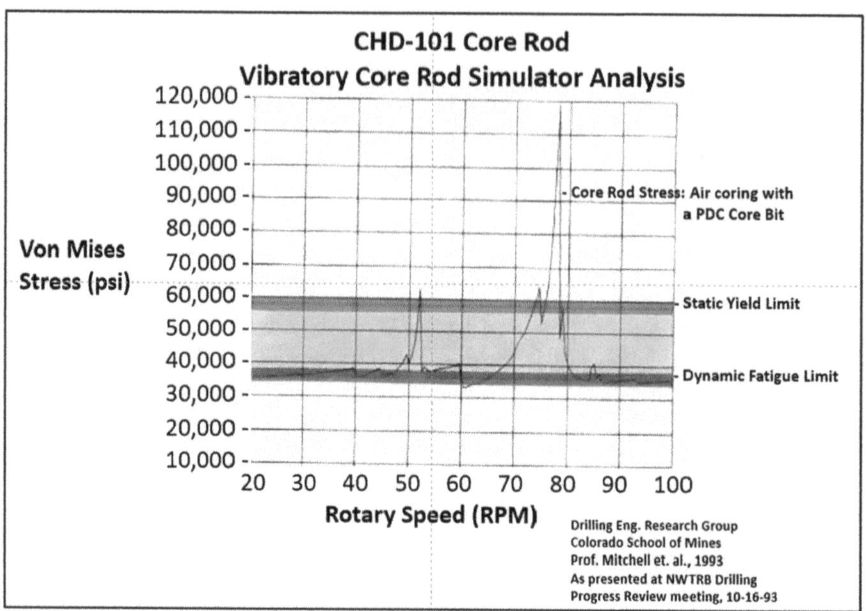

FIGURE 7.16 Vibratory core rod simulator results.

Figure 7.15) proved to be the best performer regarding penetration rate and bit life. A full range of core bits were tested as shown previously. In addition to the standard diamond impregnated bit (#5), synthetic diamond cutter bits (#1 and #6), carbonado diamond cutter bits (#2 and #4), and PDC bits (#3) were tested throughout the program (Gertsch, 1993).

Another challenge for air coring was the use of the aggressive (greater depth of cut per rotation) PDC core bits. The resulting core rod vibration occasionally led to failure of the pin connection in the core barrel reaming shell (#7 in Figure 7.15). As a result, studies were initiated to better understand the vibration issue to see if mitigations could possibly be developed to minimize these failures. A study was initiated of the system that led to development of a Core Rod Vibration Simulator capable of analyzing the dynamics of the coring system at depth with the various types of core rod used. Figure 7.16 shows an example of the output of the simulator for a particular coring depth and core bit used.

The specialized drilling/sampling system used at Yucca Mountain was built by Lang Manufacturing, a subsidiary of Boart Longyear in Salt Lake City and called the LM-300 due to its maximum mast load capability of 300,000 lb (136,078 kg). Several field trials with similar smaller systems were part of the prototype development effort. Following these trials, the system in Figure 7.17 was developed. This system, along with the surface processing equipment and sample handling trailer, resulted in successful sampling and hydrologic testing of the Unsaturated Zone (Figure 7.18).

That complete Yucca Mountain Unsaturated Zone sampling system is shown in Figure 7.19. Numbered elements 8 through 11 in the figure show the circulation and cuttings transport return path. Cuttings travel through the return line (#8) and first enter the cyclone separator (#9) where they are dropped out by means of a double sliding gate valve that is used to keep a vacuum on the return line. From #9 the cuttings pass to a final bag house filter (#10) that is used to sample fine cuttings and keep dust out of the return air from entering the vacuum blower (#11). This sampling configuration allowed continuous sampling throughout drilling and coring operations.

The drilling air was supplied by dual 900 CFM compressors (#1). It was subsequently found necessary to add a condenser (#2) downstream of the compressor. As the high pressure air was cooled coming out of the compressor, a significant amount of water was condensed (surprisingly even in

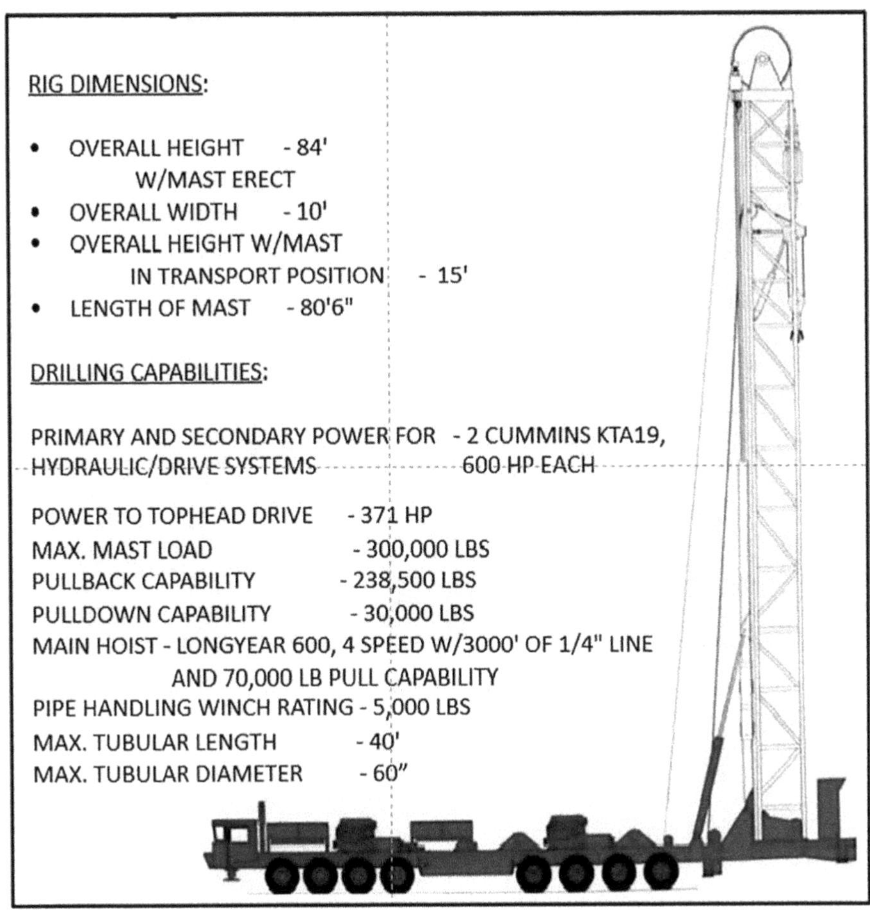

RIG DIMENSIONS:
• OVERALL HEIGHT - 84' W/MAST ERECT
• OVERALL WIDTH - 10'
• OVERALL HEIGHT W/MAST IN TRANSPORT POSITION - 15'
• LENGTH OF MAST - 80'6"
DRILLING CAPABILITIES:
PRIMARY AND SECONDARY POWER FOR HYDRAULIC/DRIVE SYSTEMS - 2 CUMMINS KTA19, 600 HP EACH
POWER TO TOPHEAD DRIVE - 371 HP
MAX. MAST LOAD - 300,000 LBS
PULLBACK CAPABILITY - 238,500 LBS
PULLDOWN CAPABILITY - 30,000 LBS
MAIN HOIST - LONGYEAR 600, 4 SPEED W/3000' OF 1/4" LINE AND 70,000 LB PULL CAPABILITY
PIPE HANDLING WINCH RATING - 5,000 LBS
MAX. TUBULAR LENGTH - 40'
MAX. TUBULAR DIAMETER - 60"

FIGURE 7.17 LM-300 specifications. *Source:* Courtesy of US Department of Energy.

Nevada's dry climate) as the air was cooled either in the condenser or downhole as the air went around any pressure restriction, such as the core bit and barrel. If the condenser were not used, there would be some question as to the water saturation found in the core and cuttings.

A core processing field lab (#4) was on site and operational during all coring and sampling operations. Personnel assigned to the lab ensured core and cuttings were processed immediately and sent to the Sample Management Facility for logging and curation. A detailed, accurate record of all core and samples was essential to validation and use for licensing of Yucca Mountain as a high-level waste repository, should a license application be initiated.

Due to the need for continuous core, a finger board (#3) was included in the rig design. This allowed stacking of dual stands of core rod for faster tripping in and out of the hole. Everything on the rig was designed to minimize pipe handling time. The top-drive along with a pipe makeup/breakout system built into the rig floor and a dual-wall pipe handling system, shown in Figure 7.19, allowed essentially total automation of all dual-wall drill pipe handling functions. This both enhanced safety and ensured the complex coring/drilling/sampling operation was completed in minimum time.

As can be noted in Figure 7.19, dual-wall pipe has no upset at the connections. As a result, this flush joint profile from surface to total depth has distinct advantages when drilling in fractured rock. As was the case in Yucca Mountain drilling, there were intervals where there was significant rock

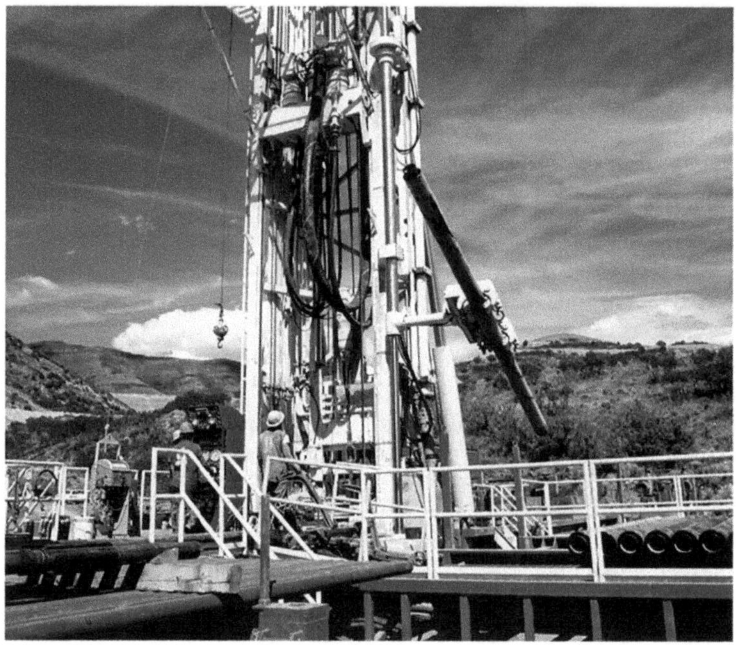

FIGURE 7.18 Yucca mountain unsaturated zone drilling/coring/sampling system. *Source:* Courtesy of US Department of Energy.

FIGURE 7.19 LM-300 dual-wall pipe handling system. *Source:* Courtesy of US Department of Energy.

FIGURE 7.20 Enhanced drill bit. *Source:* Courtesy of US Department of Energy.

breakout with a number of large pieces of hard rock falling to the bottom. However, once the loose pieces of rock in the breakout came to some stable angle of repose, they were no longer a problem since there was no continuing circulation or other destabilizing phenomena occurring at the borehole wall. However, this did present a problem initially when the rocks fell to the only upset in the string, the drill bit assembly. Once lodged between the bit and the formation wall, the pieces could only be dislodged by repeated hammering up and down with the drill string. Consequently, an "enhanced" drill bit was manufactured as shown in Figure 7.20. The feature added was tungsten carbide inserts to the top lip of the bit assembly shown in the top portion of Figure 7.20. It was found that larger inserts could be used without significant breakage. This allowed very effective back reaming when necessary and much quicker resolution of any issues due to formation instabilities. It was found that the debris in fractured zones would eventually come to a relatively stable angle of repose as long as the borehole wall remained undisturbed, as was the case with this type of drilling system.

Communication of the efficiencies of drilling and sampling with a system like the LM-300 were also made available to the geothermal industry (Rowley, 1994; Rowley et al., 1995a, 1995b, 2000). Use of dual-wall pipe and circulation of a gas drilling fluid eliminates lost circulation costs associated with geothermal drilling. Also proposed with this system was the use of a directional air hammer (Bui, 1995) in lieu of a rotary system.

Although the geothermal industry did not readily respond to the LM-300 concept, the petroleum industry did. In 2006 Encana was drilling an exploration well in south-central Washington State's Columbia River basin. The borehole was planned to a depth of 14,000 ft (4267 m); however, the upper 7800 ft (2377 m) required drilling through a fractured basalt. As a result, Encana had an LM-700 reverse circulation (RC) rig constructed to drill the basalt and avoid the lost circulation and bridging issues characteristic of conventional systems drilling through such strata (Figure 7.21). The LM-700 had a 700,000 lb (317,515 kg) hookload capacity and a mast capable of racking 100-ft stands of 6-5/8 in (16.8 cm) dual-wall drill pipe. Once casing was set through the basalt a conventional rig was brought in to complete the well to total depth.

FIGURE 7.21 LM-700 RC rig. *Source:* Courtesy of Oil and Gas Journal (2007).

7.2.3 Oil and Gas Environmental/Economic Drivers

Today's oil and gas industry continues to face pressure to reduce the environmental footprint of oil and gas exploration and production activities. Most notable in that concept of footprint is that of land disturbance. Unconventional oil and gas wells continue to require more land-use space despite the use of horizontal and extended reach drilling due to the layered heterogeneous nature of the unconventional resource/reservoir.

7.2.3.1 Footprint (DOE Microhole Initiative)

One effort to investigate what might be possible in the area of footprint reduction was the use of coiled tubing drilling in the DOE's Microhole Initiative (Perry et al., 2019). A contract award for the field demonstration of a high-efficiency hybrid coiled tubing (CT) rig designed and built by a Mr. Tom Gipson resulted in some interesting findings. The term hybrid is used because it had both coiled tubing and rotary drilling capability. In addition, the entire rig layout, including a tank for the drilling mud system, mast and drillers cabin, was packed into three custom built and very mobile tractor trailer loads. At the conclusion of the demonstration the following accomplishments were recorded (Figure 7.22):

1. Drilled 25 wells in the Niobrara in eastern Colorado and western Kansas.
   a. Drilled 330,000 ft (91,440 m) of hole in seven months
   b. Drilled and completed wells to a depth of 3000 ft (914 m) in 19 total hours
2. Nominated Colorado Oil and Gas Conservation Commission 2005 Operator of the Year.
3. Nominated World Oil Awards 2005 New Horizons.
4. Made economic development of 1 Tcf of shallow bypassed tight gas in the Niobrara formation feasible.
5. Reduced the cost of drilling wells by 25 to 38%.
6. Reduced environmental impact due to the small footprint.

FIGURE 7.22 Microhole rig field demonstration. *Source:* Courtesy of US Department of Energy.

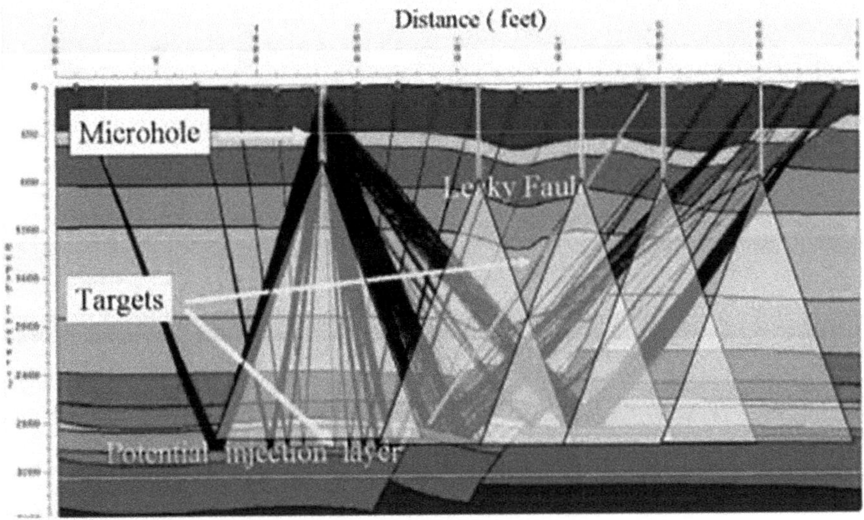

FIGURE 7.23 Downward looking VSP imaging. *Source:* Courtesy of US Department of Energy.

7.2.3.2 Imaging in E&P (Downward Looking VSP)

The Microhole coiled tubing drilling system not only offered the environmental benefit of a reduced drilling footprint for each borehole, but its ability to rapidly drill and complete boreholes, rig down, move to another location, rig up, and begin drilling in minimum time was a significant advancement, especially for a rig of this size/capability. As an example, for a 2850 ft (869 m) well the total move-in-rig-up/rig-down-move-out time was only 18% (approximately 4 hours) of total time to drill (Perry et al., 2019). This level of performance enabled another form of sampling associated with deployment of instrumentation in the subsurface for the advancement of high-resolution seismic imaging.

New advances in seismic and other imaging technologies are coming to the forefront at this writing; however, one significantly improved data acquisition system, based on a concept suggested by geophysicists at a roadmapping meeting in Albuquerque, was demonstrated during DOE's Microhole Initiative. The concept involved investigation of the potential of downward looking Vertical Seismic Profiling (VSP) applications (Figure 7.23). VSP is typically used cross-hole between wells to evaluate the vertical interval between the wells. In the downward looking application, an array of wells would be used to image much deeper than the instrumented microholes. As part of the investigation, ~6 in (15 cm) diameter holes were drilled to a depth of 1000 ft (305 m) and instrumented with geophones. The diameter of the cone that can be imaged is roughly 1/2 the distance from the bottom of the instrumented hole to the depth of the target of interest. So, for a 1000-ft (305 m) instrumented hole the imaging capability at 5000 ft (1524 m) is a circular area about 2000 ft (610 m) in diameter.

From early testing with conventional geophones the following was demonstrated:

1. Shallow Microhole VSP is capable of imaging up to 3 times (or more) shallow hole depth.
2. Up to three times better resolution than VSP in conventionally drilled boreholes (much better signal-to-noise ratio).

Some of the promising future potential benefits of these points are:

1. Permits use of microholes for low-cost, rapid VSP deployment because sensors do not need to be placed at reservoir level.
2. High resolution seismic surveys can be faster and much cheaper with permanently installed shallow, instrumented boreholes.
3. Cost-effective, permanent VSP boreholes could revolutionize complex reservoir characterization and long-term EOR monitoring.

This new imaging capability promises to allow decreasing the surface area extent of oil and gas development by optimizing well location by means of better understanding of subsurface geology/hydrology.

Another finding from the downward-looking VSP investigation was the impact of what is called the weathered zone (WZ). This phenomenon helps understanding of why there is a better signal-to-noise ratio with depth of the geophones. The WZ represents the topsoil and other loose uncompacted media that has the effect of deadening any noise from above or seismic signal from below that passes through it.

Researchers were able to quantify the degree of the insulting effect of the WZ in one field demonstration where the geophones had been set slightly shallow. Due to inclement windy weather the surface noise was very high. This provided the opportunity to actually note that the top 300–500 ft of the WZ in this location resulted in as much as a 60-db attenuation in the surface noise with depth.

Given this information, another opportunity emerged to test the potential advantage of this attenuation effect by placing much more sensitive accelerometers below the weathered zone to compare with other sensitive surface systems. That investigation is still ongoing; however early results are very encouraging. Early testing demonstrated that the accelerometers in the instrumented borehole detected a magnitude 2.6 tremor over 275 km away at a depth of 5.3 km. This is attributed to the excellent signal-to-noise ratio in this monitoring system. Many equally sensitive surface systems are not able to detect these weak signals due to the surface noise. This could have potential impacts on wide bandwidth, high-resolution seismic imaging (Figure 7.24).

7.3 OFFSHORE ENVIRONMENTAL RISK MITIGATION CHALLENGES AND TECHNOLOGIES

In the introduction of this chapter, reference was made to discussion focused on those drilling systems requiring a drilling "rig" or surface system controlling the down-hole tools. While this section does not deal with these systems per se, it does focus on the elements of the environment that influence the ability of those systems to function properly to avoid further environmental impact. Two of these major influences can be characterized as metocean and geologic uncertainty. Oil and gas

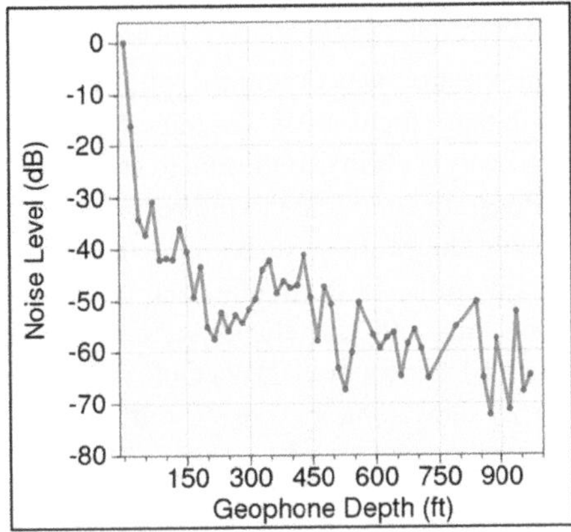

FIGURE 7.24 Surface noise vs. depth.

FIGURE 7.25 Offshore operations risks.

operations in the Gulf of Mexico can extend to water depths in excess of 10,000 ft (3048 m) water depth (Figure 7.25).

Offshore deep-water operations are characterized by significant risks of various types despite the best planning. The Deepwater Horizon risk was, in some respects, related to geologic and/or down-hole uncertainty. Chevron's Mad Dog Spar and Mars Platform were related to metocean risks associated with uncertainty in anticipated max wind speed.

Following the Deepwater Horizon blowout on April 20, 2010 and subsequent oil spill in the Gulf of Mexico, the Department of Energy's Office of Fossil Energy (DOE-FE) initiated a number of efforts to aid in the spill response [FRTG]. Following these spill response activities, DOE-FE focused its offshore funding from Title IX, Subtitle J, Section 999 of the Energy Policy Act of 2005 (hereto called Section 999) toward advancing technologies to aid in spill prevention. Other federal funding at the time was spent in spill response and evaluation of environmental impacts; whereas Section 999 was unique in that it was the only program focusing exclusively on development of spill prevention technologies.

The following details some of the highlights of technologies developed via contract with industry during this period until such time as Section 999 was repealed by the Bipartisan Budget Act of 2013, signed into law December 10, 2013 [BBA]. The graphics in this summary are taken from the final reports of the following representative projects that can be found on the National Energy Technology Laboratory's (NETL) Energy Data Exchange archive website [EDX]. The companies and researchers providing these graphics as deliverables to NETL can also be found on this website. A Complimentary Program initiated during Section 999 that continues today focuses NETL research on utilizing Big Data Analytics. That research is discussed in Section 7.3.2.

7.3.1 DOE Deepwater Research Highlights

The Offshore Research Program research was organized with four primary research focus areas as shown in Figure 7.26. Research in these focus areas was intended to provide specific products within each area that would enhance industry's ability to minimize spill risks.

The portion of the Offshore Program contracted to industry resulted in several leading-edge technologies being developed that hold the promise of both decreasing cost and enhancing industry's ability to minimize the risk of oil spills. In particular, one of the major advances for improving operations related to Surface Systems and Umbilicals were key metocean studies that improved understanding by quantifying Ultra-Deepwater (UDW) Gulf of Mexico Loop Currents. The graphics shown in Figures 7.27 and 7.28 show some of those results.

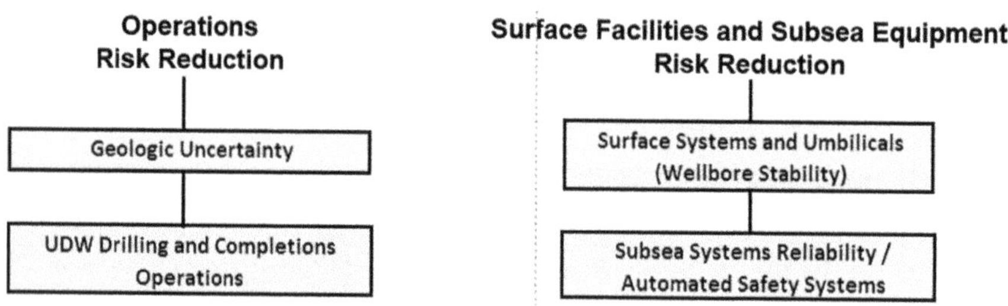

FIGURE 7.26 Offshore program research focus areas. *Source:* Courtesy of US Department of Energy.

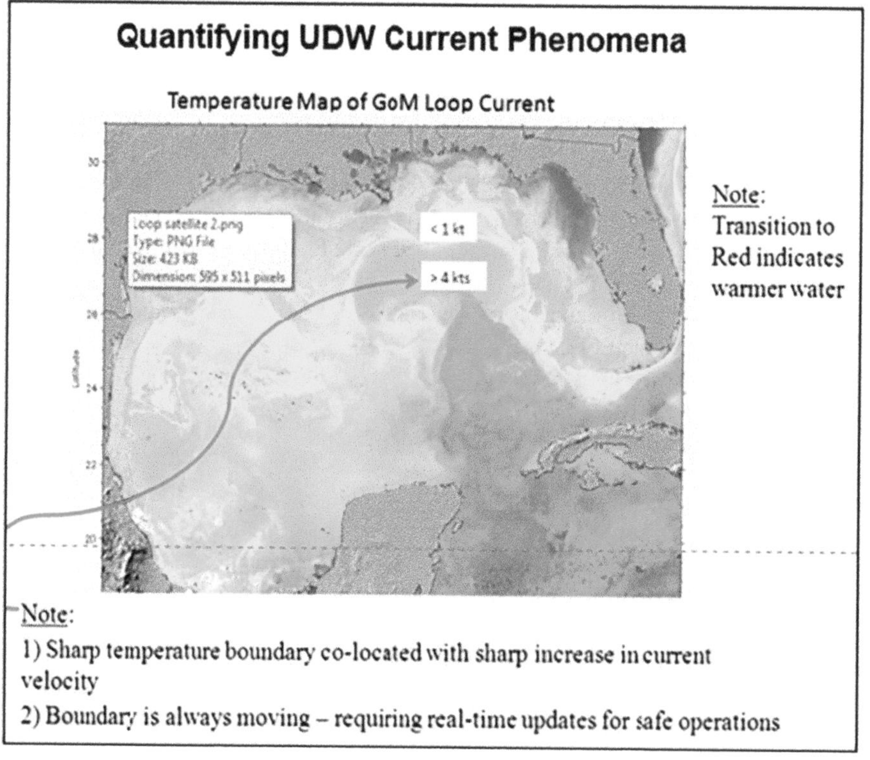

FIGURE 7.27 Project #11121-5801-01, Hi-res environmental data for enhanced UDW Ops. *Source:* Courtesy of US Department of Energy.

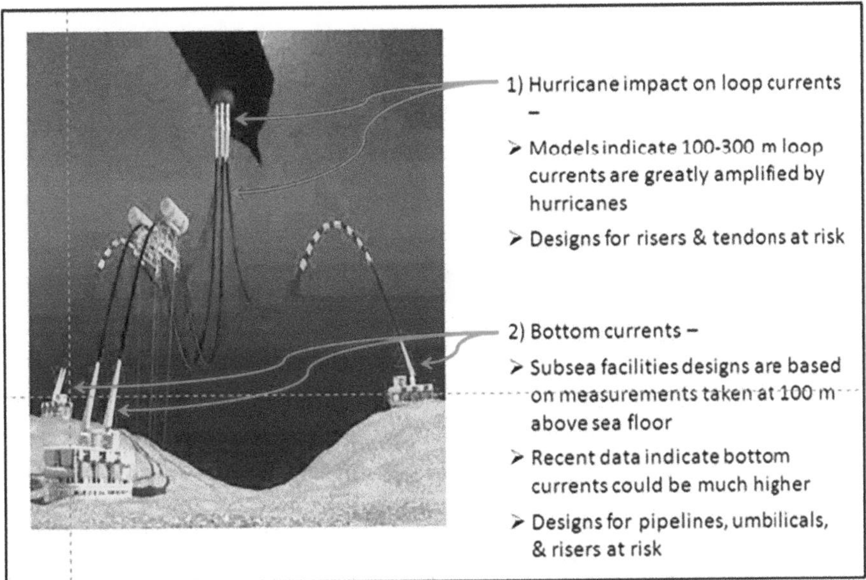

1) Hurricane impact on loop currents
-
➢ Models indicate 100-300 m loop currents are greatly amplified by hurricanes
➢ Designs for risers & tendons at risk
2) Bottom currents –
➢ Subsea facilities designs are based on measurements taken at 100 m above sea floor
➢ Recent data indicate bottom currents could be much higher
➢ Designs for pipelines, umbilicals, & risers at risk

FIGURE 7.28 Project #11121-5801-01, Hi-Res environmental data (Task 5: Measurement and modeling of near-bottom currents). *Source:* Courtesy of US Department of Energy.

Prior to these studies, industry experience was mainly from the North Sea at depths less than 2000 ft (610 m); Gulf of Mexico current and planned operations were at 10,000 ft (3048 m) and up to 12,000 ft (3658 m). This information advanced learning of UDW currents and potential system loads from a metocean joint industry project (JIP) in partnership with Chevron at the time.

In response to the new metocean information that was available, technologies as shown in Figure 7.29 were developed to improve monitoring of loop current impacts. Risers serve as an extension of the wellbore from the mud line (ocean floor) to the surface. The riser continuously responds to changes in loop current direction and velocity and are subject to what is referred to Vortex Induced Vibration [VIV] that increases cyclic loading on the riser. Although technologies have been developed (Strakes) to help reduce VIV effects, cyclic loading due to the variations in current velocity and direction are still a concern.

Another technology that was developed within the UDW Drilling and Completions focus area directly addressed a risk that was one of the potential issues with the Deepwater Horizon blowout, cement contamination. This presumably occurred when the cement mixed with oil-based mud in an extension of the borehole below the casing, causing the primary cement around the casing to have less strength following the scheduled setting time after cementing than expected.

The report noted that each well would have to be considered separately and designs developed for each case; however, there did not appear to be any "show stoppers" regarding the feasibility of potential implementation of this approach to cementing.

One project in the area of Subsea Systems Reliability addressed a major issue with resource development in the Ultra-Deepwater Offshore, that of flow assurance, Figure 7.31. One of the main safety and cost concerns was that of hydrate plugging of flow lines once the temperature of water-bearing produced fluids dropped to critical levels for hydrate development in the flow lines on the ocean floor. To better understand the phenomena associated with this problem, the Colorado School of Mines and Tulsa University were awarded a contract to develop models based on data from the University of Tulsa's Hydrate Flow Loop to improve this understanding and provide potential mitigation strategies. As noted in the study, many reservoirs that start out with no produced water can

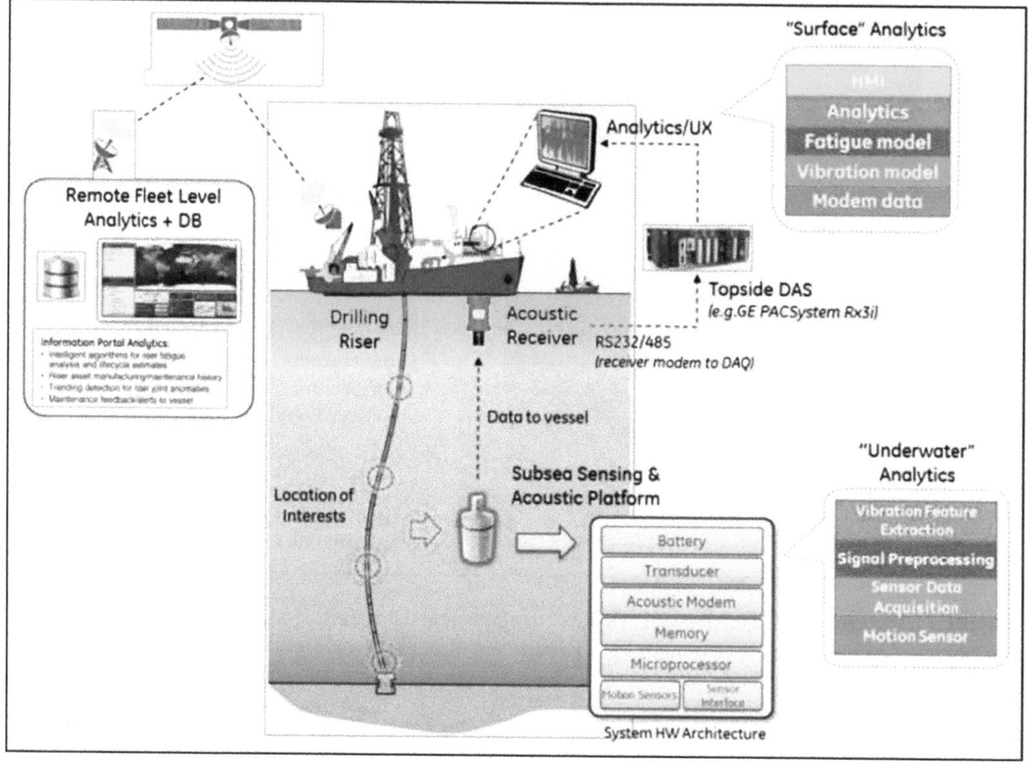

FIGURE 7.29 Project #11121-5402-01, Riser lifecycle monitoring. *Source:* Courtesy of US Department of Energy.

develop water cuts on the order of 60% over a 10-year period. This increases the cost of hydrate mitigation and the risk of hydrate plugging. Results of this research included improved predictive models and a presentation at the 2016 Offshore Technology Conference (Vijayamohan et al., 2016).

Figure 7.30a conveys conventional cement emplacement circulation with all the cement being circulated around the bottom of the casing (casing shoe) around the outside of the casing up into the annulus. This could allow any contaminated cement to remain around the annular sealing area. Figure 7.30b shows reverse circulation of the cement. In this diagram, it is assumed that any contaminated lead cement will simply be captured inside the casing. Also notable in case "b" is the use of a crossover tool at the top of the liner hanger. Various designs were considered in the report to ensure operation of this key device.

A project that made a significant impact on the design of offshore surface facilities was Project #12121-6403-01, "Development of Advanced CFD Tools for the Enhanced Prediction of Explosion Pressure Development and Deflagration Risk on Drilling and Production Facilities." This project developed a full-scale test facility which, when combined with advanced computational fluid dynamics (CFD) software, enabled determination of what levels of equipment congestion a deflagration-to-detonation transition (DDT) event can occur. The software is referred to as the Flame Acceleration Simulator (FLACS). DDT occurs when a flammable gas cloud build-up from a leak or blowout is ignited and the gas begins to burn. If equipment/facility congestion within the area of any gas cloud of a volume exceeding the designed maximum credible event (MCE), detonation or explosion is typically the result. Such an explosion, followed by fire, is considered one of the most dangerous and high-consequence events that can occur on an offshore facility.

The capability developed by this project is significant because the lack of detailed geometry information identifying congestion, in the early design phase of a facility, had the potential to result in

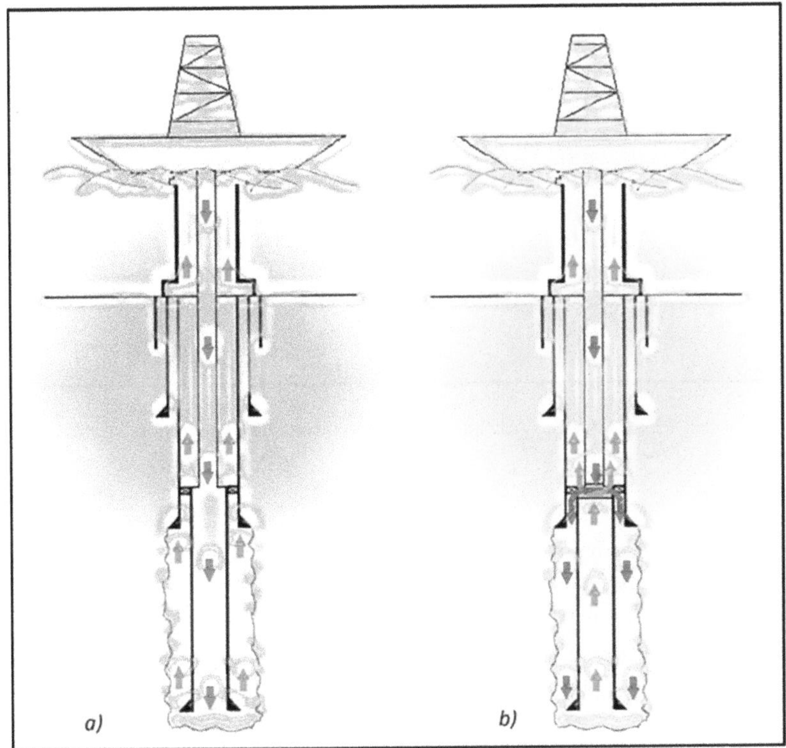

FIGURE 7.30 Project #10121-4502-01, Normal vs. reverse circulation cementing. *Source:* Courtesy of US Department of Energy.

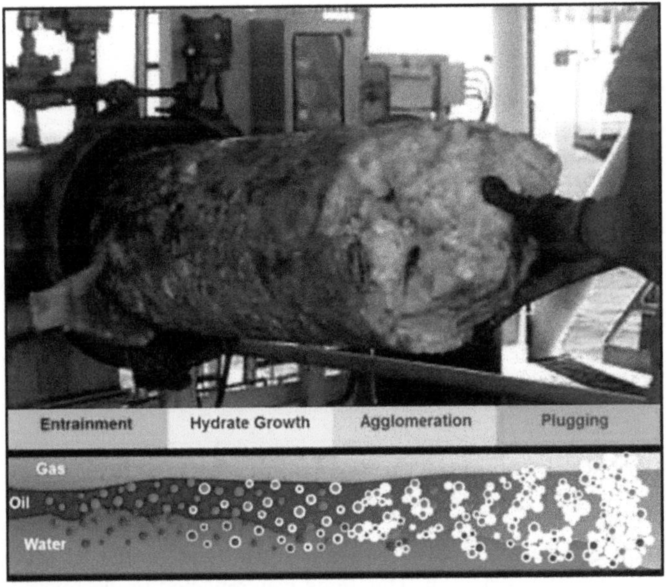

FIGURE 7.31 Project #10121-4202-01, Hydrate Modeling and Flow Loop Experiments for Water Continuous and Dispersed Systems. *Source:* Courtesy of US Department of Energy.

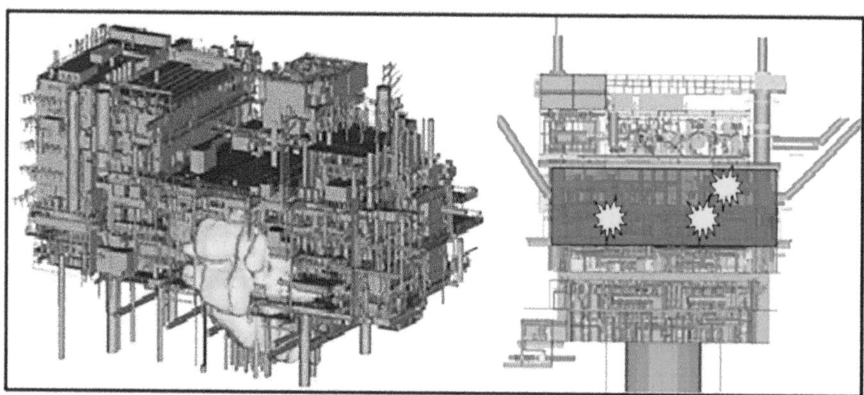

FIGURE 7.32 Project #12121-6403, "Advanced CFD tools for DDT prediction." *Source:* Courtesy of US Department of Energy.

severe underestimation of design blast loads when not accounted for in explosion studies. Operators and service companies participated in review of existing designs to ensure correct MCEs had been applied to ensure safe facility designs.

Project accomplishments are as follows:

- Preliminary design and validation of FLACS tools to predict explosion consequences was completed. Analysis of congestion GOM topsides facilities was conducted, and an interactive database tool was constructed to catalog provided platforms and other offshore installations. This database has the capability to count congestion parameters and provide comparisons between as-built installations.
- An initial test rig was designed, and analyses of pipe layout, rig brackets, and strength were conducted. Modifications to the large-scale test facilities were made, and a final rig 180 ft long and with a gross volume of over 50,000 ft^2 was built in a California facility. Large-scale DDT tests to evaluate methane, ethylene, and propane and to evaluate mitigation measures were conducted (Figure 7.32).

Inspection capabilities of subsea facilities was greatly enhanced by adding newly developed subsea LIDAR (Light Imaging, Detection, And Ranging; Project #09121-3300-06, accessed November 14, 2019) that allows unprecedented resolution of subsea equipment. This technology was combined with an autonomous underwater vehicle in this project (#10121-4903-02, "Autonomous Underwater Vehicle (AUV) Inspection Using a 3D Laser," accessed November 14, 2019) to allow high-resolution scanning of large segments of subsea facilities in a minimum time. As noted in Figure 7.33, entire platform structures can be inspected quickly. Such a capability allows cost-effective, real-time detection of flaws or damage when compared to baseline structural models previously developed.

As previously noted, projects discussed in this summary are representative highlights. The final reports and other information on these projects can be found on NETL's archive website [EDX].

7.3.2 Automation and Big Data/Analytics Impacts

7.3.2.1 Automation

The offshore oil and gas industry is moving rapidly toward automation of offshore platforms for both increased safety and cost reduction. This trend was reflected in the International Association of Drilling Contractors (IADC) Drilling Engineering Committee [DEC] Forum held September 25, 2019, titled "Drilling Data—What Is Available? How Good Is It? What Can We Do with It?" The

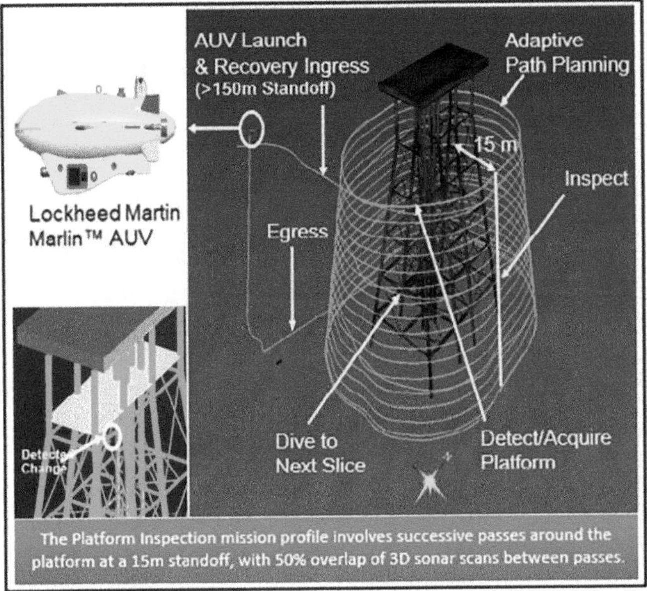

FIGURE 7.33 Project #10121-4903-02. *Source:* Courtesy of US Department of Energy.

minutes from this forum provide a good summary of many of the issues, focus areas, and industry participants involved in developing viable automation systems for the Gulf of Mexico and international offshore operations.

One of the issues noted in the DEC Forum was the need for appropriate data mining and analytical and predictive tools to underpin automated, machine-supported decision making. This is a continuing focus area of NETL's inhouse offshore research group, Research Innovation Center (RIC), as initiated in the Section 999 research. Following lessons learned from the Deepwater Horizon spill, DOE NETL pivoted research to drive and develop big data-driven models and tools in support of offshore risk reduction and prevention. Phase 1 of this effort concluded with the Offshore Risk Modeling suite's development in 2016. Phase 2 of this effort is enhancing these models and tools to integrate machine learning and adaptive, near-real time predictions of geohazards, infrastructure maintenance forecasting, operational design, and decision-support needs amongst other applications. An example of an advanced DOE R&D project that could afford opportunities for automation is discussed next.

7.3.2.2 Big Data, Models, and Analytics for Offshore Systems

This section describes some novel, big data tools and models custom-built to address offshore operational risks and decision-support needs. The information generated from aspects of these models can be used to decrease both geologic and metocean risk in offshore operations, including pre-drill planning and operational design decisions that increase risks to operations.

For example, Figure 7.34 represents NETL's Offshore Risk Modeling [ORM] suite that compiles data- and built science-driven tools and crosscuts multiple offshore environments in an attempt to provide users with a clear picture of the full system for more informed decision making.

An online video is available that shows the workings and types of graphical output information provided by the ORM at the weblink in the references [ORM Video]. In addition, the following is offered as a more detailed description of the ORM suite.

The ORM suite is comprised of eight innovative science- and data-driven computational tools and models designed to predict, prepare for, and prevent future oil spills. The data-driven models

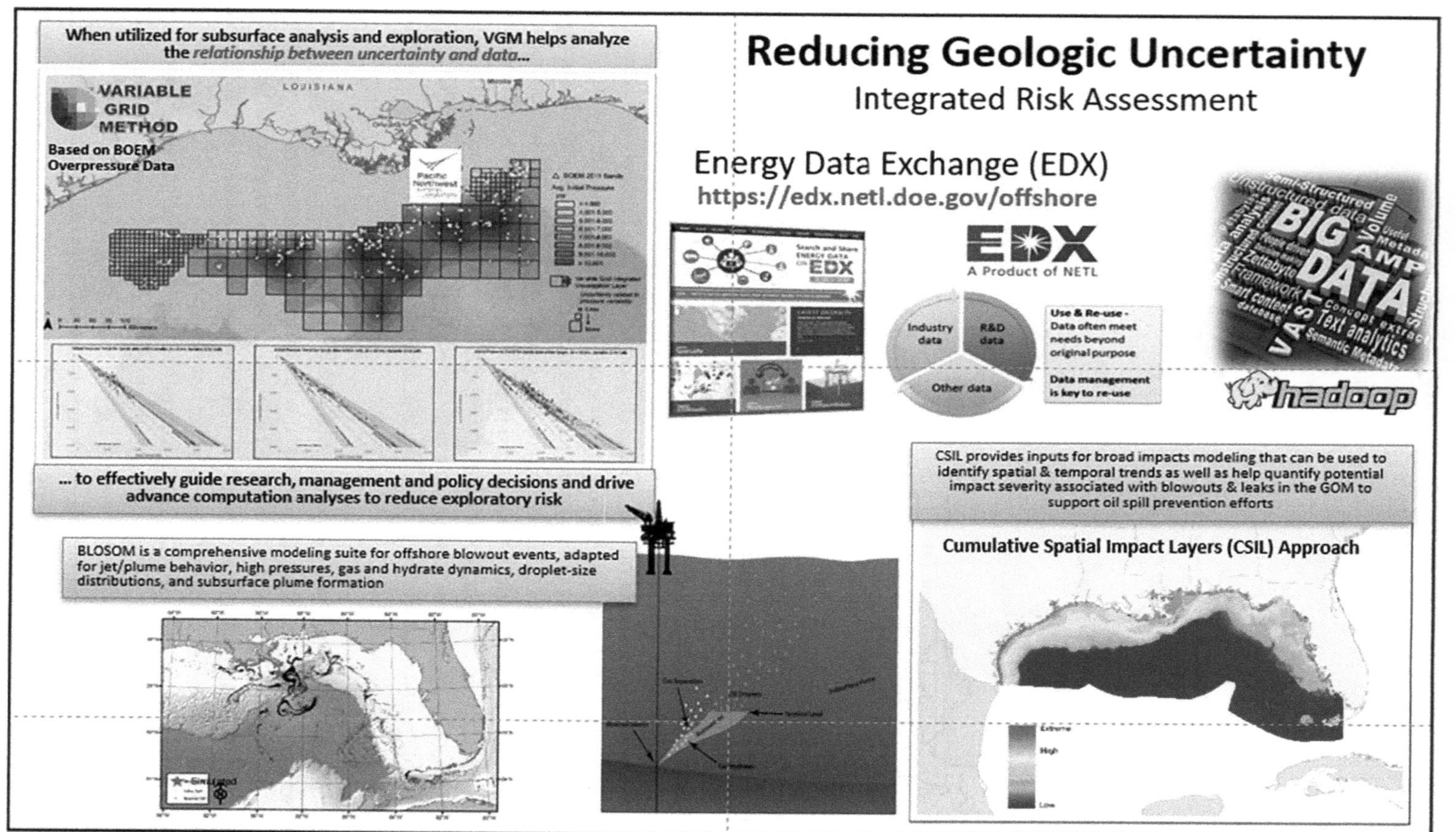

FIGURE 7.34 The Offshore Risk Modeling Suite (ORM). *Source:* Courtesy of US Department of Energy.

and tools in the suite offer a novel approach that spans the full engineered and natural offshore system from the subsurface, through the water column and to the coast. The ORM suite holistically assesses the offshore system to mitigate key knowledge and technology gaps and better predict hazards, risks, and inefficiencies for specific locations, such as metocean and subsurface conditions for predrill planning. During the six years preceding 2020, the ORM suite has been validated and adopted for use by domestic and international regulators, researchers, academia, and industry.

The ORM suite addresses lessons learned from anthropogenic and natural disasters, including Hurricanes Rita and Katrina and the 2010 Deepwater Horizon oil spill in the Gulf of Mexico. When the Deepwater Horizon disaster occurred—releasing over 130 million gallons of oil into the Gulf of Mexico over 87 days—off-the-shelf tools were inadequate for identifying risk and predicting behaviors of engineered or natural offshore systems.

The ORM suite provides a comprehensive framework for future predictions, analyses, and visualizations to better inform offshore hydrocarbon operations. Specifically, the suite incorporates disparate data with system-wide, science-driven models and tools to improve predictions using advanced spatio-temporal, metocean, subsurface, and spill modeling capabilities in a manner not previously employed. It provides a secure, coordinated system for inter-agency/entity assessment and evaluation, offering users an open-source suite for simulating real-world and hypothetical scenarios. In addition, the ORM suite is a "one-stop shop" for data, tools, and models to predict hazards and mitigate challenges that help solve many of the offshore energy industry's most demanding and complex issues.

Tools and models of the ORM suite can be used individually or synergistically, offering users the flexibility to evaluate their area or item of interest within the context of the full offshore system. Whether utilizing one component or a combination of tools and models, users benefit from the power of the ORM suite's innovative big-data computing infrastructure to evaluate potential risks, improve safety, and support spill prevention. Additionally, the ORM suite allows the user to efficiently utilize millions of data points from numerous sources in proven tools and models, thereby allowing rapid predictions at a range of spatial and temporal scales.

Researchers at NETL have applied the ORM suite to evaluate and reduce risks associated with offshore hydrocarbon operations. Applications include the analysis and prediction of subsurface geologic properties such as pressure, temperature, porosity, and permeability in offshore regions of low data density and resolution. Interest has been expressed by a number of federal agencies for applying the suite to predict the fate and transport of offshore oil spills and blowouts from a range of natural and anthropogenic sources, including seeps, pipelines, wellbores, and tanker spills by industry, regulators, and researchers worldwide.

7.3.2.2.1 Use-Case Examples

The tools and models in the ORM suite are applications that can be deployed directly on a personal device, such as a desktop computer, laptop, or tablet, or run through a virtual cloud-computing framework. The use of online computing resources offers greater technical efficiency, as the models and tools are maintained on one server while data are hosted and/or streamed real-time from authoritative sources. The use of hosted and streamed resources enables the ORM suite to use gigabytes to petabytes of data depending on the scale of the analyses being performed. Tools and models of the ORM suite can be used independently or together to provide a wider scope of knowledge surrounding the subsurface, water column, and shore. The six primary ORM suite analytical components include the following tools and models:

- **The *Climatological Isolation and Attraction Model©*** (CIAM) applies mathematical theories of dynamical systems and metocean data—including real-time ocean current patterns—to determine where oil and other particles in the ocean (e.g. debris, hazardous waste, plankton, etc.) are likely to be attracted or repulsed. CIAM offers a novel and efficient way to summarize big ocean current and wind data to determine the ultimate destination of ocean particles. A manuscript for CIAM was published in the prestigious journal *Nature* (Duran, Beron-Vera, and Olascoaga, 2018).

- **The *Blowout Spill Occurrence Model*™** (BLOSOM) is an open-source, comprehensive model that predicts the fate and transport of oil following offshore blowout and spill events. BLOSOM is the first open-source oil spill and blowout model in 4-D (latitude, longitude, depth, and time) that has been compared to and validated against traditional and industry-applied spill models (Duran et al., 2018; Socolofsky et al., 2015). Built upon a flexible framework, BLOSOM consists of several modules to help visualize, characterize, and simulate spills, including behaviors in high-pressure environments, gas and hydrate dynamics, as well as how oil particles move throughout the water column.
- ***Cumulative Spatial Impact Layers*™** (CSIL) is a spatial tool that rapidly identifies and quantifies potential socioeconomic and environmental risk. The CSIL tool is capable of handling multiple disparate datasets, measuring data density and producing multivariable layers that identify vulnerabilities within a given area. The CSIL tool has been adapted to integrate BLOSOM simulation outputs and summarize the potential risks or response availability associated with hydrocarbon events over time. A manuscript on the CSIL tool has recently been accepted for publication with the international journal *Transactions in GIS* (Romeo et al., 2019).
- ***Spatially Weighted Impact Model*™** (SWIM) explores relationships among oil spill simulations, response availability, and potential risk to inform decision making. SWIM enables users to apply weights to rank and compare different scenarios and come up with varying plans of action in the event of a spill under different conditions.
- ***Subsurface Trend Analysis*™** (STA) is a multi-resolution method that predicts and characterizes subsurface properties leveraging geologic expertise and advanced spatio-temporal statistical methods. Built over five years of research, the STA has been applied for reserves calculations, exploration and resource identification, geohazard prediction, drilling safety, and improved well design. A manuscript for the STA method has recently been accepted for publication with the journal *AAPG Interpretation* (Rose, Bauer, and Mark-Moser, 2020).
- ***Variable Grid Method*©** (VGM) communicates the uncertainty in data and modeled results, which is critical to efficiently communicate results in an intuitive manner. The VGM tool provides the flexibility to use different data types and uncertainty qualifications, preserves overall trends and patterns observed within the data and enables users to customize the analysis and final product to meet their needs. VGM was invited to be part of a special issue for the international GIS journal *Transactions in GIS* (Bauer and Rose, 2015).

In addition to these analytical tools and models, the ORM suite offers applications to support data acquisition, curation, and collaboration critical for offshore oil spill prevention. Although public, these data are often disparate and considered "dark" due to difficulties in resurrecting these resources for use. Once acquired, stakeholders often encounter issues to securely share and visualize data. The following applications of the ORM address these data challenges.

7.3.2.2.2 Capabilities

As discussed earlier, the ORM suite represents over six years of development, innovation, and validation resulting in a robust suite of advanced data computing-driven tools and models focused on offshore spill prevention. It has and can be applied to improve safe and efficient operations in offshore systems for a range of stakeholder needs. Examples of how the ORM suite has been used by NETL and other end-users include:

- **Simulating 4-D oil spill and blowout scenarios**: BLOSOM can model multiple hypothetical and historic oil spill simulations at various locations and times. CIAM can enhance and validate these simulations, improving the prediction of the fate of oil spill particles. Both BLOSOM and CIAM pull big metocean data via EDX, which can be visualized through GeoCube, a custom web-based mapping application.

- **Identifying critical subsurface characteristics**: Various subsurface conditions can negatively impact hydrocarbon exploration safety and costs. The STA can be used to define potential hazards based on geologic data and expertise to improve predictions of subsurface characteristics that may impact drilling, such as areas of overpressure. The uncertainty associated with these predictions can be quantified and visualized using the VGM. Together, the STA and VGM help decision makers pinpoint areas where overpressure might be present and make confident decisions regarding mitigation strategies that ensure safe operating conditions.
- **Evaluating response preparedness, supporting efficient operational decisions**: CSIL can be used to summarize and visualize oil spill fate and transport from BLOSOM with the spatial and temporal distribution of socioeconomic and environmental variables. Overlapping these data with CSIL helps identify gaps in response infrastructure readiness. SWIM can compare and rank multiple scenarios to strategically identify regions where additional prevention equipment is needed to improve oil spill preparedness.
- **Assessing offshore infrastructure integrity**: Data from EDX and GeoCube have been leveraged to identify and visualize historic pipeline and platform failure incidents. CSIL summarized metocean data (e.g., wave height, current velocity, wind speed) to statistically evaluate extreme environmental conditions' effect on infrastructure integrity over time.

7.3.2.3 Adapting for Machine-Driven, Real-Time Decision Support

The benefit of this big-data driven suite of tools and capabilities in the ORM is that it can plug and play with commercial resources, and also take advantage of machine learning capabilities, such as knowledge graphs, natural language processing, etc. to provide the end-user with real-time forecasting and analyses. The ORM has been leveraged in an ongoing project that is combining metocean and infrastructure risk modeling with machine learning insights to provide an adaptive, near real-time forecasting tool for predrilling, operational planning, and risk assessment [ORM].

Strong metocean conditions (the combined wind, wave, climate, etc. conditions found at a certain ocean locations) as well as natural shifts in the ocean seafloor can have a significant effect on the success and longevity of offshore infrastructure. These conditions can impact safety and cost during oil and gas exploration and production activities. Changes in the ocean environment, like mudslides and stress from subsea currents, can be spurred by strong weather events (e.g., hurricanes) or the natural fluctuations in the system over time. These events have been linked to billions of dollars in damage to offshore infrastructure and major environmental impacts. The ever-shifting offshore environment introduces a currently unquantified level of risk and vulnerabilities to offshore energy.

This project is leveraging historic observations of metocean and bathymetric data to train, test, and develop a validated smart technology that can identify current hazardous metocean and bathymetric conditions, as well as forecast changes and potential vulnerabilities that may impact existing or future offshore infrastructure. The end result, expected public in late 2020, will be a Smart Tool to advance the current state of knowledge of ocean conditions, offering insights to improve infrastructure longevity, support the identification of shallow hazards and offshore enhanced oil recovery innovations, thus decreasing cost and improving safety during oil and gas exploration and production.

This project will result in a smart tool plug-in within the ORM GeoCube application on EDX. Specifically, several key analytical results will be made available as interactive map layers. The tool will be built off of advanced, science-based analytics and big data, with near real-time assessments of current hazards and the support of forecasting areas with higher vulnerabilities. This information is critical to help reduce potential risks to daily oil and gas activities, as well as other offshore energy development (i.e., carbon storage, offshore wind/wave energy, etc.) in the future.

7.4 SUMMARY/CONCLUSIONS

From the descriptions provided in this chapter, it could be concluded that environmental drilling primarily involves acquisition of in situ data in support of improved understanding of subsurface geologic and hydrologic processes that potentially impact our environment. In many cases this understanding is best developed with adequate data to support an effective modeling effort. What becomes interesting is that the line between drilling and environmental drilling/sampling becomes less distinct as the need for real-time down-hole information increases for drilling operations, especially in challenging environments like the deep-water offshore. Examples of this capability are discussed in the other drilling chapters in this book within the categories of Measurement While Drilling (MWD) and Logging while Drilling (LWD) technologies.

Offshore Deepwater drilling is unique in the respect that it now includes the added risk of the metocean environment potentially impacting industry's ability to maintain control of operations in the down-hole geologic/hydrologic system (Figure 7.25). As can be noted in projects such as Project #11121-5402-01 (Riser Lifecycle Monitoring), instrumentation installed for monitoring the condition of the riser also allows monitoring of subsurface currents. This sort of data can be fed directly into models monitoring seasonally varying trends in the Gulf of Mexico Loop Current [Loop Current]. As noted in the discussion of the ORM, this type of real-time data can be very important in case there is need to mitigate unintended events such as an offshore oil spill.

Drilling is going to continue to be essential to an improved understanding of the subsurface. Also, as has been shown in development of these technologies, advancement of drilling technologies will likely still be needed to enable continued resource development and environmental sustainability as the industry is faced with the challenges of drilling more complex wells to greater depths, operating at hard-to-access locations, operating at extreme temperatures, and autonomous drilling in hazardous environments and/or environmentally sensitive areas.

ACKNOWLEDGMENTS

The Department of Energy, Office of Civilian Radioactive Waste funded the work in Section 7.2.2.2 related to the Yucca Mountain Project (YMP). Although YMP was closed in 2011, additional information can be found at the link in the references [YMP]. Most of the rest of the research reported in this Chapter was funded by the Department of Energy Office of Fossil Energy (DOE-FE) and managed by the National Energy Technology Laboratory (NETL). The author would like to thank DOE-FE and NETL in addition to those researchers and project managers involved for working together to achieve the high level of success noted in the projects discussed in this chapter. Also, the authors would like to thank Alfred William (Bill) Eustes III, Petroleum Engineering Department at the Colorado School of Mines, Golden, CO; and Gary Covatch, Petroleum Engineer, the National Energy Technology Laboratory (NETL), US Department of Energy's (DOE), Morgantown, WV, for reviewing this chapter and providing valuable technical comments and suggestions.

REFERENCES

Barratt, T. (2012). Exploration Drilling: Hunting Cost Efficiency. *Mining World*, 9(2), 34–40.

Bauer, J. and Rose, K. (2015). Variable Grid Method: An Intuitive Approach for Simultaneously Quantifying and Visualizing Spatial Data and Uncertainty. *Transactions in GIS*, 19, 377–397.

[BBA] Bipartisan Budget Act of 2013. (2019). In *Wikipedia*. http://www.gpo.gov/fdsys/pkg/BILLS-113hjres59enr/pdf/BILLS-113hjres59enr.pdf, accessed October 29, 2019.

Bui, H. (April 1995). Steerable Percussion Air Drilling System, Proceedings of the Natural Gas R&D Contractors Review Meeting, Baton Rouge, Louisiana, US DOE Contract #DE-AC21-92MC28182. https://www.osti.gov/servlets/purl/10128298, accessed on October 14, 2019.

Carson, R. (September 1962). *Silent Spring*, New York: Houghton Mifflin Harcourt Publishing Company, ISBN 978-0-618-24906-0.

[DEC International Association of Drilling Engineers, Drilling Engineering Committee, (September 252019). Automation Forum, "Drilling Data—What Is Available? How Good Is It? What Can We Do with It," https://www.iadc.org/wp-content/uploads/2019/10/25-September-2019-DEC-Tech-Forum-Minutes.pdf, accessed on November 11, 2019.

Duran, R., Beron-Vera, F.J., and Olascoaga, M.J. (2018). Extracting Quasi-Steady Lagrangian Transport Patterns from the Circulation: An Application to the Gulf of Mexico. *Science Reports*, 8, 5218, doi:10.1038/s41598-018-23121-y, accessed November 22, 2019.

[EDX] NETL. (2019). Energy Data Exchange archive website, https://edx.netl.doe.gov/group/prior-section-999-funded-offshore, accessed November 9, 2019.

Fortin, J. (2019). Description of Drilling Methods, https://www.rgc.ca/?page=page&id=92, accessed on September 16, 2019.

[FRTG] Flow Rate Technical Group Deepwater Horizon Blowout Flow Estimates. (2019). https://en.wikipedia.org/wiki/Flow_Rate_Technical_Group, accessed December 23, 2019.

Gertsch, R.E., (1993, September). "The Selection and Use of Coring and Reaming Bits at Yucca Mountain", Technical Report for The Comparative Bit Testing Program, Colorado School of Mines, Science Applications International Corporation DOE contract #39-930015-39

Hafner, F. (2017a, January 12). *Seven tips for getting the most out of sonic drilling*. Boart Longyear. https://www.boartlongyear.com/insite/getting-the-most-out-of-sonic-drilling/, accessed November 21, 2019.

Hafner, F. (2017b, February 26). *How sonic drilling works*. Boart Longyear. https://www.boartlongyear.com/insite/sonic-drilling-works/, accessed November 21, 2019.

Hafner, F. (2017c, April 11). *The top 5 advantages of sonic drilling over conventional drilling*. Boart Longyear. https://www.boartlongyear.com/insite/top-5-advantages-sonic-drilling-conventional-drilling/, accessed November 21, 2019.

Ingolfson. (2007). Construction drill auger [jpg]. Retrieved from https://en.wikipedia.org/wiki/Auger_(drill)

Kennedy, D.C. *The Evolution of Instrumentation for Environmental Monitoring*", *National Environmental Monitoring Conference*, San Antonio, TX, August 4–8, 2013, Abstract and Presentation accessed September 14, 2019.

[Loop Current] The Gulf of Mexico Loop Current. (2019). https://coastwatch.noaa.gov/cw/stories/emilys-post/the-gulf-of-mexico-loop-current.html, accessed on December 24, 2019.

[LTBT] Limited Test Ban Treaty. (2019). https://2009-2017.state.gov/t/avc/trty/199116.htm, accessed on September 22, 2019.

Lucon, P.A. (2013). Resonance: The Science Behind The Art Of Sonic Drilling, Internet Edition. https://scholarworks.montana.edu/xmlui/bitstream/handle/1/3483/LuconP0513.pdf?sequence=1, accessed on November 15, 2019.

Lucon, P.A., "Automatic Control of Oscillatory Penetration Apparatus", US patent 8,925,648, May 26, 2009. Internet Edition. https://www.osti.gov/doepatents/biblio/1167031, accessed on November 15, 2019.

[NAP] National Academies Press. (2019). Appendix B: Yucca Mountain: Ground-Water Flow. https://www.nap.edu/read/2013/chapter/10, accessed December 27, 2019.

[NAS] National Academies of Sciences, Engineering, and Medicine. (2019). *Manual on Subsurface Investigations*. Washington, DC: The National Academies Press. https://doi.org/10.17226/25379.

National Driller-1. (2019). https://www.nationaldriller.com/drilling-history, accessed on September 12, 2019.

National Driller-2 (2013). https://www.nationaldriller.com/articles/84736-auger-drilling-of-monitoring-wells, accessed on September 15, 2019.

ODEX (2019). *Odex drilling*. http://ontrackdrilling.com/geotechnical/odex-drilling/, accessed on December 21, 2019.

Oil and Gas Journal. (June 11, 2007). *Custom-built rig uses reverse-circulation pipe to drill basalt*. https://www.ogj.com/general-interest/companies/article/17227887/custombuilt-rig-uses-reversecirculation-pipe-to-drill-basalt, accessed on October 28, 2019.

[ORM] Offshore Risk Modeling Suite (2019). *NETL Advanced offshore research portfolio*. https://edx.netl.doe.gov/offshore/portfolio-items/risk-modeling-suite/, accessed on November 19, 2019.

[ORM Video] National Energy Technology Laboratory(2019). *Offshore Risk Modeling (ORM) Suite*. https://www.youtube.com/watch?v=iN1REKcfXxQ&feature=youtu.be, accessed on December 23, 2019.

Perry, K., Batarseh, S., Gowell, S., and Hayes, T. (2019). "Microhole Coiled Tubing Rig (MCTR) Technology", Final Technical Report for US DOE Contract #DE-FC26-05NT15482, https://www.osti.gov/biblio/888550, accessed on September 22, 2019.

Piddock, C., (2009), *Rachel Carson: A Voice for the Natural World*, Pleasantville, NY: Gareth Stevens Publishing, 87–97, ISBN-10:1-4339-0058-0, ISBN-13: 978-1-4339-0058-7.

Profaizer, S. (2006). Mechanical drill head [jpg]. Retrieved from https://commons.wikimedia.org/wiki/File:Drilling_mechanical-drill-head.jpg

Romeo, L., Nelson, J., Wingo, P., Bauer, J., Justman, D. and Rose, K. (2019). Cumulative Spatial Impact Layers: A Novel Multivariate Spatio-Temporal Analytical Summarization Tool. *Transactions in GIS*, 23(5), pp. 908–936.

Rose, K., Bauer, J.R., and Mark-Moser, M., (2020). A Systematic, Science-Driven Approach for Predicting Subsurface Properties. *Interpretation*, 8(1), pp. 167–181, https://doi.org/10.1190/INT-2019-0019.1

Rowe, P. (2019). "Blind Shaft Drilling—The State of The Art", Technical Report, DOE Contract #AC08-89NV10630, www.osti.gov/servlets/purl/10169243, accessed on September 24, 2019.

Rowley, J.C. (January 1994). Design Concept for an Advanced Geothermal Drilling System. *SME/ETCE Drilling Technology Symposium Proceedings, PD-Vol. 54*, New Orleans, LA, 239–247.

Rowley, J.C., Saito, S., and Long R. (May 1995a). An Advanced Geothermal Drilling System. *1995 World Geothermal Congress*, Florence, Italy, 1379–1383.

Rowley, J.C., Saito, S., and Long, R. (October 1995b). An Advanced Geothermal Drilling System, *Geothermal Resource Council Annual Meeting*, Reno, NV, Vol. 19, pp. 123–128.

Rowley, J.C., Saito, S., and Long, R. (May–June 2000). *Advanced Drilling System for Drilling Geothermal Wells—An Estimate of Cost Savings. Proceedings World Geothermal Congress 2000*, Kyushu—Tohoku, Japan, pp. 2399–2404.

[SCPR-12] (2019), Yucca Mountain Site Characterization Progress Report Number 12, https://www.nrc.gov/docs/ML0224/ML022460384.pdf, accessed December 12, 2019.

Socolofsky, S.A., Adams, E.E., Boufadel, M.C., Aman, Z.M., Johansen, Ø., Konkel, W.J., Lindo, D., Madsen, M.N., North, E.W., Paris, C.B. and Rasmussen, D., Reed, M., Ronningen, P., Sim, L., Uhrenholdt, T., Anderson, K., Cooper, C., and Nedwed, T. (2015). Intercomparison of oil spill prediction models for accidental blowout scenarios with and without subsea chemical dispersant injection. *Marine Pollution Bulletin*, 96(1–2), pp. 110–126.

Thamir, F., Thordarson, W., Rouseau, J., Long, R., and Cunningham, D. (1998). Drilling, Logging, and Testing Information from Borehole UE-25 UZ#16, Yucca Mountain, Nevada; US Geological Survey Open-File Report 97-596, 49 p.

US Department of Energy. (August 1995). *Site Characterization Progress Report: Yucca Mountain, Nevada*, Number 12, DOE/RW-0477, Office of Civilian Radioactive Waste Management, Washington, DC.

US Department of Energy. (May 2001). *Yucca Mountain Science and Engineering Report—Technical Information Supporting Site Recommendation Consideration*, DOE/RW-0539, Office of Civilian Radioactive Waste Management, Washington, DC.

Vijayamohan, P., Majid, A., Chaudhari, P., Sloan, E.D., Sum, A.K., Koh, C.A., Dellacase, E., and Volk, M. (2016). Hydrate Modeling & Flow Loop Experiments for Water Continuous & Partially Dispersed Systems, OTC 25307, *Proceedings of The Offshore Technology Conference*, Houston, TX, May 2–5, 2016.

[VIV], (2019), Vortex Induced Vibration, as discussed in the example at https://www.rigzone.com/training/insight.asp?insight_id=359&c_id=17, accessed November 11, 2019.

Warrington, D.C. (1997), "Closed Form Solution of The Wave Equation for Piles" Internet Edition. Retrieved from https://ceprofs.civil.tamu.edu/llowery/Things/piledriv/warrington thesis.pdf, accessed on September 29, 2019.

Whitfield, M.S., Thordarson, W., Ammermeister, D., and Warner, J. (2020). Drilling and Geohydrologic Data for Test Hole USW UZ-1, Yucca Mountain, Nevada. US Geological Survey Open-File Report 90-354, 44 p.

Yucca Mountain Project. (2019). http://yuccamountainproject.org/, accessed December 26, 2019.

8 Drilling Automation

Richard Meehan
Drilling Automation at Schlumberger Limited, Sugar Land, TX

CONTENTS

8.1 OILFIELD DRILLING INTRODUCTION

In the oil industry wells are constructed primarily to access oil and gas, and to control its production. There is a wide range of activities associated with this endeavor, but in this chapter, we will concentrate on those that are central to the drilling of the wellbore itself. Many excellent and detailed texts have been written on the drilling and well construction process [1–3], and the reader is recommended to consult these for a more comprehensive explanation.

There are several ways to categorize the various types of drilling. Normally field development will start with an exploration phase. In this phase wells are drilled primarily to obtain data about the subsurface, and to determine the probability of hydrocarbons being present in commercially relevant quantities. These wells can provide a wealth of geological data which is then analyzed to determine the suitability of the location for further investigation and development. The data acquired are useful not only for indicating the production potential of the reservoir, but also for providing information about potential challenges in drilling through the overburden above the reservoir.

If the results from the exploration drilling phase are promising, an appraisal phase will follow. In this phase, extra information is sought to more accurately determine the potential of the field. More data is gathered, much of it by means of drilling additional wells. Once again, this data is useful for both establishing the potential of the reservoir and highlighting well construction issues that will need to be addressed in the following development phase.

Assuming the appraisal phase indicates that commercially viable reserves are present, the next step is development. Here the goal is to exploit the reservoir in a safe and efficient manner by constructing wells of the appropriate design. In this phase it may be that many tens or hundreds of wells will be drilled over the course of several tens of years.

In this progression of activity from exploration, through appraisal, to development, the basic processes in well construction are common, and much of the equipment used is generally similar. There are, however, significant differences in terms of uncertainty, economics, and risk that have a large impact on the barriers to process automation. In the exploration phase the largest uncertainty usually concerns the details of the geological setting. In such situations, execution choices are often made based on experience gained in other environments—experience which does not always translate appropriately. During the appraisal phase more local information becomes available as more wells are drilled and the drilling team typically starts to ascend the learning curve. By the time the development phase is reached the geological setting and its associated constraints[1] are usually well understood, any drilling risks[2] are evaluated, and mitigation strategies are put in place. Now the focus is on efficiency of design and operations, but there are still risks—the subsurface is never completely predictable, the pressure regimes within the formations from which hydrocarbons are produced may change significantly if production is taking place during field development—and drilling teams can become complacent.

Another way to characterize types of drilling activity is to consider the environment. Obviously drilling on land, in a location with good transport infrastructure, moderate climate, and a local economy that can supply the appropriate goods and services, is a very different prospect to drilling in a remote location with little or no infrastructure and a lack of available equipment, materials, and labor. Drilling on land can be challenging; drilling offshore even more so. Then there is the architecture of the well to consider. Vertical wells are generally easier than deviated wells. Extreme deviation, so-called extended reach wells, can present even bigger challenges. As with exploration, appraisal, and development wells, the basic processes and equipment used for these different environments and well types can be similar; however the costs and risks can be very different. These differences present a challenge for process automation.

8.2 THE WELL CONSTRUCTION PROCESS

Well construction proceeds in stages, the number of which is often determined based on subsurface data such as formation pressures and mechanical properties. After site preparation the first stage is usually to install the surface conductor. This is large diameter steel pipe (perhaps 90 cm in diameter) which is driven into the ground for several tens of meters. This conductor prevents the surface formations, which may not be well consolidated, from collapsing while the construction process continues.

With the conductor installed, the drilling rig can begin to drill the next section of the well. This section, smaller in diameter than the conductor, may be drilled down to several hundred meters. Another steel pipe, called casing, of a smaller outside diameter than the inside diameter of the conductor, is placed in this new hole. Cement is then pumped down the inside of the casing. The cement exits the bottom of the casing and returns to surface through the annular space between the formation and the casing. The cement in the annular space sets and provides a seal between the formation and the casing, preventing formation fluids from rising up to the surface through the annulus.

The next phase is to install the wellhead on top of the casing string. The wellhead supports the weights of further casing strings that will be run into the well, and provides a structure to install a set of valves, known as the Blowout Prevention stack, or BOP stack. The BOP stack should be installed before the well encounters any permeable hydrocarbon-bearing formations. These valves allow the crew to apply a seal to the wellbore, preventing formation fluids escaping during subsequent operations.

With the BOP stack in place, further hole sections can be drilled, cased, and cemented. Each hole section is necessarily smaller in diameter than the previous one.

If the purpose of the well is to produce hydrocarbons, then a production casing string will be set across the reservoir section. Various strategies are available to produce the oil or gas from the well, most involve installing a tubing string—a small diameter pipe—through which the oil or gas will flow to the surface. The annular space between the inside of the casing and the production tubing is sealed by placing a flexible packer at the end of the tubing string. This isolates the producing section of the wellbore from the space inside the casing strings. When the packer is set the BOP valves are removed, and a set of production control valves are installed instead to allow the rate of hydrocarbon production to be optimized.

When the system is ready to produce, it is necessary to allow the hydrocarbons to flow into the wellbore. To do this, perforations must be made through the production casing and the cement annulus to allow formation fluids to enter. This is often achieved by running small explosive charges on a wireline down the production tubing and into the production casing. When detonated these charges blow holes in the production casing and cement and allow the influx of formation fluids.

This description is a simplified version of the various phases of well construction, and many variants exist. However, the basic approach is generally similar—construct the well by drilling a hole, install and cement casing, repeat this process until the desired depth is reached, install production equipment, establish flow from the reservoir into the well, and use surface valves to control production. Sometimes it is necessary to install active systems to pump the hydrocarbons out of the well. Sometimes the reservoir section is left uncased. Complex completions systems with active downhole valves are used in some locations. Hydraulic fracturing may be used to create flowpaths in low permeability reservoirs. The completions strategy is usually dictated by the geology of the reservoir and the economics of producing in that location.

8.2.1 Rig Systems

Most drilling rigs have the same basic structure and utilize similar sets of equipment (Figure 8.1). There are five main systems: the hoisting system, the rotary system, the circulating system, the well control system, and the power system. There is also usually some infrastructure that allows monitoring of operations; this requires collecting and collating information from the five systems previously mentioned and presenting that information to the crew on the rig and to decision makers that may be remote from the rig location.

Because drilling operations are often conducted in remote locations, and because the energy demands in well construction are significant, most rigs have their own power generation equipment.

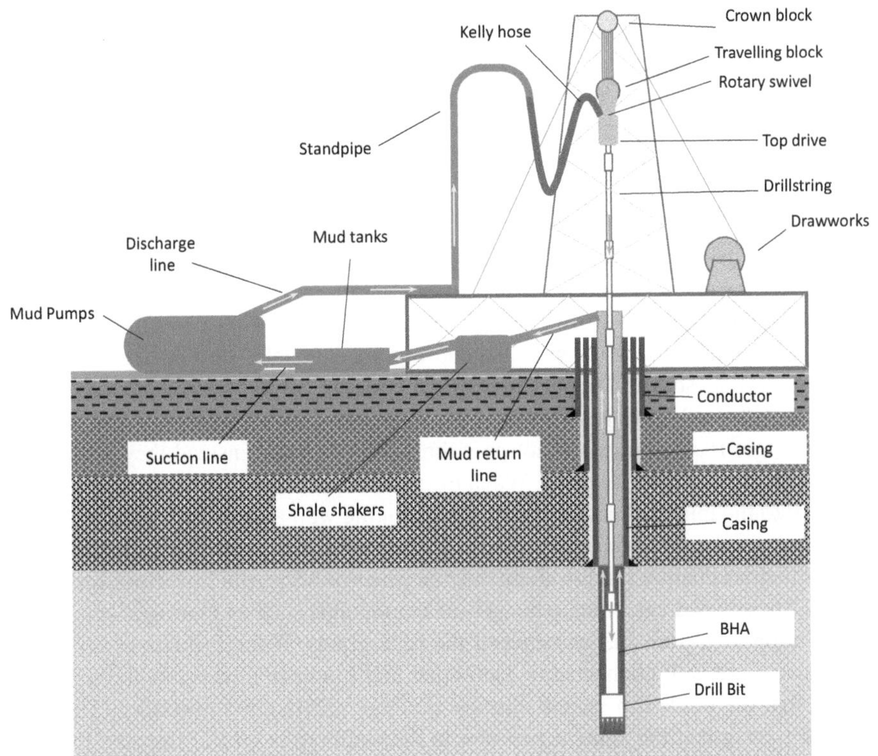

FIGURE 8.1 Rig layout.

These are usually diesel-driven generators, although some larger offshore installations have gas turbine generators. The electricity generated is used to power the major equipment systems, and to provide electrical power for the general infrastructure on the rig and associated rig site. Automated management of the power generation is normal on most modern rigs.

The hoisting system is used to move drillpipe and casing into and out of the wellbore. It is a large pulley system which comprises a rotating drum, around which is wrapped a steel cable. This rotating drum is referred to as the drawworks. Attached near the top of the rig mast is a set of pulley sheaves (the crown block) and below this hangs another set of sheaves (the traveling block). The wire rope from the drawworks is threaded through the crown block and traveling block sheaves, and the end is anchored to the rig floor. When the drawworks drum rotates, the traveling block can move up and down. This up-and-down movement is what is used to run equipment into and out of the wellbore.

In order to drill the well it is usually required to be able to rotate the drillpipe. In modern rigs the rotary system is called a top drive. This is a large electric motor which is suspended below the traveling block. The top drive is connected directly to the drillpipe and can provide the required torque and rpm to rotate the drillstring and power the drillbit to drill the hole. Today most rig top drives use AC motors with VFD controllers.

The circulating system is used to pump fluids into the well. This drilling fluid, usually called drilling mud, gives the crew a way to ensure that the pressure in the wellbore exceeds that of the fluid pressure in the formations being drilled, thus preventing an uncontrolled release of formation fluids. The mud also provides the essential function of carrying the drilled rock cuttings out of the hole. The mud is mixed in a set of tanks on the surface. It is usually a mixture of some base fluid (oil or water) with the addition of various weighting agents and additives. Some other essential functions of the drilling fluid are to help maintain the stability of the wellbore, provide cooling and lubrication to the drillbit and reduce friction between the drillstring and the formation or casing. Drilling fluid is also used to provide hydraulic power to various tools in the drilling assembly.

The drilling mud is prepared in surface tanks and then pumped into the top of the drillstring by means of positive displacement pumps. Most rigs have at least two pumps and many have three or more. These pumps are generally capable of supplying several cubic meters of drilling mud per minute at pressures up to several hundred bar. The discharge line from these pumps is connected to a pipe called a standpipe. The end of the standpipe is usually around 15 m above the rig floor. Attached to this end is a flexible hose (the Kelly hose) which connects the fluid path to a rotary swivel that sits above the top drive. The swivel allows the mud to be pumped into the top of the rotating drillstring through the top drive. Hence mud can be pumped into the drillstring while the block is moving up or down, and the drillstring is rotating.

The mud flows down the center of the drillstring, out through the nozzles in the drillbit and up the annulus formed between the drillstring and the wellbore. The mud carries the rock cuttings as it flows. Upon reaching the surface the mud and cuttings are channeled through a mud return line to a device known as a shale shaker. This is essentially a vibrating table with a mesh screen on top. The liquid part of the returned mud flows through the mesh into tanks where any remaining solids can settle out. The larger rock cuttings do not pass through the mesh—instead they fall off the end of the shaker and are collected for disposal. When the smaller particles have settled out of the mud, the fluid is returned to the mud tanks to be circulated down the well again.

The well control system is designed to prevent the uncontrolled release of formation fluids. Most subsurface formations contain fluids—these may be water or brine, oil, or various forms of gaseous hydrocarbons. These gases can contain dangerous compounds such as hydrogen sulfide. The fluids are generally trapped in pore spaces between the rock grains. If the formations are permeable—i.e. the pore spaces are well connected to each other and fluid can flow easily through them—then there is a risk that the fluids can escape into the wellbore and rise to the surface. This is normally prevented by ensuring that the annular pressure in the wellbore is greater than the formation pressure. Since the formation pressures are not always known ahead of time it sometimes happens that the annular pressure is insufficient to prevent the formation fluids from entering the wellbore. When this happens these formation fluids start to rise up the wellbore, displacing the drilling mud. This phenomenon, known as taking a kick, requires a swift response. Upon detecting an influx, the rig crew will activate the blowout preventer valves contained in the BOP stack. These valves will close around the drillpipe and seal off the annulus from the surface. If a kick happens when there is no drillpipe in the hole another set of valves can close off the wellbore completely. Once the wellbore is closed the kick can be dealt with by pumping higher density mud through the drillstring while slowly bleeding off pressure at the wellhead by means of a choke valve. Eventually the density of the mud in the wellbore should be sufficient to prevent any additional influx of formation fluid. Then the annulus can be opened, and drilling can continue.

8.2.2 The Process

As described previously there are three basic processes involved in well construction. These are drilling the hole, casing the hole, and cementing the casing. Of these three processes, the first two are more relevant when it comes to automation.

Drilling the hole requires that an appropriate drilling assembly is prepared and lowered into the hole. This drilling assembly can be composed of many different parts but is generally divided into the Bottom Hole Assembly (BHA) and the drillstring.

At the bottom of the BHA is the drillbit. This is the cutting tool which removes rock by scraping, crushing, gouging, or grinding against the bottom of the hole. There are many drillbit designs, and the appropriate choice of which to use is usually informed by the type of rock that needs to be drilled, the type of drilling mud that may be used, and the desired performance in terms of rate of penetration, directional tendency, durability, and cost. The bit contains flow passages that allow the circulation of drilling fluid from inside the drillpipe into the wellbore. The flow passages are terminated by nozzles which direct the flow over the bit to ensure it is kept cool and that the rock cuttings are removed from the hole bottom.

The bit is attached to heavy steel pipes known as drill collars. These provide weight to push the bit into the rock and provide some stiffness to help with directional control. The drill collars are necessarily of smaller diameter than the bit to allow the passage of mud and cuttings along the annulus. The BHA may also contain components known as stabilizers. These are short sections of similar diameter to the bit and they provide centralization of the BHA in the hole and add stability to the rotating assembly. These components have flow channels to enable the passage of mud and cuttings.

Various measuring instruments may also be included in the BHA. These include survey instruments to measure the inclination and azimuth of the borehole, pressure, temperature, and vibration sensors, and instruments that measure the properties of the rock formations being drilled. Most of these instruments are mounted in specially adapted drill collars. In order to use these measurements to make real-time decisions, most BHAs will contain a Measurement While Drilling (MWD) component. This is usually mounted in a drill collar and has the ability to transmit the sensor information to the surface, either by causing pressure variations in the mud flow inside the drillpipe that can be detected at the standpipe, or by emitting electromagnetic waves into the formation that can be received by special detectors placed on the ground at the surface.

The BHA can also contain active components. The most common active component is a positive displacement mud motor. This device converts the hydraulic energy of the mud flowing through the drillstring into rotation energy at the bit. This can allow the bit to rotate even if the drillpipe is not rotated at the surface. These motors are often used in drilling directional wells and enable effective control of the direction of the wellbore. Another common component is a tool known as a Rotary Steerable System (RSS). There are several popular designs, all of which have active components. The RSS enables effective steering of the well even when rotating the drillstring from surface.

The top of the BHA is connected to the drillpipe. Drillpipe is made from seamless metal tubes, typically around 10 meters long. Each end of the pipe has threaded connections, known as tooljoints, allowing the drillpipes to be joined together to form a long continuous string. The purpose of the drillpipe is to provide a means of circulating fluid to the bottom of the wellbore, to transit rotation and torque to the bit from the top drive, and to suspend the BHA in the wellbore.

Although drillpipe normally comes in 10-meter sections, standard practice today is to assemble three pipes together to form what is called a stand. This stand, approximately 30 meters long, is a convenient length for rig operations. The drilling process proceeds by first lowering the bit and BHA towards the bottom of the hole. Since the drillpipe is made up into stands, this process, known as Tripping in Hole (TIH), is composed of a number of repeated operations. The drillstring, attached to the traveling block, is lowered until the traveling block is near the rig floor. The drillstring is then suspended from the rig floor by clamping it in a set of steel slips. The traveling block is disconnected from the drillpipe and hoisted to the top of the mast. Another stand of drillpipe is then positioned so that its lower tooljoint can be connected to the tooljoint at the top end of the drillstring. The tooljoints are screwed together. The traveling block is then latched on to the top of this new stand and is hoisted a little higher. This has the effect of suspending the weight of the whole drillstring. The slips are removed, and the traveling block is lowered towards the rig floor, thus moving the bit and BHA closer to the bottom of the hole. This process is repeated until the bit is close to the bottom.

When the bit is close to bottom, drilling ahead can begin. To do this the drilling mud is prepared, and the top drive is connected to the top of the drillstring. The pumps are switched on and circulation established. Rotation is turned on and the bit is slowly lowered towards the bottom of the hole. When it touches the bottom, the torque increases as the bit begins to remove the rock. Drilling progresses until the top drive reaches the rig floor. Then rotation is turned off, circulation is stopped, and the pipe is set in slips as before. The top drive is disconnected from the top of the drillpipe, and the traveling block is hoisted to the top of the mast. A new stand is positioned and connected to the top of the drillpipe in slips. The top drive is connected to the top of this stand, the block is hoisted slightly, and the slips removed. The pumps are switched on and circulation is established, the top drive rotation is turned on, and the bit lowered towards the bottom of the hole and drilling recommences. This sequence continues until the desired hole depth is reached. When this happens, the next operation is Tripping out of Hole (TOOH). This is essentially the reverse operation to TIH. The block, still connected to the drillstring, is hoisted to the top of the mast, and the drillstring set in slips. The tooljoint directly above the slips is unscrewed, freeing the stand hanging from the traveling block. This stand is racked back in the mast, the traveling block is lowered to the drill floor and latches on to the top of the drillstring. The block is hoisted slightly, slips are removed, and then the block is hoisted to the top of the mast. The process repeats until the BHA is out of the hole.

Once the BHA is out of the hole, and assuming the hole has reached the appropriate depth, the next operation is to run casing. Casing is similar to drillstring in the sense that it is composed of sections of metal pipe with threaded connections at each end. The diameter of casing is larger than that of drillpipe, and the threaded connections have a different form. Drillpipe is designed to be run in and out of the hole many times, whereas casing is designed to be run in only once, and then cemented in place. The process of running casing in the hole is similar to tripping drillpipe in the hole. The traveling block picks up a section of casing and runs it in hole. When the block is close to the rig floor, the casing is set in slips, the block disconnected and hoisted up to the top of the mast, and a new piece of casing is positioned to connect to the top of the casing held in slips. The connection is made up, the block connects to the new casing section, hoists it slightly to allow removal of the slips, then runs it in hole until the block is once more at the rig floor. The process is repeated until the casing has reached the desired depth.

During drilling and tripping operations, the crew must pay careful attention. For the drilling process, the major parameters that can be controlled from surface are the flow rate of the drilling mud, the rotation speed of the top drive, and the motion of the traveling block. These parameters all influence how the bit interacts with the rock, and how the drillstring interacts with the wellbore. Increasing mud flowrate or rotation speed increases the amount of energy being transferred to the drillstring, BHA, and wellbore. When the bit is on bottom drilling, increasing the speed at which the block moves down increases the weight pressing the bit into the formation. This weight is usually referred to as Weight on Bit (WOB). This increase in WOB normally means that the teeth on the bit penetrate further into the rock and hence more torque is needed to maintain the bit speed. If the teeth are penetrating further, more rock is removed with every rotation and the overall rate of penetration (ROP) increases. The requirement for increased torque means the top drive must work harder to maintain a constant speed.

The basic relationship between WOB, torque, rotation speed, and ROP is not dissimilar to many other types of drilling—for example using a drill press to drill a hole in a piece of steel or other material. The challenges for the oil industry come from the facts that:

- The bit is connected to the surface by a very long flexible shaft—the drillstring. A typical drillstring may be 0.125 meters in diameter and more than 5000 meters long.
- This long shaft is encased in a hole that is usually only between 1.5 to 2.5 times that of the shaft diameter. The drillstring interacts with the walls of the borehole along its entire length.
- The properties of the material being drilled (the rock) are unknown and often unpredictable.
- The rock cuttings must be transported along the narrow annulus between the rotating drillstring and the wellbore, to the surface, as drilling progresses.
- Measurements of what is happening at the bit are difficult to make, and the transmission of these measurements to surface usually involves a considerable delay—of the order of several seconds to several tens of seconds.

The challenges do not end with trying to identify and apply the correct combination of drilling parameters. Simply tripping the BHA in and out of the hole, or tripping the casing in, can be very difficult.

8.3 AUTOMATION PERSPECTIVE

In many industrial settings, automation is widespread. Automation generally leads to improved efficiency, consistency, and quality. It helps to improve safety by removing the need for people to work closely with powerful machinery or in dangerous environments. Automation is ideal for repeatable processes carried out in controlled environments, where the properties of any input material can be guaranteed to be fit for purpose.

Across manufacturing industry, some processes are only partly automated. Those parts of the process that are predictable and require precise movement and force control tend to be automated first. The parts of the process that require judgment by humans, or that have a dependency on unpredictable input material or environmental conditions, are often still carried out by people. This partial automation can still bring enormous benefits in terms of efficiency and productivity.

In many settings, equipment, rather than process, is automated. In the construction industry many large pieces of earth-moving equipment are automated in the sense that closed loop control systems are used to carry out the commands of the equipment operator. Such equipment is extremely versatile and can be used in many different processes. For some industries, such as open cast mining, the current trend is for more and more automation of individual pieces of equipment. These automated machines are then controlled by an overall process automation layer which orchestrates how the machines work together. This development of systems which are composed of many semi-autonomous parts is likely to become more prevalent.

8.4 CURRENT STATE—OFFSHORE AND LAND

In the oilfield, some automation exists on many rigs today. Indeed, over the last few years there has been an explosion of interest in bringing industrial automation techniques to the oilfield (see [4]). Activities that are repeatable, with the appropriate measurements available in a timely manner, are suitable for automation. In the brief description of rig operations earlier, the first places automation is available is in the manipulation of the various components of the drillstring on surface, and on some rigs, the preparation of the various types of drilling fluid. These activities are characterized by some of the features seen in industrial automation scenarios: the component parts (joints or stands of drillpipe, or containers of chemicals) are standardized, the machines that are manipulating them have access to the appropriate sensor measurements, and the environment (the rig) is to some extent controlled. Today these automated activities are still generally driven by a human operator who oversees the process and can (and often must) make interventions to ensure operations are successful. Most offshore rigs today have automated or semi-automated pipe handling equipment and fluid preparation equipment. On land these systems are less common. This is partly due to cost, but also due to the fact that offshore rigs are more like factories in that the equipment, once installed and set up, is usually left in place. The entire rig may move to many new locations, but the infrastructure on the rig moves with it. On land, moving a rig from one location to another usually involves dismantling much of the rig site structure, putting it on trucks, and re-assembling at the new location. Most automated systems that manipulate large pieces of equipment need careful setup and tuning. The repeated assembly and dis-assembly of such systems requires time and expensive resources, and this is a barrier to widespread adoption.

For the actual drilling and tripping, automation systems do exist, but in general they follow the concept of automating the equipment rather than automating the well construction process. On a modern rig the hoisting system, rotating system, and circulating system are typically controlled in a closed loop manner. For example, when hoisting the block, the driller can set up the acceleration rate and final speed required. He or she can set limits on the amount of force that should be used, and define the behavior of the system if these limits are exceeded—should it slow down, stop, reverse, etc. He or she can do the same for the rotation system and the fluid system. These systems are generally coupled together to ensure safe operation on the rig. For example, the system will not allow the pumps to be turned on if the mud valve in the top drive shaft is closed, the hoisting system will automatically slow down the block as it approaches the top of the mast or the rig floor, coordination of movement between the pipe handling systems and the block is implemented to prevent equipment collisions and damage [5].

When on bottom drilling, the driller can set up the surface control system to adjust the flow, rotation, and block movement to try to achieve the appropriate performance in terms of rate of penetration. The driller can define limits for drillstring torque, standpipe pressure, and weight on bit. The control challenge here is that the downhole conditions, the interaction of the drillstring with the wellbore, and the condition of the wellbore itself, need to be inferred from surface measurements. Although a limited set of downhole measurements is available, the latency involved in the transmission of those measurements to surface means they cannot easily be used in a closed loop fashion. To get an idea of the system response time to a change in input at the surface, consider a drillstring 3000 meters long. If the driller makes a change to the rotation speed at surface, it takes just over 1 second for that speed change to reach the bit. If the driller changes the speed of the block, the resultant axial force change on the bit will happen approximately 0.6 seconds later. For a flow change the mud pressure wave reaches the bit perhaps 3 seconds later. The typical downhole data acquisition system can sense these changes as they happen, but transmission to the surface is heavily bandwidth limited. Typical commercially available rates are of the order of 10 to 20 bits per second. Since the transmission mechanism is usually pressure waves in the mud flow inside the drillpipe, the transit time for the information from downhole to surface in our example is of the order of 3 seconds. The downhole tools can collect a wealth of information about the physical response of the BHA and the formation properties, which means there is competition for which data should be sent to the surface. This implies that the available bandwidth cannot only be used for transmitting information relevant to closed loop control.

In recent years methods to overcome these bandwidth and latency limitations have been investigated. Perhaps the most promising of these is the introduction of “wired” drillpipe. This is specially manufactured drillpipe which allows transmission of information along wires incorporated into the drillpipe structure. Couplers at each tooljoint allow the transmission to pass from one joint to another. At the surface a special sub provides wireless connectivity to the top of the drillstring. This methodology effectively enables wired communication from surface to downhole. Bandwidth limitations are now in the hundreds of kilobits per second and latency is reduced, potentially to millisecond levels. This technique provides an opportunity to develop effective closed loop systems for controlling the detailed dynamics of the BHA. Now commercially available, wired drillpipe systems may signify a major step in drilling automation [6,7]. Of course, challenges still exist. Low latency signals are available, but robust physics models of cause and effect need to be developed, to enable effective control. When there are several kilometers of flexible steel pipe between the place where energy is input and the location of the desired response, prediction can be difficult.

Well construction has many practical challenges. Some of these involve the uncertainty of the geological environment and the practical difficulties associated with controlling a system that has a length scale of several kilometers. There are, however, additional challenges that need to be considered for any proposed technical solutions to automate well construction operations. Perhaps one of the biggest of these is organizational.

Most well construction projects are collaborative in nature. Several different organizations are involved, each with their own drivers and priorities. Since the wells must be drilled somewhere, there is always the involvement of a regulatory body, sometimes more than one. A typical arrangement is as follows. The overall process is controlled by an oil company (or a partnership between two or more oil companies). The lead oil company, usually referred to as the Operator, engages the services of several different specialist companies to carry out the overall process. These may comprise: a Drilling Contractor company that supplies the rig and crew; a drilling fluids company that may supply the fluids, some of the fluids handling equipment and, a Mud Engineer; a directional drilling company that supplies some BHA components and directional expertise on the rig (a Directional Driller or DD); a measurement while drilling (MWD) or logging while drilling (LWD) company that supplies sensor subs for the BHA and interpretation expertise; a Bit company that supplies the drillbits; a casing running service; a cementing company; a completions service supplier; and so on.

The key decision makers on the rig are usually:

- The Operator representative, known as the Company Man. The Company Man is responsible for delivery of the well.
- The Drilling Contractor manager of the rig, known as the Toolpusher.
- The Driller, employed by the Drilling Contractor.
- The Directional Driller (DD) employed by the directional services company.
- The Mud Engineer, employed by the drilling fluids company.

These key people usually have a wealth of expertise and experience between them and make crucial execution decisions as the well is constructed.

Of course, before drilling begins there is a planning phase where many technical decisions are made. These usually cover the well design, equipment to be used, processes and procedures to follow, mitigation strategies in the event of execution problems, and contingency plans for when things go badly awry. Planning is a collaborative effort and involves technical and process experts, from both the Operator and the many service companies involved. Many of the design and execution processes and procedures are based on decades of experience, informed by problems encountered in past drilling campaigns. A large part of these procedures may be dictated by regulatory requirements, usually based on health, safety, and environmental laws in the local jurisdiction. In addition to the regulatory requirements, some procedures are defined in the contractual arrangements between the different organizations involved. It is worth noting that these contracts can make it difficult to align priorities and to ensure the decisions made on the rig optimize well construction.

There is one further layer of difficulty in decision making. The oil and gas industry has a long tradition of "getting it done." This mentality can lead to exceptional performance, but it can also lead to poor quality outcomes or even catastrophic results. It encourages crews to take responsibility but also encourages them to try to solve problems on their own which may mean operating without all the available information or expertise.

8.5 THE AUTOMATION WORKFLOW

The well construction process can be decomposed into three parts as shown in Figure 8.2. The first part concerns planning, policies, and procedures. Wells are planned as part of the Operator's strategy to exploit the reservoir. The goals and objectives of the well are defined, and these are considered in the light of cost, regulation, and HSE concerns. The well is designed to achieve the engineering objectives, within a cost budget, and with reference to available equipment and resources. Well designers build on experience to try to ensure continuous improvement. The architecture of the well and the procedures used to construct it are inextricably linked. Many Operator procedures have been developed over decades and include (implicitly and explicitly) lessons learned from previous drilling campaigns. These procedures also incorporate any required regulatory steps. The procedures are well documented and Operator and service company personnel are usually well versed in their execution (Figure 8.2).

This first part of the process entails building and using models of the world—engineering models, financial models, etc. Reasoning about goals, objectives, and constraints is deliberative, and generally slow. The reasoning process is not tied to the execution timeline, but the well planning timeline.

The third part of the process concerns the interaction of the drilling equipment with the environment, both surface and downhole. Here the rules of physics apply. The equipment is often under

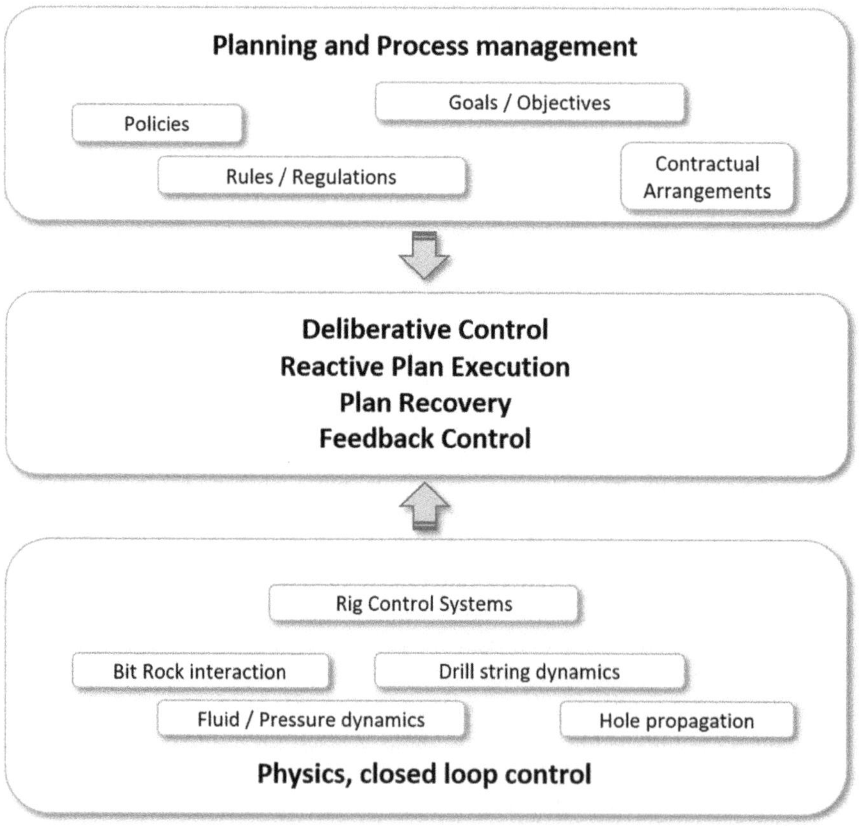

FIGURE 8.2 The automation workflow.

PLC (Programmable Logic Controller) control, supervised by the rig crew. When good quality models of the physics exist, for example the movement of the equipment at surface, these models are incorporated into the control loops. When such models do not exist, for example the dynamics of the interaction between the BHA and the rock, then the closed loop controls use simple assumptions and the rig crew oversees the performance of the system [8–10].

The second part of the process, going from the high-level plans, policies, and procedures to the detailed operation of the equipment, is where most of the uncertainty inherent in well construction is dealt with. This is an area of deliberative control mixed with reactive plan execution. Higher level policies and constraints are interpreted in the light of current conditions, and decisions are taken to execute more detailed closed loop procedures.

An example of this is "drilling the curve" [11]. Most wells drilled today have relatively complicated trajectories. In the current unconventional market in the US these wells are typically vertical down to some depth—perhaps 3 kilometers, then extend horizontally for another 1 to 3 kilometers. The transition between the vertical and horizontal sections, the curve, may be of the order of 300 to 500 meters in length. During planning, care is taken to ensure the equipment chosen (BHA, bit, etc.) is suitable for the task of drilling the curve, and various engineering calculations are carried out to verify that this is the case.

During execution the DD is responsible for ensuring the trajectory is drilled in such a way that the technical objectives of the plan are met. Given the uncertainty in formation properties, the steering response of the BHA cannot be predicted exactly, so the DD must pay attention to the real-time measurements and has to be proficient at predicting the likely future behavior of the drilling assembly. To help with this, he or she has access to engineering software routines that can evaluate the outcome of potential decisions. Since this is the execution phase, there are time pressures. It is risky to stop drilling since the risk of the deterioration of the wellbore increases with time. The DD must evaluate the current situation, decide on a course of action, persuade the other decision makers of the merit of this decision, and execute. For each steering decision, feedback comes after some distance has been drilled. This distance may be of the order of 10 meters or so. The DD must then re-evaluate the situation and make the next decision.

There are several key points in how this is done:

- The DD must understand the nature of the goals and constraints. These are not always explicitly represented in the plans and procedures—there is an assumption that the DD "knows" what is expected.
- Although engineering software tools are available, the uncertainty of the formation properties and the lack of appropriate real-time measurements of the situation means that these tools are often not sufficiently accurate to determine to the required level of accuracy the likely consequences of any particular steering decision.
- The quality of the decision rests heavily on the ability of the DD to accurately understand the "state of the world" as it relates to the steering decision. With insufficient measurements available, this understanding is often based on previous experience, with similar wells, BHAs, etc.
- Once a decision has been arrived at, the DD may need to convince others that the decision is correct, before execution. This means the DD may have to rely on his or her powers of persuasion. This "social" requirement is extremely important. Factors such as the DD's experience, reputation, confidence, and relationships with the other decision makers become critical.

This cycle of evaluation of the situation, comparison of progress against goals, rapid modeling of scenarios, decision making, and execution is characteristic of many of the processes during the reactive plan execution phase.

Automation of this phase of well construction is a challenging area. Advances in sensors and telemetry of information from downhole to surface, coupled with improved models of the physics of interaction between the wellbore and the drilling assembly, and improved calibration techniques for these models, should lead to better closed loop control of the drilling and tripping processes. This will increase efficiency by improving the consistency and repeatability of some operations. It will also reduce the pressure on the rig crew by reducing the need to constantly supervise the equipment. In the directional drilling example, this should make it easier to implement the decision made by the DD. However, the decision still needs to be made. Automation of this process is a different challenge. To achieve this several steps are required:

- A model of the world needs to be constructed. This model needs to represent the key features of the environment that will affect the process. It needs to be constructed in such a way that it can be related to the goals and constraints of the process.
- A system to reason about this world model must be implemented. It needs to be able to represent goals, constraints, actions, and outcomes. This system needs to be able to decide on a course of action and establish a plan for how it should be executed.
- Once the plan is created it needs to be translated into a set of concrete instructions that can be executed by the closed loop systems, or a combination of closed loop systems and systems monitored and controlled by the rig crew. The plan needs to be scrutable, understandable by humans—this is required to build the appropriate level of trust in the system.
- Well construction is typically an under-instrumented process with humans using sight and sound to compensate for the lack of sensor measurements. Combined with the inherent uncertainty in the environment, the lack of sufficient data to correctly monitor the state of the world will often mean the plan will fail. The system needs to be able to recover gracefully from plan failure.
- Finally, the system needs to be able to refine its model of the world, and its representation of goals, constraints, actions, and outcomes based on the success or failure of the plans.

8.6 THE CHALLENGES AHEAD

Many methods for automated generation and execution of plans have been developed over the last decades. Such techniques have been extensively explored in the field of autonomous vehicles, and in other domains. In the oil industry, attention is turning to this methodology and trial systems have been implemented.

As experience is gained several key questions will need to be answered. The first and most important of these is the validation and verification challenge—how do we test such systems, including the veracity of the inference and the completeness of the state of the world description. Testing on simulators is essential, but simulators do not always capture the nature of the uncertainty of the environment in real life. Deliberative decision techniques that issue automation commands to closed loop machinery control systems need to be tested with real hardware in the loop. This hardware, at a minimum, needs to include the actual PLCs used to control the equipment, and ideally the equipment itself. This requires testing on real rigs and in real wellbores—an expensive prospect.

Decision making in a group setting, as on the rig site, is already complicated. Adding an automated decision-making system that may issue instructions to people increases the complexity. Scrutability of automated systems is extremely important. The people involved need to be able to understand the reasoning behind the decision, to build trust, and to be able to contribute their own experience and expertise to the group. Scrutability is also important for regulatory reasons. As with any safety critical systems, it should always be possible to ascertain why the system behaved in a particular way.

Having automated systems involved in decision making changes the roles and responsibilities of the people involved. Because people have enormous flexibility in their capacity to react to changes and make decisions on the spot, many important aspects of current plans are not made explicit. Much work needs to be done to ensure these aspects are captured in the description of the world and that the automated reasoning system takes them into account [12–15]. This will change the role of the planning engineers. During execution, decisions that were once taken by individuals, such as the DD or Company Man, will now be automated. In the current legal and commercial landscape, the automation tool itself cannot be the accountable party. The industry needs to foster the appropriate relationships between the human users and the automation systems in place. This is an opportunity to rethink the distribution of tasks and expertise, which will lead to changes in the way rigs are crewed and operated.

Finally, the impact of automation on business models may be significant [16,17]. As discussed earlier, well construction is a collaborative venture, and the nature of the collaboration is mediated by a mixture of market forces, contractual relationships, and supply and demand of skills and services. Automation will change this balance and new business models will emerge.

ACKNOWLEDGMENTS

I would like to thank Gokturk Tunc, Richard Harmer, and Rodrigo Gallo Covarrubias for their comments and inputs, and for the time and effort spent reviewing the content of this chapter.

NOTES

1. In this context constraints are usually a set of limitations and restrictions that should be observed during execution of the well construction process. They can be of different forms, for example operational ranges of parameters, defined sequences of activities, standard operating procedures, etc.
2. Risk is an uncertain event or condition that, if it manifests, has an effect (generally a negative one) on one or more well objectives such as scope, schedule, cost, quality, or safety.

REFERENCES

[1] Bourgoyne Jr., A.T., Millheim, K.K., Chenevert, M.E., and Young, Jr., F.S., *Applied Drilling Engineering*, SPE, Richardson, TX (1991).

[2] Gatlin, C., *Petroleum Engineering—Drilling and Well Completions*, PrenticeHall, Inc., Englewood Cliffs, NJ (1960).

[3] Mitchell, B., *Advanced Oilwell Drilling Engineering Handbook*, Mitchell Engineering, Houston, TX (1993).

[4] Drilling Systems Automation Roadmap Initiative https://dsaroadmap.org/

[5] Abrahamsen, E., Bergerud, R., Kluge, R., and King, M., Breakthrough in Drilling Automation Saves Rig Time and Safeguards Against Human Error, Society of Petroleum Engineers, SPE-177825-MS. (2015).

[6] Trichel, D.K., Isbell, M., Brown, B., Flash, M., McRay, M., Nieto, J., and Fonseca, I., Using Wired Drill Pipe, High-Speed Downhole Data, and Closed Loop Drilling Automation Technology to Drive Performance Improvement Across Multiple Wells in the Bakken, Society of Petroleum Engineers, SPE-178870-MS, SPE. (2016).

[7] Pink, T., Cuku, D., Pink, S., Chittoor, V., Goins, A., and Facker, B., Hanford, R., World First Closed Loop Downhole Automation Combined with Process Automation System Provides Integrated Drilling Automation in the Permian Basin SPE-184694-MS, SPE/IADC Drilling Conference and Exhibition, March 14–16, 2017.

[8] Carpenter, C., Automation-Adoption Approach Maps Human/System Interaction, Society of Petroleum Engineers. doi:10.2118/0219-0065-JPT (2019).

[9] Cayeux, E., Daireaux, B., Saadallah, N., and Alyaev, S., Toward Seamless Interoperability Between Real-Time Drilling Management and Control Applications, Society of Petroleum Engineers. doi:10.2118/194110-MS (2019).

[10] Abughaban, M., Alshaarawi, A., Meng, C., Ji, G., and Guo, W., *Optimization of Drilling Performance Based on an Intelligent Drilling Advisory System, International Petroleum Technology Conference*. doi:10.2523/IPTC-19269-MS (2019).

[11] Wylie, R., McClard, K., and de Wardt, J., Automating Directional Drilling: Technology Adoption Staircase Mapping Levels of Human Interaction, Society of Petroleum Engineers, SPE-191408-MS, SPE. (2018)

[12] Pastusek, P., Payette, G., Shor, R., Cayeux, E., Aarsnes, U.J., Hedengren, J., Menand, S., Macpherson, J., Gandikota, R., Behounek, M., Harmer, R., Detournay, E., Illerhaus, R., and Liu, Y., Creating Open Source Models, Test Cases, and Data for Oilfield Drilling Challenges, Society of Petroleum Engineers, SPE-194082-MS, SPE. (2019).

[13] Brackel, H.U., Macpherson, J., Mieting, R., and Wassermann, I., An Open Approach to Drilling Systems Automation, Society of Petroleum Engineers, SPE-191939-MS, SPE. (2018).

[14] Solvi, L.-J. R., Revheim, O., Schaefer, S., and Schutte, F.J., An Electronic Rig Action Plan—Information Carrier Equally Applicable to the Driller and the Automation Platform, Society of Petroleum Engineers, SPE-195959-MS. (2019).

[15] Isbell, M.R., Groover, A.C., Farrow, B., and Hasler, D., What Drilling Automation Can Teach Us about Drilling Wells, Society of Petroleum Engineers, SPE-195818-MS, SPE. (2019).

[16] Israel, R., Farthing, J., Walker, H., Gallo Covarrubias, R., Bryant, J., and Vahle, C., Development to Delivery—A Collaborative Approach to Implementing Drilling Automation, SPE-184695-MS, SPE/IADC Drilling Conference and Exhibition, March 14–16, 2017, The Hague, The Netherlands. https://doi.org/10.2118/184695-MS

[17] Iversen, F.P., Thorogood, J.L., Macpherson, J.D., and Macmillan, R.A., Business Models and KPIs as Drivers for Drilling Automation, Society of Petroleum Engineers, SPE-181047-MS. (2016).

9 Specialized Drilling Techniques for Medical Applications

Yoseph Bar-Cohen, Hyeong Jae Lee, Mircea Badescu, Stewart Sherrit, Xiaoqi Bao, and
Jet Propulsion Laboratory (JPL)/California Institute of Technology (Caltech), Pasadena, CA
Yoseph Shalev
Placidus Medical, Milwaukee, WI

CONTENTS

9.1 INTRODUCTION

Increasingly, artificial components including screws and fixtures are inserted into human bodies as part of medical surgery or treatment and these processes involve the use of drilling. Drilling bones is done for such applications as joining fractured or broken bones where screws and braces are used to attach and support bone sections. An example of a drill bit that is used in orthopedic bone drilling is shown in Figure 9.1 and an example of using screws for fusing vertebrae in the back is shown in Figure 9.2. In addition, drilling is done as part of the replacement of hip and knee joints with artificial components and an example of a replaced hip radiograph is shown in Figure 9.3. In teeth, drilling is done for removal of decayed parts in preparation for filling cavities as well as for root canal

FIGURE 9.1 Example of a drill bit that is used in orthopedic bone drilling. *Source:* Courtesy SterileBits, Inc.

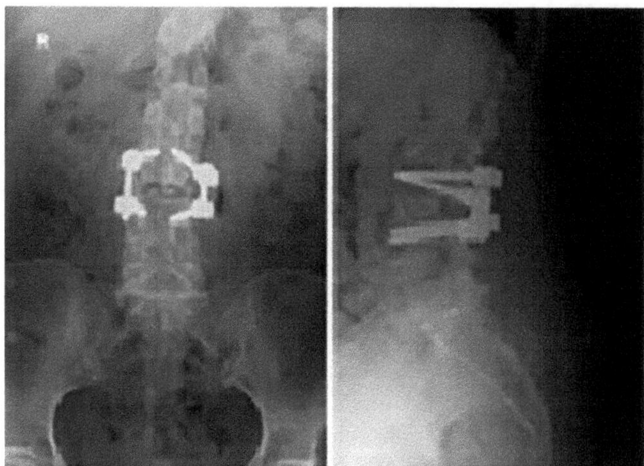

FIGURE 9.2 X-ray images that were taken from back vertebrae showing a fusion made using screws and support. *Source:* Courtesy Rina Eshet, Raanana, Israel.

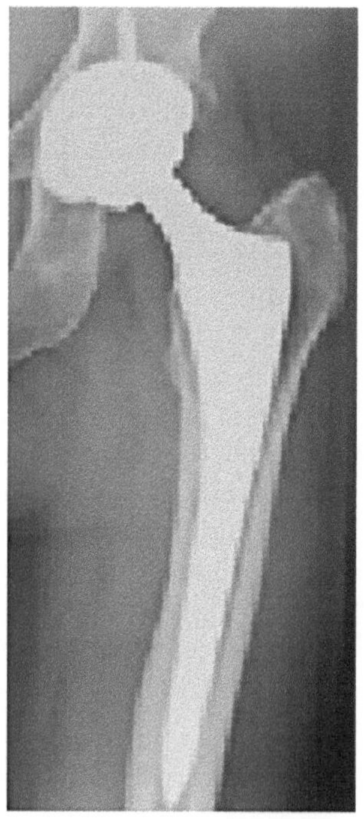

FIGURE 9.3 Radiographed hip replacement showing artificial components that are installed after drilling the related bones.

treatment. It is interesting to note that the sound from the drill provides surgeons a guide for the drilling process (Praamsma et al., 2008). Since drilled bones are mostly covered by tissues, orthopedic surgeons are seeking to rely on various sensory cues for guidance including the sound pitch that is produced. This is particularly essential since they need to stop the drilling quickly once the full bone thickness is traversed. Thus, potential injury to soft tissue parts is prevented including damage to the arteries, veins, and nerves. Moreover, piezoelectric actuated drilling is done in various hard parts of the human body including the bones and teeth as well as for breaking large kidney stones.

Examples of drilling via piezoelectric drills for opening blocked arteries and breaking kidney stones as well as high-speed rotary dental drilling are covered in this chapter. In reviewing piezoelectric actuated drilling, the authors covered the development work that have been done jointly with other investigators (Bar-Cohen et al., 2005; Bar-Cohen et al., 2016; Sherrit et al., 2019). These drills include the Ultrasonic/Sonic Driller/Corer (USDC) (Bar-Cohen et al., 2005) and the constrained-mass mechanisms (Sherrit et al., 2019).

The USDC technology was initially developed for planetary sampling in cooperation between some of the authors and engineers from Cybersonics (Sherrit et al. 1999, 2000; Bar-Cohen et al., 2005; Bar-Cohen et al., 2018). This mechanism employs an ultrasonic horn that drives a free-mass onto a floating bit or probe and the impact delivers a large stress pulse to the tip of the probe. This method of delivering stress at a probe tip has found utility also in medical applications and, specifically, in the "CyberWand" tool for fragmenting large kidney stones (marketed by Olympus).

The constrained-mass mechanism is based on a concept of probe actuation that has been investigated by the coauthors as a means of fracturing or penetrating kidney stones and occlusions in blocked arteries with a better control of the impact energy and frequency (Sherrit et al., 2019). The mechanism is potentially applicable to drilling various other hard parts in the human body including the bones and teeth. The concept enables transferring ultrasonic vibrations and impacts onto a firmly attached (fixed) probe and it has been theoretically modeled, simulated, produced in prototypes and tested. In their related studies, the authors used their extensive expertise with the development of piezoelectric actuators and drilling mechanisms for potential future planetary exploration missions (Bar-Cohen and Zacny, 2009; Bao et al., 2003; Bar-Cohen et al., 2016; Bar-Cohen et al., 2018; Sherrit et al., 2019). The approach addressed the need to use low axial forces and holding torques, lightweight hardware, producing low heating of the tool with ability to efficiently duty cycle the used ultrasonic power. These percussive mechanisms have been enhanced by augmenting rotation of the drilling bits.

The example of dental drilling is covered in this chapter in order to include a solely rotary mechanism that is driven at extremely high speeds using pneumatic actuation.

9.2 LITHOTRIPSY USING PIEZOELECTRIC ACTUATION

In relation to lithotripsy, kidney stones can be divided into small and large. The small ones are routinely fractured by extracorporeal shock wave lithotripsy where the stones in the kidney and ureter are treated by high energy shock waves. The waves are transmitted through the body and they fragment the stones into pieces that are as small as grains of sand. The fragmented stones are made sufficiently small to pass through the urethra along with the urine. The treatment is done while the patient is monitored under X-ray or ultrasonic imaging to allow locating the stones and precisely directing the shock waves. For treating large kidney stones, more invasive mechanisms are used and they include the CyberWand Dual Lithotripsy System, which is partially based on the Ultrasonic/Sonic Driller/Corer (USDC) mechanism and is widely used (Bar-Cohen et al., 2005; Bar-Cohen et al., 2018). This mechanism employs piezoelectric actuated hammering onto a free-mass that generates low frequency impacts onto a probe transmitting stress pulses to the probe tip. The probe, which is effectively a thin rod, is inserted through the urinary track and is placed in contact with the individual stones to fragment them.

To enhance this lithotripsy-related technology, the authors investigated alternatives to the use of free-mass, and introduced the constrained-striker configuration (Bar-Cohen et al., 2016; Sherrit et al., 2019). The constrained-striker is excited at a reduced frequency from the one that is generated by the piezoelectric ultrasonic actuator. The actuator strikes a rigidly attached (fixed) probe head with a larger stroke and it creates much larger stress pulses. The general concept of the related design is illustrated in Figure 9.4, where the striker is constrained by a membrane support allowing for the excitation of sub-harmonic vibrations. In addition, there is a clear annulus or central path region for providing passage for tubing to pump or vacuum fluids or small solid pieces of broken stones. Moreover, a vibratory actuated probe is used that can be removed from the driving horn and replaced. When implemented medically, the replacement of the probe in each new procedure would be part of the preparations for the process application.

The advantage of the constrained-striker design is its ability to allow for effective conversion of high-frequency, high-force vibration of the horn tip into low-frequency, high-displacement hammering blows while eliminating lateral motions. These impacts create a large stress in the target and it enables efficient drilling (Sherrit et al., 2019). A finite element model was developed to simulate the operation of this type of lithotripsy system. The model is able to predict the displacement amplitudes of the probe with different types of flexure configurations and tip loads. The software traces the translation movements of the piezoelectric actuator and the constrained-striker as well as the vibration of the probes as a function of time. Within a reasonable accuracy, it predicts the time and location of the striker/horn or striker/bit collision. It also calculates the changes of the variables as time evolves. The movements and vibration due to the impact are recorded along with the impact momentum and time. The software then proceeds to predict the next impact.

Finite element modeling results of the constrained-striker configuration analysis are shown in Figure 9.5. These results are showing the time-dependent displacement and velocity amplitudes for the horn and striker, respectively, when the horn vibrates at 20 kHz with displacement boundary condition of ~10 μm on the horn tip. Note that the striker vibrates at reduced frequencies that can be in the range of 100 Hz to 1 kHz but with higher velocities and thus effective momentum transfer is made to the probe and the target.

An example of time-dependent FE analysis is shown in Figure 9.6 presenting the equivalent stress distribution on the target material (kidney stone) as a function of time. The properties of the kidney stone model, BegoStone (15:3), that was used are summarized in Table 9.1. This material has high compressive strength but low tensile strength-simulating materials that are damaged by brittle failure. The brittle failure model assumes, according to Rankine failure criterion, that the material is elastic in compressive loading and a crack forms when the principal tensile stress exceeds the tensile strength. Before failure, the material is linearly elastic in tension, as shown in Figure 9.7. Note that when the target material is subjected to high cyclic loading, cracks develop with time, and

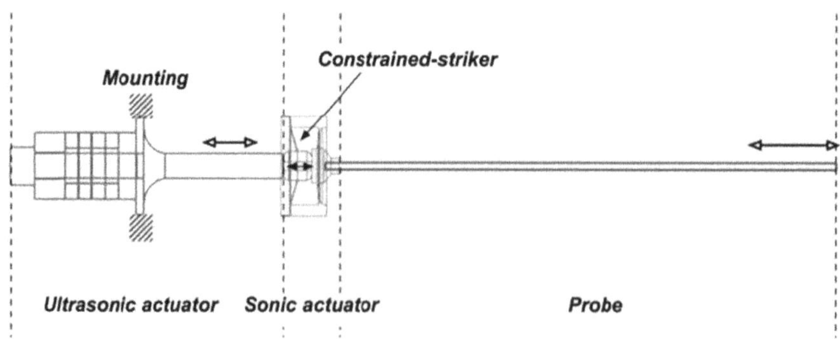

FIGURE 9.4 Design and description of the constrained-striker configuration.

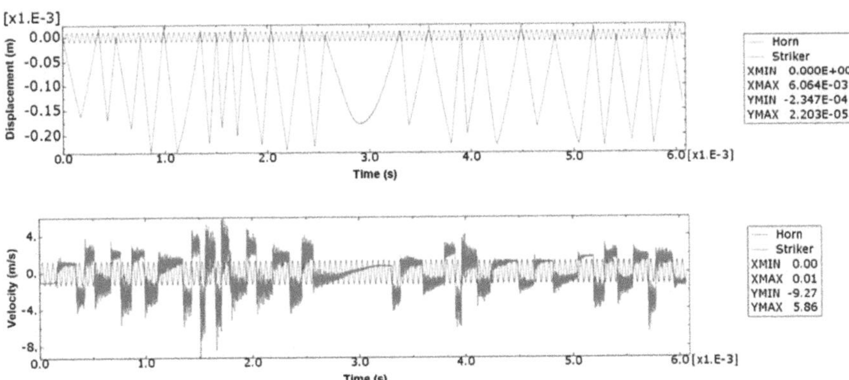

FIGURE 9.5 Time-dependent displacement and velocity amplitudes for the horn and striker, respectively.

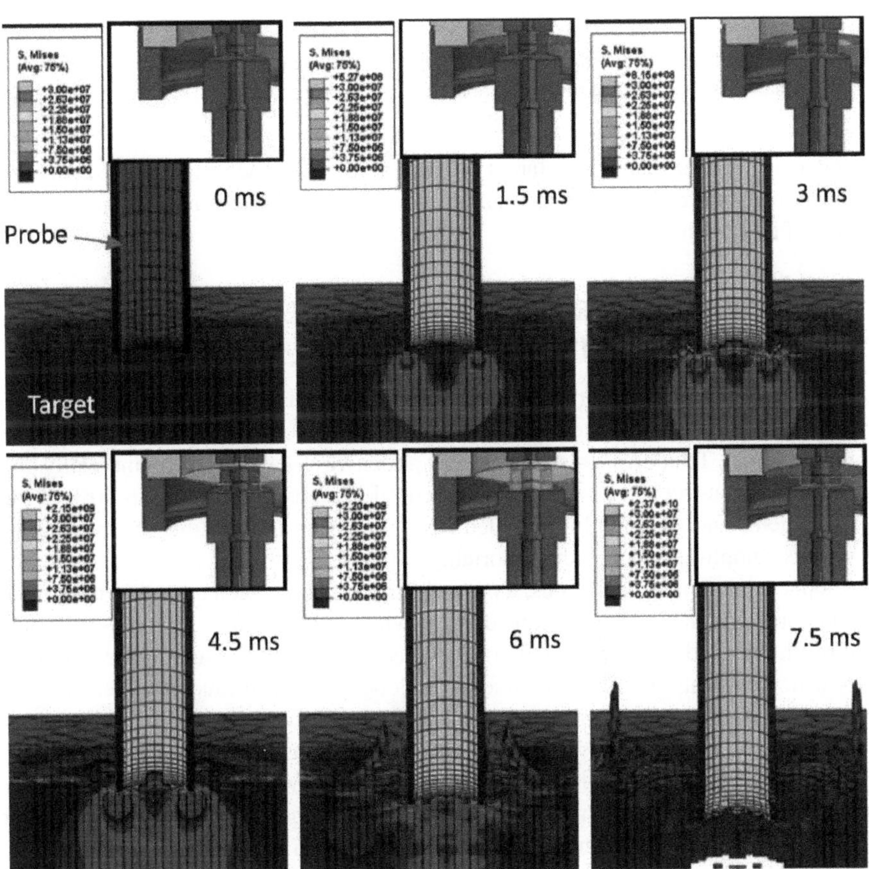

FIGURE 9.6 Equivalent stress distribution in the target material as a function of time. The time interval graphic inset shows the corresponding reaction level along the horn and striker cross section.

TABLE 9.1
The Properties of a Typical Kidney Stone Model. Young's Modulus (YM), Shear Modulus (SM), and Tensile Strength (TS) are Identified in the Table

Material	σ	ρ (kg/m³)	YM (GPa)	SM (GPa)	V (m/s)	TS (MPa)
BegoStone(15:3)	0.27	1995	27.4	11.78	4142	16.3(dry)/7.12(Wet)

Source: Esch et al. (2010).

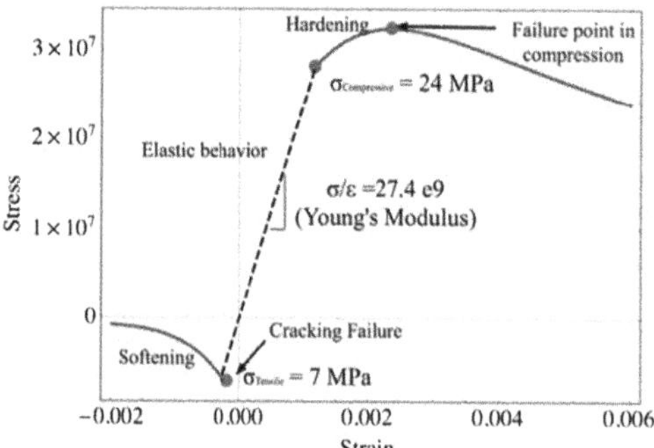

Young's modulus:	27.4 GPa
Poisson's ratio:	0.27
Density:	1995 kg/m3
Compressive initial yield stress:	24 MPa
Compressive ultimate stress:	32 MPa
Tensile failure stress:	7 MPa

FIGURE 9.7 The response of soft BegoStone (15:3) to uniaxial loading in compression and tension. The main failure mechanisms are cracking in tension and crushing in compression.

significant damage is induced onto the target material after 8 ms. The results of the analysis indicate that the constrained-striker configuration can offer high-impact forces on the target, which enables the breakage of kidney stones.

9.3 PERCUTANEOUS INTERVENTION OF PERIPHERAL ARTERIES AND CORONARIES CHRONIC OCCLUSION

Chronic total occlusion of peripheral or coronary arteries is difficult to treat successfully, mostly because these types of occluded arteries are filled with hard calcified plaques that standard medical wires fail to penetrate. To address this issue, stiffer glider wires have been commercially developed. However, even when using these wires by experienced operators, only 50% to 70% of the procedures are completed successfully. In the failed cases, the effort to penetrate the occlusion tends to result in serious complications including mortality.

To develop a penetrator of occlusions, a piezoelectric actuated mechanism has been investigated by the coauthors. The device was designed for driving drilling-wires via high-frequency vibrations. Test results showed that ultrasonic horn-based actuators generate high power, high-frequency (>10 kHz) wire tip vibrations. However, due to the long length and much smaller diameter of the drilling-wires compared to the horn (~6.5 mm and 0.35 mm for horn and wire diameter, respectively), unwanted bending motions can be excited that increase the power losses and decreases the efficiency of the device operation. In an attempt to improve the performance of the actuator, reduce the fabrication cost, and miniaturize its size, flextensional and cymbal actuator alternatives to the ultrasonic horn actuators were considered. Compared to the horn-based actuators, the flextensional actuators were thought to have general advantages, such as easy fabrication, compactness, low weight, and large stroke. Even though test results showed that flextensional-based actuators provided a satisfactory vibration amplitude and frequency, one of the issues that was encountered has been the significant noise generation due to their relatively low-resonant frequencies operation in the sonic range (1–6 kHz). In addition, the drilling performance of flextensional-based actuators was found to be inferior to that of horn-based drilling-wire systems. Due to the identified limitations, the research investigating the use of flextensional actuators was redirected to focusing on the use of horn-based actuators.

To improve the driving performance of ultrasonic horn-based drilling-wire systems, efforts have been made to enhance the vibration coupling between the piezoelectric horn actuator and the drilling-wire (Bar-Cohen et al., 2016). With optimized ultrasonic horn and horn adapter configurations, the drilling-wire was able to penetrate different types of rocks, such as brick, concrete, and limestone as well as a sample that simulates clogged artery. The test results have shown that it took 2 min to penetrate 7-mm thickness of brick with an input power of 1 to 2 W. For consistency of the terminology used in this chapter, Table 9.2 lists the related definitions.

As part of the development task, the authors investigated the effects of the excitation frequency and found that even at frequencies as high as 34 kHz the horn actuator did not damp in water and it exhibited performance that is similar as in air. Based on the experimental results, it was concluded that horn-based actuators are good candidates for driving guidewire probes. Thus, vibratory actuation of drilling-wires has been established, allowing for treating chronic occlusions towards completely removing plaques (Bar-Cohen et al., 2016) and keeping the arteries open following an angioplasty procedure and stent placement. Intra-Vascular Ultrasound (IVUS) showed post drilling of a central channel without damage to the vessel wall and repeated IVUS post stenting have shown excellent lumen with complete apposition of the stent (Figure 9.8).

Schematic representation of the developed mechanism is depicted in Figure 9.9. This mechanism is intended to address the serious effect of chronic total occlusion associated with peripheral vascular disease (PVD) that may include physical disabilities and increased probability of mortality. The drilled hole through the occlusion provides a path for other medical means of expanding the hole and the possibility of removing the rest of the occlusion. These means may include balloon angioplasty or thoracotomy. Their use is followed with placement of a stent. In this reported study, the authors tested the use of commercial guidewires with 0.014 and 0.018 inch diameters.

9.3.1 Modeling the Interaction between Drilling-Wire and Tissue

Atherosclerotic tissue consists of various materials, such as fibrous tissues, fatty deposits, thrombus, and calcification. Examples of atherosclerotic tissue are shown in Wong et al. (2012) indicating

TABLE 9.2
The Terminology Used to Describe the Wire-related Mechanisms of Drilling Vascular Systems

Term	Definition
Drilling-wire	A rigid low attenuation wire used for drilling occlusions. For an operation, the wire is connected directly to the actuator and is pushed against the occlusion for its drilling.
Guiding-element	A cone, ball, or other solid element that is mounted at the tip of the drilling-wire. The element has a diameter that is larger than the wire and it is used to center the wire along the artery and minimize potential damage to the wall.
Guiding-sleeve	The sleeve within which the drilling-wire is inserted and is used to guide the drilling-wire along the sleeve and serves as a conduit for delivering fluids such as medications or cooling saline.
Guidewire	A guiding-sleeve with inserted drilling-wire.
Smart-Wire system	The presented occlusion penetration system that consists of a piezoelectric actuated mechanism of delivering percussive action into drilling-wires. It can be augmented by rotation from a motor and the drilling-wire path is guided by a sleeve.

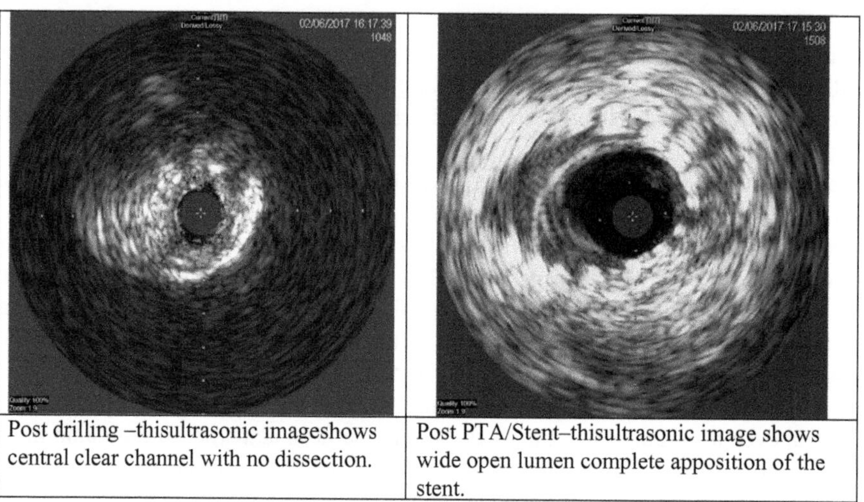
Post drilling –thisultrasonic imageshows central clear channel with no dissection.
Post PTA/Stent–thisultrasonic image shows wide open lumen complete apposition of the stent.

FIGURE 9.8 Intra-Vascular Ultrasound (IVUS) showing the results post drilling and post insertion of a stent.

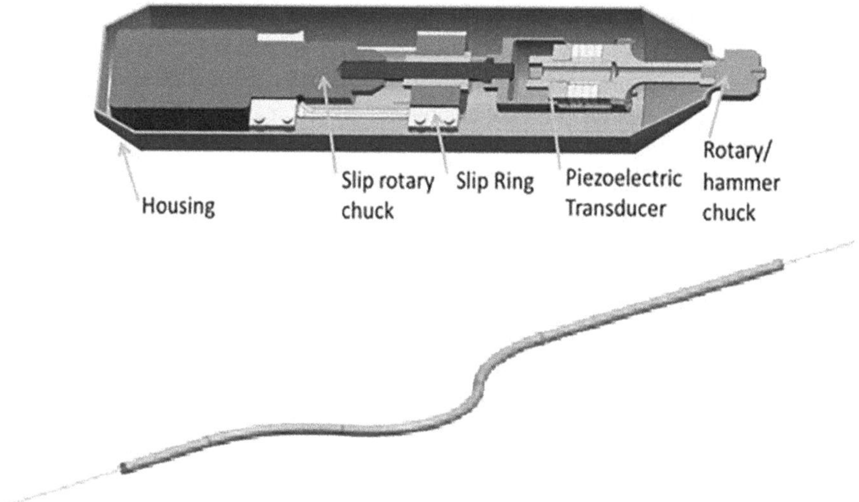
Housing
Slip rotary chuck
Slip Ring
Piezoelectric Transducer
Rotary/ hammer chuck

FIGURE 9.9 Illustration of the vibratory actuator with rotary capability (top) that is being developed to drive a drilling-wire (bottom).

that multiple modeling parameters are required to simulate the mechanical response of calcified plaques under dynamic cyclic loading from piezoelectric actuators. The mechanical behavior of the individual layers of the artery is highly nonlinear. Therefore, the artery wall and plaque were modeled using a 5-parameter third-order Mooney–Rivlin hyperelastic constitutive equation. This has been found to be adequate to describe the nonlinear stress-strain relationship of elastic arterial tissue. The calcified plaque was assumed to be linear isotropic elastic material with Young Modulus 2.7 MPa and Poisson's ratio 0.49, with the density of 1450 kg/m^3 (Loree, Grodzinsky et al. 1994; Lally, Dolan et al. 2005; Schiavone, Zhao et al. 2013; Karanasiou, Sakellarios et al. 2014). In the analysis, the drilling-wire is assumed to have a circular cross section with a diameter of 0.014 inch and a total length of 300 cm, the wire is made of steel and it has Young's Modulus of 200 GPa and

TABLE 9.3
Material Properties of the Artery Wall, Plaque and Calcified Plaque

	Linear elastic			Mooney-Rivlin Hyperelastic Constants (kPa)				
	ρ	Y (MPa)	σ	C10	C01	C20	C11	C30
Artery wall	1066			18.9	2.75	590.42	857.18	0
Plaque	1450			–495.96	506.61	1193.53	3637.8	4737.25
Calcified plaque	1450	2.7	0.49					

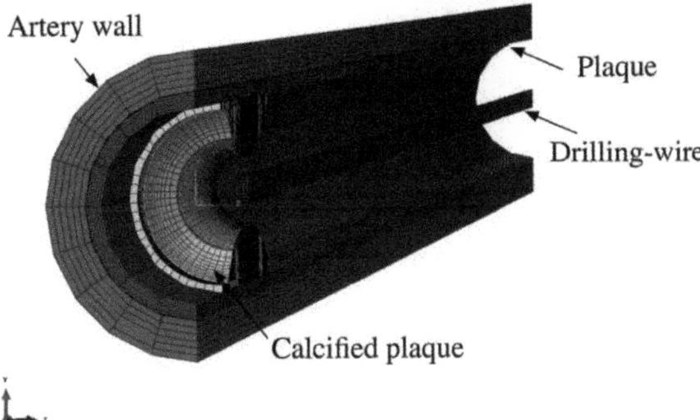

FIGURE 9.10 The finite element mesh showing the cross section of the drilling-wire, plaque, calcified plaque, and artery wall. The green, red, yellow, and blue colors represent the arterial wall, plaque, calcified plaque, and the wire, respectively.

Poisson's ratio of 0.3. The material properties of the artery wall, plaque, and calcified plaque that were used in the model are summarized in Table 9.3. Eight-node brick elements (C3D8I) were used to model the wire as these elements can overcome the issue of shear locking and are suitable for contact with bending model. The finite element model is shown in Figure 9.10 and it consists of the artery, a calcified plaque, and a stiff drilling-wire at their initial condition. The plaque is located inside the arterial wall, and the calcified plaque is lying between the plaque and the drilling-wire. The thicknesses of plaque and the calcified plaque are assumed 0.2 mm and 0.75 mm, respectively. The arterial segment is 100 mm in length, while the length of the calcified plaque is 9.5 mm.

To simulate the oscillation motions, the kinematic coupling constraint was used, where a number of nodes from one end of the drilling-wire are constrained to the rigid body motion of a reference node. A periodic displacement of 20 μm with ~20 kHz was then applied to the reference point of the wire. Contact constraints were used to contact separate surfaces of artery wall, plaque, and calcified plaque. The step time was chosen to be 1 ms with 1-μs time interval. Explicit dynamic analysis was applied to simulate the whole process. In all the simulations, the ends of the artery wall were fully constrained in order to prevent rigid body motion and Figure 9.11 shows the boundary and load conditions of this model.

The selected explicit dynamic analysis results are shown in Figure 9.12 and Figure 9.13 exhibiting damaged calcified plaque through its thickness at 0.75 ms and 0.805 ms. Note that the drilling-wire shows enough force on the target by way of ultrasonic percussion from piezoelectric elements, removing the elements of calcified plaque. However, as can be seen in the displacement plots, the amplitudes of vibration of the wire tip are found to be much more dominant in the lateral direction (U2) compared to those along the axial direction (U1).

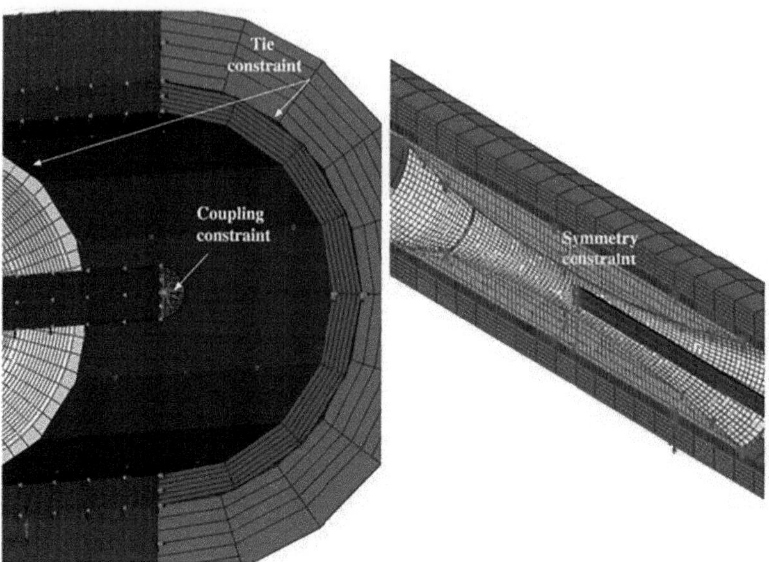

FIGURE 9.11 Boundary and loading conditions.

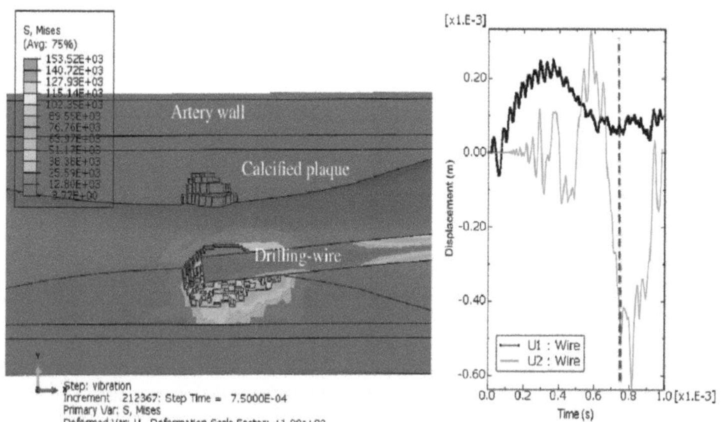

FIGURE 9.12 On the left, the stress contour of calcified plaque and the drilling-wire at 0.75 ms (left) is shown. On the right, the displacements of the wire tip as a function of time is shown. U1 and U2 represent the amplitudes in the X and Y directions, respectively.

9.3.2 Prototype Fabrication and Testing of Piezoelectric Horn Actuator

Besides the performance optimization of the piezoelectric actuator through FE analysis, the authors also investigated the system-level design. An initial concept of the drilling-wire system in a portable configuration has been considered with the goal of enabling operators or surgeons to manipulate the guiding-sleeve while holding the drilling device. However, such a manipulation was found to be difficult to control when using a sleeve that consists of a very long (300 cm) and thin (0.014 or 0.018 in) wire. Therefore, the driving components were combined into a table mount unit as shown

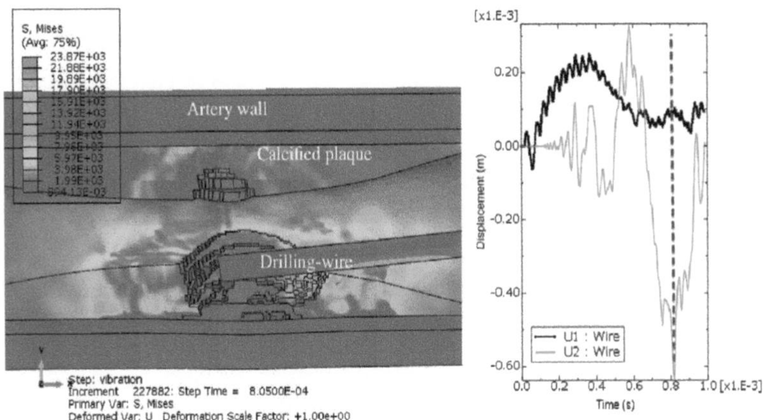

FIGURE 9.13 On the left, the stress contour of calcified plaque and the drilling-wire at 0.805 ms is shown. On the right, the displacements of the wire tip as a function of time is shown. Here too, U1 and U2 represent the amplitudes in the X and Y directions, respectively.

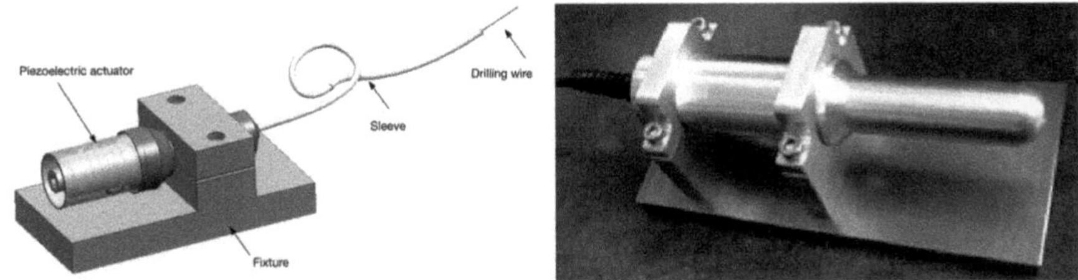

FIGURE 9.14 Table mount unit containing the driving components (actuator, sleeve, drilling-wire, fixture, etc.). Only the guiding-sleeve with the drilling-wire are articulated by the user/doctor who pushes them along the artery to open calcified plaques.

in Figure 9.14, where only the guiding-sleeve with the drilling-wire are articulated by the user/doctor who pushes them along the artery to open calcified plaques. This design gives better manual manipulation control of the drilling-wire system. The fabricated assembly is shown in Figure 9.14 (right).

To allow for easy attachment of the wire with the sleeve to the actuator, an interface can be considered as shown in Figure 9.15. In addition, as shown in this figure, a sleeve section that operates as an extender can be used for fine adjustment of the guidewire along the artery. Moreover, a fluid access port is designed as part of the mechanism to remove pulverized plaque or deliver drugs. The wire is inserted into the sleeve and the interface adjustment components as well as the wire are attached to the tip of the horn. The sleeve interface is then attached to the actuator housing and a knob on the housing is used to clamp the wire to enable coarse adjustment. The fluid access port is then attached to a pump.

During the cadaver superficial femoral artery (SFA) trials, the feasibility was tested using the actuator and the drilling-wire, named PlacidWire, breadboard to canalize long segments of calcified SFA (Peripheral arterial calcification) in a scenario that mimics actual intervention in a patient (Figure 9.16). For this purpose, a silicon model was created that included abdominal aorta and entire

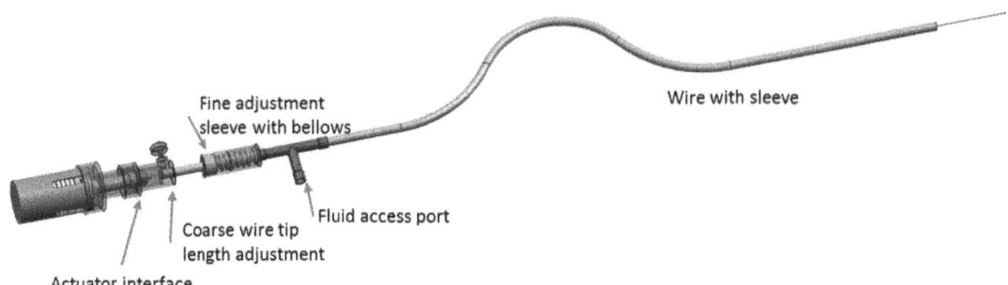

FIGURE 9.15 Smart wire unit with piezoelectric actuator, sleeve and wire couplings, and adjustments and fluid access port.

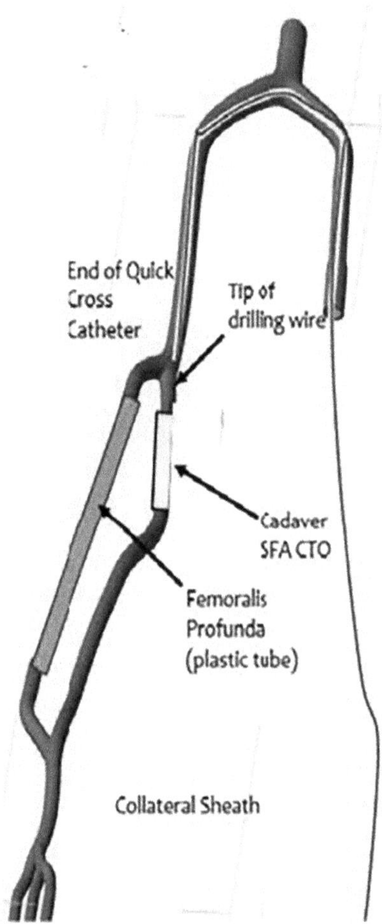

FIGURE 9.16 A scenario that mimics actual intervention in a patient where the feasibility of using the drilling-wire (also called PlacidWire) and its actuator were tested to determine the ability to canalize a long segment of peripheral arterial calcification (SFA).

legs arterial circulation. A window was made to allow inserting a long cadaver SFA and the results of the experiment were:

1. No difficulties to navigate the wire to the front of the occlusion.
2. It took about 60 s to drill an entire 12-cm occlusion.
3. It took a total of about 20 min to complete angioplasty and stenting of the long segments of chronically occluded and calcified SFA. In contrast, using current techniques and equipment

usually takes about 3–4 hr with long radiation exposure and usage of large amount of contrast material that can damage the kidneys.

At its current state, the authors believe the technology is ready for testing towards a commercial prototype of the actuator and drilling-wire, first in a silicon model and then in real patients with peripheral vascular disease (PVD). In the future, experiments will be considered where the use of the developed system will be tested in coronary arteries with chronic total occlusion.

9.4 DENTAL DRILLS

Dental drills are important tools that are used in the treatment of teeth. They are used to remove decayed material in teeth prior to filling the cleaned cavity (Figure 9.17). The use of the dental drills started in the 18th century and demonstrated a major improvement in the ability to repair teeth. The use of pneumatic power to drive dental drills was first implemented by George F. Green in 1868, and he followed this innovation in 1875 with his electrically powered dental drill (NYU Dentistry, 2019). Improvements that have been made over the years have enabled production of dental drills that work faster and more accurately than ever before and with much less pain to the patients. The air turbine powered dental drills were first introduced in 1911 with significant enhancement implemented in 1953. Today, these type of drills are the leading form of dental treatment and they have been developed with enormous speeds of up to hundreds of thousands of rotations per minute (RPM).

A full view of a dental hand-piece is shown in Figure 9.18a while an assortment of drill bits (called burs) are shown in Figure 9.18b. The drill hand-piece has a chuck to hold the burs via easy snap on and off and the burs are made of light and corrosion resistant materials. Tubing through the hand-piece (can be seen on the left of the hand-piece in Figure 9.18a) air and water are applied to perform cleaning, cooling, and driving the turbine of the drill. Moreover, the turbine consists of ceramic or metal bearings and a central spindle, and pressurized air at the level of about 35 psi is fed via a fine tube into the turbine. Unloaded burs can spin at speeds of hundreds of thousands of RPM; while under the torque load of a tooth it spins at slower speeds that are of the order of tens of thousands of RPM. For washing out the cuttings and cooling the bur, a fine spray mist of air and water are applied through the hand-piece and a separate vacuum tube is used to remove the accumulating wet cuttings and saliva. The hand-piece and the internal bearing need to operate accurately and with stable rotation over the extremely high number of revolutions that are involved in drilling a tooth. Also, the lubrication and the mechanism have to sustain many repeated sterilization applications in an autoclave with heated steam.

a. A dental drill hand-piece. On the left, the entry for air and water are shown and, on the right, the chuck for clamping the bur is located.
b. An assortment of burs (dental drill bits).

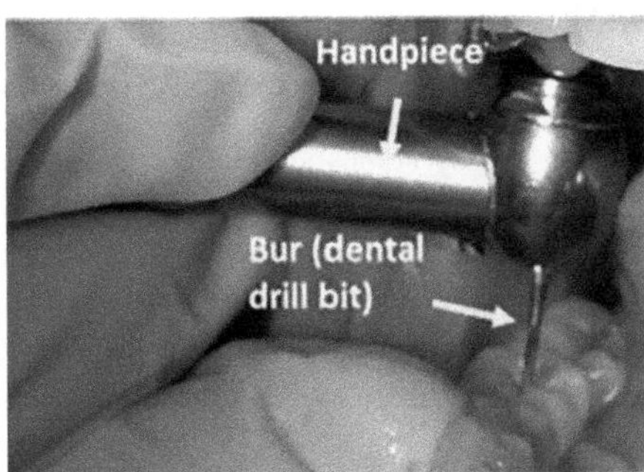

FIGURE 9.17 A dental drill is shown removing decayed tooth material. Courtesy Alan Grant, Dentist, Long Beach, California.

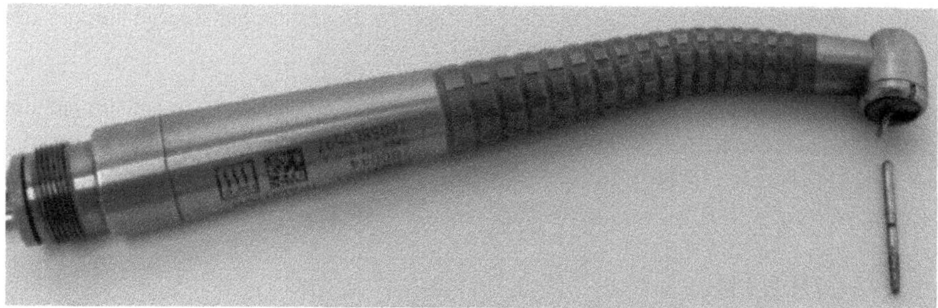

a. A dental drill hand-piece. On the left, the entry for air and water are shown and, on the right, the chuck for clamping the bur is located.

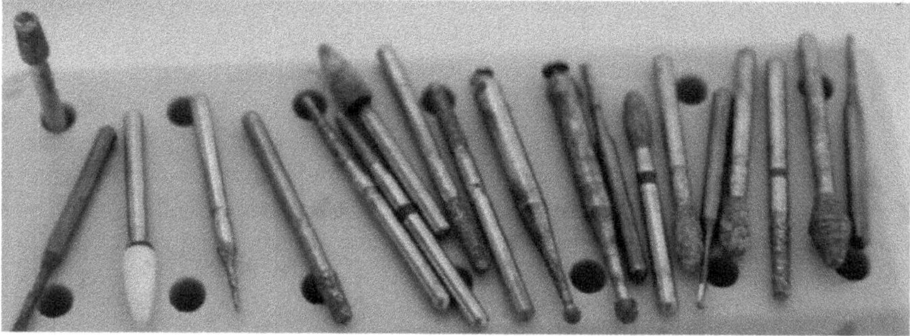

b. An assortment of burs (dental drill bits)

FIGURE 9.18 Photos of (a) full view of a dental hand-piece and (b) assortment of dental drill bits. This photo was taken with permission from the dentist Martin Galstyan at his dentist's office of South Pasadena dental clinic, California.

Typical burs (see assortment example in Figure 9.18b) have a diameter of 0.8–1 mm and they have an end effector (cutting surface) section that is about 4 mm long. Burs are generally made of stainless steel with a tungsten carbide coating or entirely of tungsten carbide, but sometimes they are coated with diamond or other abrasive materials. The burs need to be able to penetrate the tooth enamel, which is a crystalline material made of calcium hydroxyapatite, and it is the hardest material in the human body. Starting to drill on the surface of a tooth can cause "drill-walk" and undesirable extensive damage to the enamel. To prevent such damage, the bur is applied in light strokes onto the surface to microscopically mill the surface without stalling the bur. Dental drilling is a difficult task due to the limited operation space in the mouth, the obstructed view and light as well as the presence of various fluids containing debris. It is even further challenged by the need to drill while observing through mirror.

9.5 PERFORMANCE TESTS

9.5.1 Drilling Kidney Stone Simulants Using a Constrained-Striker-Based Actuator

Using the FE modeling results and kidney stones simulant, experiments were conducted to test the concept and corroborate the predictions. Snapshots of the performed drilling tests taken every 10 milliseconds are shown in Figure 9.19, where the damage evolution and fracture of the target material, soft BegoStone (15:6), can be clearly seen. In order to constrain the environment of in situ kidney stones and simulate operation conditions, the drill was tested on hard surface (light color) and soft foam (dark color) substrates. In both cases, the target kidney stones simulant was broken in less than 5 sec.

The time to break the soft (15:6) and the hard (15:3) BegoStones is shown in Figure 9.20. Although the time to drill through the sample varied relative to the location on the target object, they were generally broken in less than 40 sec for both the soft and the hard BegoStones.

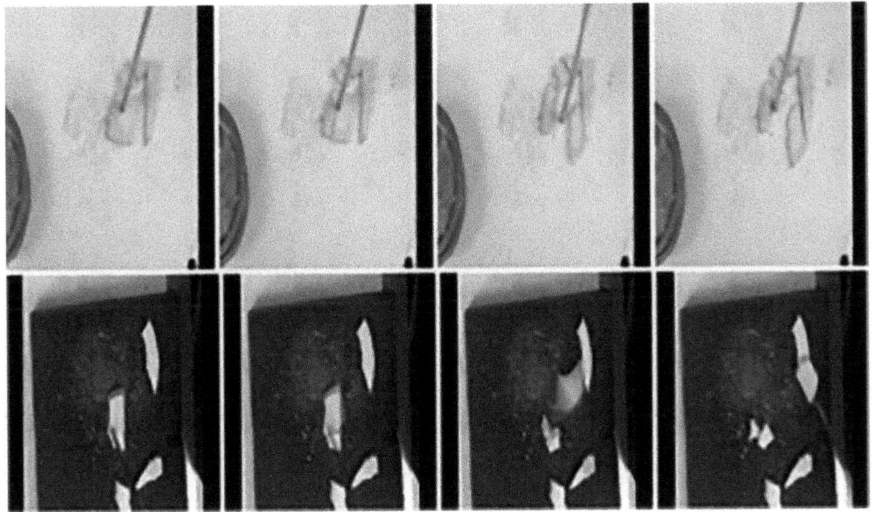

FIGURE 9.19 Snapshots of kidney stone model BegoStone (15:6) drilling test every 10 ms. The drill was tested on hard surface (light color) as well as on soft foam (dark color).

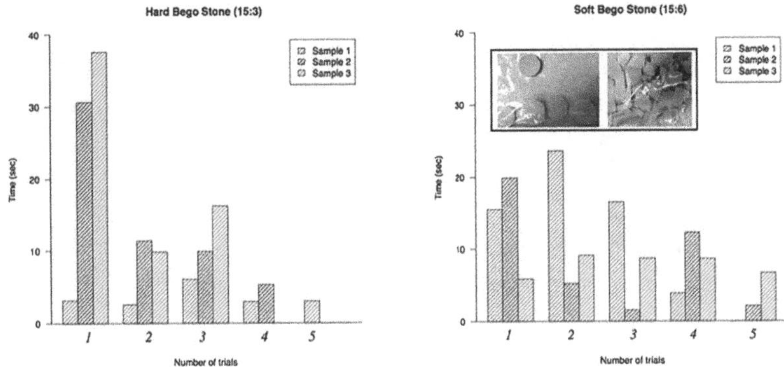

FIGURE 9.20 Time to break the soft (15:6) and the hard (15:3) BegoStones. The dimensions of the tested disk samples are diameter: 33–34 mm and thickness: 9–10 mm. The inserted pictures show the BegoStones before and after the tests.

9.5.2 Drilling Various Rocks Using the PlacidWire System

The performance of the ultrasonic horn-based PlacidWire system with drilling-wire has been investigated penetrating various type of rocks. The rocks included brick (compressive strength ~10 MPa), limestone (compressive strength ~25 MPa), and concrete (compressive strength ~30 MPa), and it was found that the wire was able to penetrate all these tested materials. Photographs that show the breadboard wire drilling a brick and concrete appear in Figure 9.21.

Snapshots showing the drilling of samples that simulate a clogged artery using the developed ultrasonic horn actuator are shown in Figure 9.22. The drilling-wire penetrated the simulated clogged artery in less than 1 min. Using the developed PlacidWire system, the rate of penetration has been tested by measuring the time to penetrate a 5-mm-thick brick (compressive strength ~10 MPa). The result is shown in Figure 9.23, where the photo was taken after 2 min of operation. The experimental results have shown that the rate of penetration of this actuator system (under an applied power of ~1.5 W) is around 0.058 mm/s. Figure 9.24 shows performance test of the horn-based

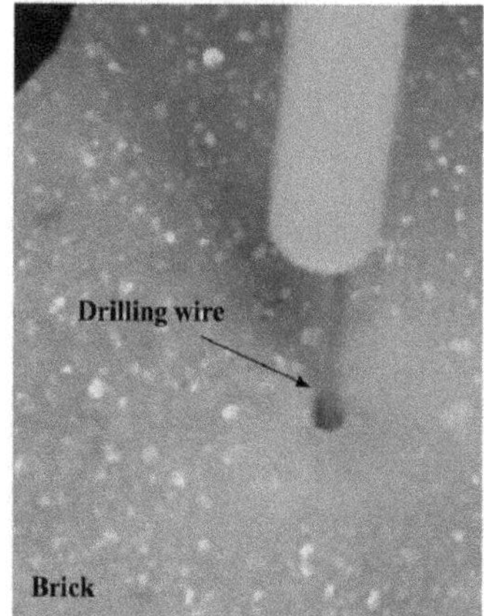

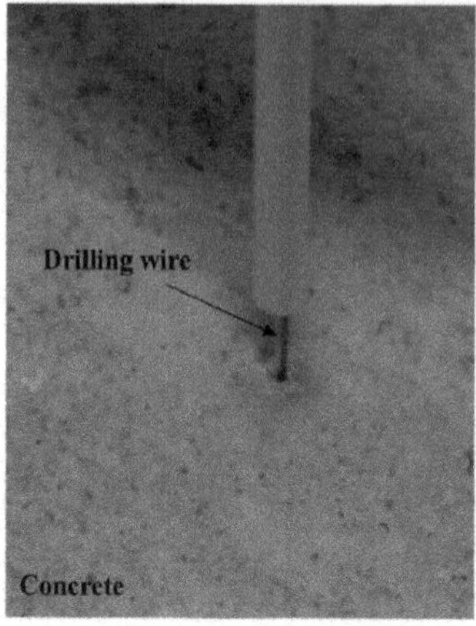

FIGURE 9.21 Snapshots showing drilling a brick (left) and concrete (right).

FIGURE 9.22 Snapshots of drilling a sample that simulates a clogged artery using the ultrasonic horn actuator.

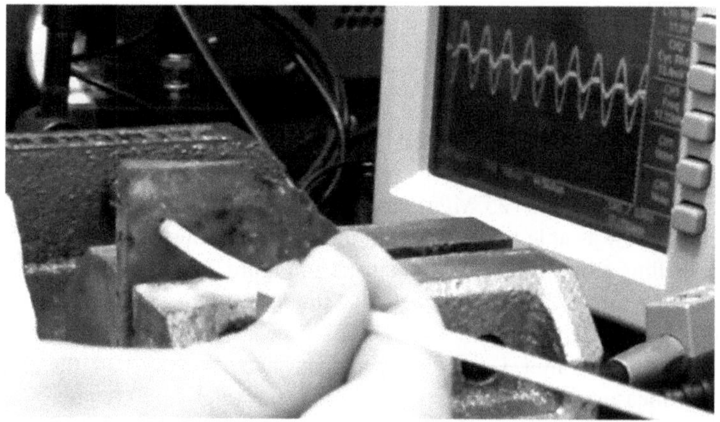

FIGURE 9.23 Drilling at 17.8-kHz drive frequency using the 0.018-inch drilling-wire. The drilling-wire was driven at ~1.5 W, and the 5.0-mm-thick brick (compressive strength >10 MPa) was completely drilled after 2 min.

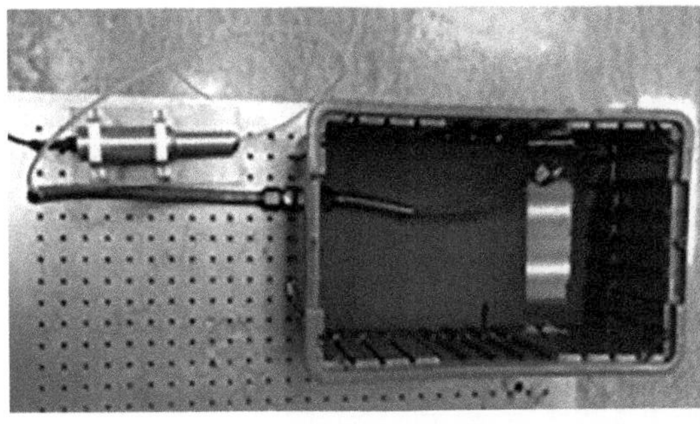
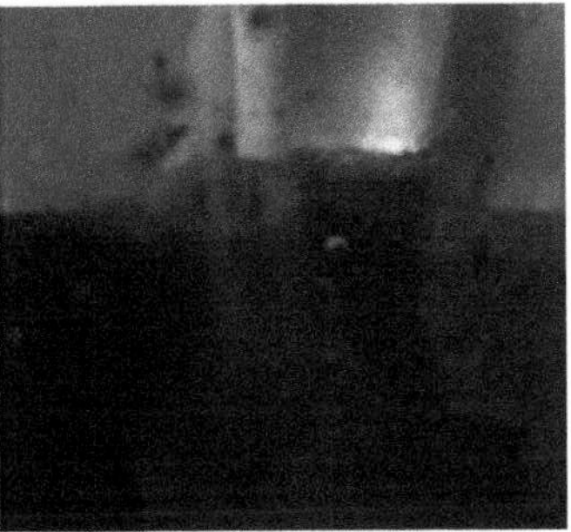

FIGURE 9.24 Snapshots showing the drilling in water.

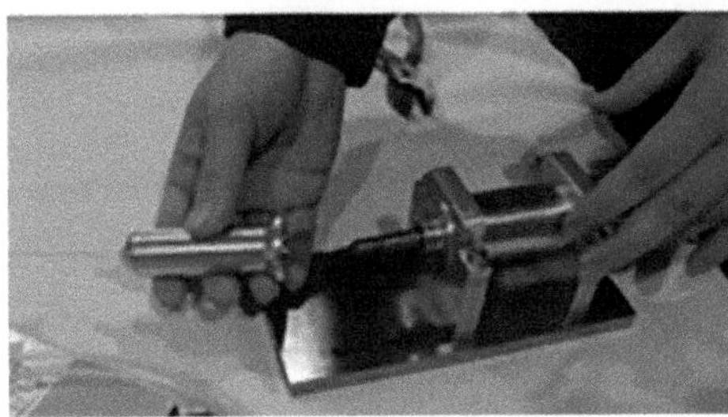
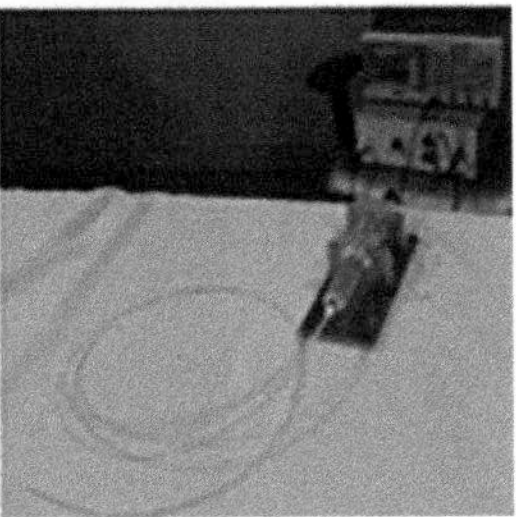

FIGURE 9.25 The actuator and the connected wire (left) as well as the full breadboard view (right).

actuator and drilling-wire that was immersed in water. It was noted that even at high frequency (34 kHz) the vibrations have not been damped by the water immersion and the recorded performance has been similar as in air.

9.5.3 Drilling Occluded Arteries

The performance of the developed horn-based breadboard, called PlacidWire, was tested drilling occluded arteries of cadavers and its operation demonstration has been successful. The system was assembled prior to the test and the connected components are shown in Figure 9.25 (left). The overall view of the breadboard is shown in Figure 9.25 (right) where the drilling-wire is covered with a guiding-sleeve and, in the back of the photo, the amplifier and function generator can be seen.

To determine potential thermal effect of the drilling on the tested artery, the temperature was measured before and after the drilling test (Figure 9.26). It is important to note that the temperature did not change from the 19°C initially measured value, as was expected given the very short time of the drilling and the low power levels that are consumed by the actuator. The actual drilling of the occlusion in a sample of an artery is shown photographically in Figure 9.27. Two positions of holding the artery segment were made: (1) attaching the artery to a Styrofoam support using pins, and (2) holding the artery segment manually in the hand of the cardiologist while drilling allowing for the cardiologist to get a feel for the action that is involved.

Based on feedback from the cardiologist who conducted the tests, the pathology tests of the drilled occlusions confirmed that no thermal damage took place, the occlusions were penetrated, and the fragments were quite small. As for the size of the fragments that were observed by the pathologist, they were small enough to be removable by a circulating fluid in the drilling-wire sleeve.

The test results showed that using the 0.46 mm (0.018″) diameter drilling-wire driven at 17.8 kHz under 1 W of input power, the occlusion in the artery samples was penetrated in less than 1 min. As

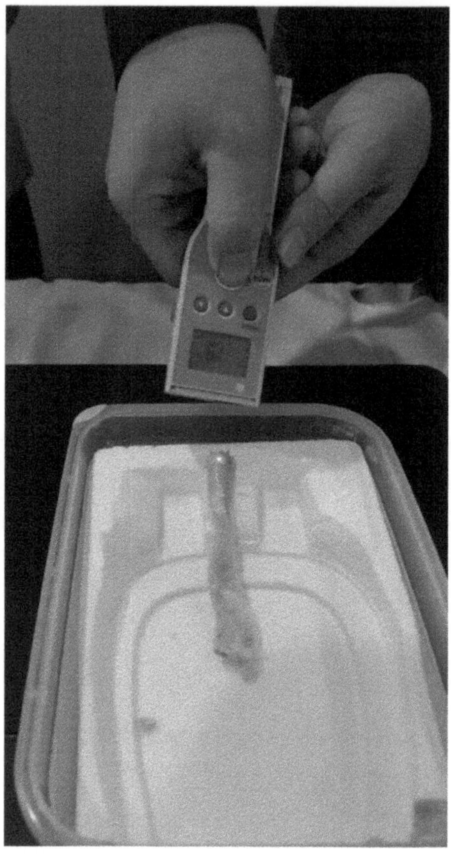

FIGURE 9.26 Using a thermal-sensor, the temperature of the artery was measured before and after the drilling tests.

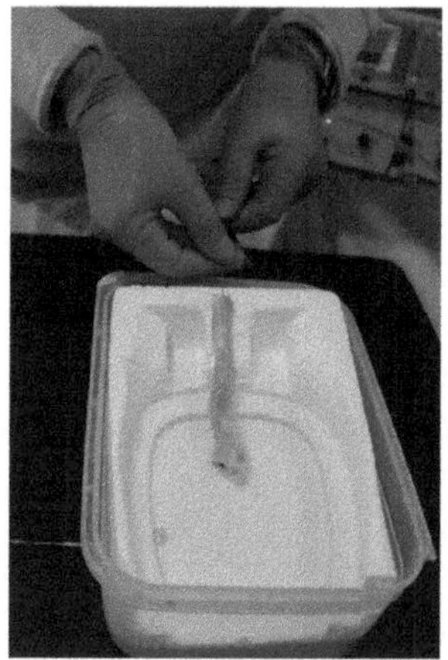

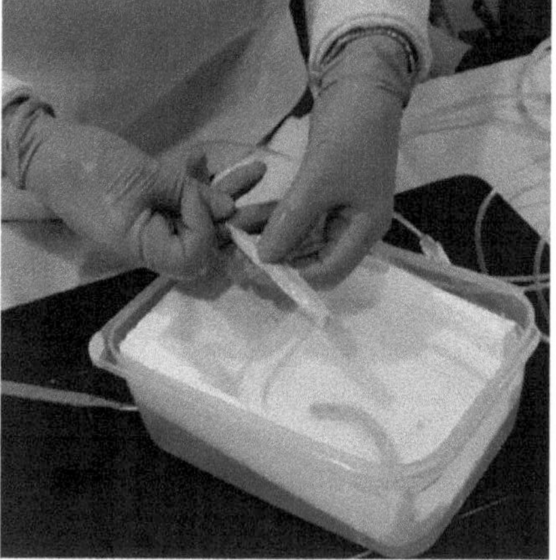

FIGURE 9.27 Penetrating an occluded artery sample using the drilling-wire that is activated by the piezoelectric actuator.

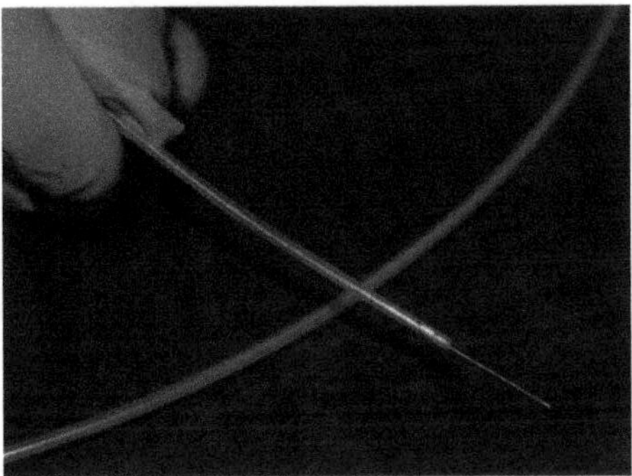

FIGURE 9.28 A view of the sleeved wire with the metal tip shown to stick out.

expected from the cardiologist experience, once the first hard part of the occlusion was penetrated, the drilling-wire was able to advance without a need for vibratory enhancement, i.e., performing piezoelectric actuations.

Since the obtained results in the first test were quite fast and surpassed the expectations of the cardiologist, the test was repeated a few months later in order to confirm the performance. Three artery segments were used in the second experiment. After testing them manually by pushing a drilling-wire through to verify the hardness of the occlusion—two of the arteries had penetrable occlusions that made them inappropriate for the test. The third one was confirmed to have a clogged artery and it was served as an acceptable sample for the test. This artery was placed in a container that was filled with water and the artery segment was placed on a Styrofoam support. The temperature of the sample was measured as 22°C and the sleeved wire was activated by the developed piezo-driver. The sleeve was brought to cover the metal wire insert and had the tip of the insert extended about 1 cm from the tip of the sleeve (Figure 9.28). The sleeved wires that were tested included those listed next along with the results of the test:

1. A Confianza wire was used having 0.36 mm (0.014″) diameter wire, which has 0.23 mm (0.009″) diameter tip and is 180 cm long. Upon activation, the Confianza failed at the brazing connection.
2. An AV-18 control wire was used as a replacement and was found to be too short for the selected sleeve.
3. A Cordis SV-5 type wire was used to do the drilling demonstration. This wire has 0.46 mm (0.018″) diameter, 300 cm long, having 4 cm spring tip. This time, the sleeve extension section next to the brazing was held manually to avoid undesirable bending and the wire did not break at the connection. It took approximately 1 min to penetrate the occlusion and there was no change in temperature during the test.

Again, in this test, the pathology test of the penetrated occlusion in the cadaver's artery did not find any heat damage or any perforations of the artery wall (Figure 9.29).

9.6 SUMMARY

Drilling mechanisms are increasingly being developed for medical tools to perform surgical procedures that involve penetrating bones in the human body, fracturing kidney stones, drilling occlusions as well as removing dental decay in teeth. Drilling bones is usually done for such applications as joining fractured or broken bones where screws and braces are used to attach and support bone sections. This chapter covered several examples of mechanisms that have been developed in recent years

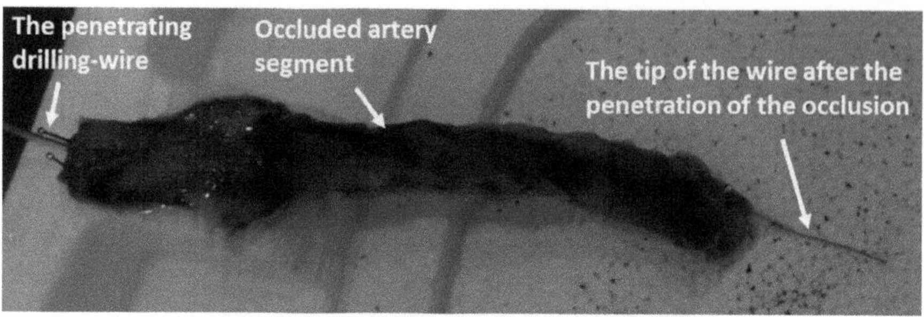

FIGURE 9.29 An occluded artery segment and the drill wire entered and exited the artery after penetrating the occlusion.

using percussive piezoelectric actuation as well as high-speed pneumatic rotary drills. Piezoelectric-based actuators have significant potential for medical applications including lithotripsy for large stones in the kidney as well as occlusion-penetrating tools for blocked arteries. The piezoelectric actuation using a free-mass is currently being medically used to break large kidney stones and it is considered one of the most effective mechanisms for this purpose. This drill has been originally developed by some of the authors of this chapter jointly with engineers from Cybersonics. To advance the capability that was originally developed with the use of the free-mass mechanism, the authors introduced the constrained-mass mechanism and it has been shown to be highly effective in breaking simulated kidney stones. Drilling teeth for filling cavities, root canal treatments, and shaping teeth for crowns is widely practiced by dentists throughout the world and is another example of highly effective use of drilling technology for medical applications.

ACKNOWLEDGMENTS

Some of the research reported in this chapter was conducted at the Jet Propulsion Laboratory (JPL), California Institute of Technology, under a contract with the National Aeronautics and Space Administration (NASA). The authors would like to thank Mark Mewissen, Interventional Radiologist, SLH Milwaukee, WI; Ron Waxman Cardiology Director, Cardiovascular Research and Advanced Education, MedStar Heart and Vascular Institute, MedStar Washington Hospital Center, Washington, DC; and Craig Ford, President, SterileBits, Inc., for reviewing this chapter and providing valuable technical comments and suggestions. In addition, the authors would like to thank Joshua Leavitt from Placidus LLC and Sujat Sukthankar, Vice President, Research & Development, Boston Scientific for supporting the reported studies.

REFERENCES

Bao X, Y. Bar-Cohen, Z. Cheng, B.P. Dolgin, S. Sherrit, D.S. Pal, S. Du, and T. Peterson, (Sept. 2003), "Modeling and Computer Simulation of the Ultrasonic/Sonic Driller/Corer (USDC)," *IEEE Transactions on Ultrasonics, Ferroelectrics and Frequency Control*, 50, 1147–1160.

Bar-Cohen Y., and K. Zacny (Eds.), (July 2009), "*Drilling in Extreme Environments—Penetration and Sampling on Earth and Other Planets*," Wiley-VCH, Hoboken, NJ, 827, ISBN-10: 3527408525, ISBN-13: 9783527408528

Bar-Cohen Y., S. Sherrit, B. Dolgin, T. Peterson, D. Pal, J. Kroh, and R. Krahe, (March 8, 2005), "Smart-ultrasonic/sonic driller/corer," U.S. Patent No. 6,863,136.

Bar-Cohen Y., S. Sherrit, H. J. Lee, M. Badescu, X. Bao, Y. Shalev, J. Leavitt, and S. Shalev, (2016), "Placid-Wire—Mechanism of Penetrating Blocking/Occlusion in arteries," NTR Docket No. 50075, Submitted on Feb. 24, 2016. Provisional Patent Application No. 62/303,989 was filed by Caltech CIT 7465-P on March 4, 2016.

Bar-Cohen Y., S. Sherrit, M. Badescu, H.J. Lee, and X. Bao, (May 2018), "Drilling Mechanisms Using Piezoelectric Actuators Developed at Jet Propulsion Laboratory," Chapter 6 in V. Badescu and K. Zacny (Eds.), *Outer Solar System. Prospective Energy and Material Resources*, Springer-Verlag, Heidelberg, Germany, ISBN 978-3-319-73845-1, 181–259.

Esch E., W.N. Simmons, G. Sankin, H.F. Cocks, G.M. Preminger, and P. Zhong, (2010). "A simple method for fabricating artificial kidney stones of different physical properties," *Urological Research*, 38(4), 315–319.

Karanasiou G.S., A.I. Sakellarios, E.E. Tripoliti, E.G.M. Petrakis, M.E. Zervakis, F. Migliavacca, G. Dubini, E. Dordoni, L.K. Michalis, and D.I. Fotiadis, (2014), "Modeling of stent implantation in a human stenotic artery." *XIII Mediterranean Conference on Medical and Biological Engineering and Computing 2013*, Springer, Seville, Spain.

Lally C., F. Dolan, and P.J. Prendergast, (2005). "Cardiovascular stent design and vessel stresses: a finite element analysis," *Journal of Biomechanics* 38(8), 1574–1581.

Loree H.M., A.J. Grodzinsky, S.Y. Park, L.J. Gibson, and R.T. Lee, (1994). "Static circumferential tangential modulus of human atherosclerotic tissue," *Journal of Biomechanics* 27(2), 195–204.

NYU Dentistry website, (visited on March 22, 2019), "Technology in Dentistry, Through the Ages," https://dental.nyu.edu/aboutus/history-of-nyucd/technology-in-dentistry-through-the-ages.html

Praamsma M., H. Carnahan, D. Backstein, C.J.H. Veillette, D. Gonzalez, and A. Dubrowski, (2008), "Drilling sounds are used by surgeons and intermediate residents, but not novice orthopedic trainees, to guide drilling motions," *Canadian Journal of Surgery*, 51(6), 442–446. PMCID: PMC2592586, PMID: 19057732. https://www.ncbi.nlm.nih.gov/pmc/articles/PMC2592586/

Schiavone A., L.G. Zhao, and A.A. Abdel-Wahab, (2013), "Dynamic simulation of stent deployment–effects of design, material and coating," *Journal of Physics: Conference Series, IOP Publishing*.

Sherrit S., B.P. Dolgin, Y. Bar-Cohen, D. Pal, J. Kroh, and T. Peterson, (October 1999), "Modeling of horns for sonic/ultrasonic applications," *Ultrasonics Symposium, 1999. Proceedings. 1999 IEEE*, Vol. 1, 647–651. IEEE.

Sherrit S., Y. Bar-Cohen, and X. Bao, (2000), "*Ultrasonic Materials, Actuators and Motors (USM)*," *Automation, Miniature Robotics and Sensors for Nondestructive Evaluation and Testing*, edited by Y. Bar-Cohen, TONE Series, ASNT, Columbus, OH, 215–228.

Sherrit S., X. Bao, Y. Bar-Cohen, M. Badescu, and H. J. Lee, (2019), "Dual Frequency Ultrasonic and Sonic Actuator with Constrained Impact Mass," Patent application 16/367,075 has been filed on March 27, 2019.

Wong, K.K., P. Thavornpattanapong, and J. Tu, (2012), "Effect of calcification on the mechanical stability of plaque based on a three-dimensional carotid bifurcation model," *BMC Cardiovascular Disorders* 12(1), 7.

Index

H

I

J

K

L

M

N

O

P

R

Milton Keynes UK
Ingram Content Group UK Ltd.
UKHW050449071024
449327UK00014B/309